The Alkali–Silica Reaction in Concrete

in Concrete

Edited by

R. N. SWAMY
Department of Mechanical and Process Engineering
University of Sheffield

CRC Press
Taylor & Francis Group
Boca Raton London New York

CRC Press is an imprint of the
Taylor & Francis Group, an informa business
A TAYLOR & FRANCIS BOOK

CRC Press
Taylor & Francis Group
6000 Broken Sound Parkway NW, Suite 300
Boca Raton, FL 33487-2742

First issued in paperback 2019

© 1992 by Taylor & Francis Group, LLC
CRC Press is an imprint of Taylor & Francis Group, an Informa business

No claim to original U.S. Government works

ISBN-13: 978-0-216-92691-2 (hbk)
ISBN-13: 978-0-367-86545-0 (pbk)

British Library Cataloguing in Publication Data

Alkali–silica reaction in concrete.
I. Swamy, R.N.
620.1

ISBN 0-216-92691-2

Library of Congress Cataloging-in-Publication Data

The Alkali–silica reaction in concrete / edited by R. N.
Swamy.
  p.   cm.
Includes bibliographical references.
ISBN 0-442-30299-1
1. Concrete—Deterioration.   2. Alkali–aggregate reactions.
3. Concrete—Deterioration—Case studies.   I. Swamy, R.N.
TA440.A4567 1990
620.1'3623—dc20   89–70439
                                                      CIP

Phototypesetting by BPCC AUP (Glasgow) Ltd

Visit the Taylor & Francis Web site at
http://www.taylorandfrancis.com

and the CRC Press Web site at
http://www.crcpress.com

Foreword

At first sight, and when looked at superficially, the problem of alkali–silica reaction (ASR) may appear to be similar to the other deterioration processes that cause damage to concrete with which researchers and engineers are already familiar. Like these other problems, ASR is a time-dependent phenomenon—it occurs during the service life of a structure (often after several decades of good or satisfactory performance), gives visible external warnings of internal damage, and generally appears to be harmless to the useful life of a structure. Although the actual number of structures affected by ASR is few and far between, experience has shown that, when considered in the context of concrete construction globally, the effects of ASR can be very profound and severely detrimental to concrete structures, affecting strength, stiffness, serviceability, safety, and stability.

The phenomenon of ASR is complex, and there are many interacting and interdependent parameters that influence its occurrence. Even now there is probably no single test that can guarantee that a structure will never be affected by ASR. In several respects, therefore, ASR is an unusual and frightening phenomenon, not widespread or likely to be so, but quite devastating when it does occur. What makes this process of deterioration more unusual and complex than others is that ASR is much more difficult to recognise, identify and monitor. Further, it often provides a means through which other deterioration mechanisms can develop and operate, and indeed structures affected by environmental or structural deterioration other than ASR may provide favourable conditions for the rekindling of a dormant ASR problem. There is probably no country which has not experienced ASR in some form or other, and the likelihood of the problem diminishing in the future is not hopeful given the progressive depletion of deposits of so-called safe unreactive aggregates. It might even be more realistic to say that there are probably no aggregates that are completely unreactive, although they may remain so for a very long time.

There are some similarities between the problem of ASR and the high alumina cement (HAC) problem. Both are the result of internal chemical reactions, with the HAC problem arising from the nature of the cement whilst in ASR the aggregates are responsible for the expansive reactions. Both types of reactions are highly susceptible to environmental conditions, and the severity of the problem is very much influenced by adverse ambient temperature and humidity. Both are time-dependent, occur in structures already stressed and cracked, and superimpose additional stresses to both concrete and steel. It is crucial to note that the effects of both HAC and ASR are heightened by local conditions: temperature, humidity, reactive silica concentrations, concrete alkali concentrations, and the variable material sorptivity and diffusivity. Like HAC, ASR problems often result in non-uniform, non-homogeneous forms of distress. But here the similarities end: whilst 'conversion' and consequent loss of strength are inherent in the use of HAC cement, there is only limited guidance to anticipate where and when ASR may show up as a potential cause of material and structural distress. The solution to these problems does not lie either in banning the use of HAC or stopping the use of aggregates in concrete. Such actions are merely expedient and strike at the very roots of scientific and engineering solutions to problems.

It is important to realise that real concrete structures behave very differently from laboratory studies. In practice, the method of concrete placement, variable compaction, moisture/thermal movements, and above all the micro-climatic conditions of heat and moisture, all lead to surprisingly varying and variable local concentrations of reactive silica and concrete alkalinity arising from the highly unpredictable, *in-situ* material porosity and per-meability characteristics. Under these conditions the chemical and physical components of the alkali–silica reaction become interactive, interdependent and synergistic—and they become almost inseparable. This is what is so unique about ASR, and puts it in a totally different category from other deterioration processes in concrete. Thus there may be no visible evidence of alkali–silica reactivity for a long time if the quantities of the reactants and moisture available at a given site are insufficient to initiate the reactivity. On the other hand, the amounts of susceptible silica or alkali concentration may be inadequate because of lack of availability of capillary water, which is required for both the transportation of the reactive ions and the swelling of the gel. All these could be changed dramatically, albeit in a slow but steady fashion over many years and decades, by a steady change to more favourable circumstances.

It is also important to realise that laboratory studies are not irrelevant. We need both types of investigations, at microstructural and engineering level, but we need to apply the results with engineering judgement and a good under-standing of material properties and engineering performance. The greatest danger lies in believing that laboratory studies are the sole basis for *in-situ* performance; worse still, is to assume that micromechanics and micro-structure provide a universal panacea for engineering problems. What we need are engineering solutions based on real microstructural perspectives— if we are able to do this, problems posed by material distress, whether due to the presence of reactive cements, reactive aggregates or reactive fibres, can be controlled through appropriate modification of the concrete constituents and through structural design and detailing.

It will be clear that a problem of this nature cannot be satisfactorily tackled by a single author, nor can it be based on either laboratory studies or field observations alone. Alkali–silica rections involve the physics, chemistry and engineering of materials and structures. There is much to learn and understand from both laboratory testing and field assessments. By necessity and various constraints, laboratory investigations are often designed to provoke uniform and homogeneous reactions under conditions not readily prevalent in actual structures, but they give an insight into deterioration processes that no amount of field evaluation can provide. However, they cannot blindly be extrapolated to engineering solutions without an understanding of and feel for structural behaviour. Equally, field evaluations can be very misleading if not adequately supported by intelligent and relevant microstructural examinations. It is not unknown for concrete structures to be diagnosed as suffering from forms of concrete deterioration other than ASR, when the real culprit is alkali–silica reactivity. The dependence of one on the other cannot be overemphasised.

This book is different from other books on ASR. Because of the complex nature of ASR, the limitations of both laboratory research and field evaluations, and the very wide experience gained in the diagnosis, assessment and rectification of structures affected by ASR, we have attempted to pull together and synthesise a wide range of knowledge and expertise of the phenomenon of ASR. The first four chapters are devoted to the basics of ASR, while the remaining seven chapters describe the experience gained globally in tackling the problem of ASR from microstructure and micromechanics and their extension to material and structural implications as observed in real structures. The editor is fortunate in having secured the services of internationally-recognised experts in compiling this volume. Their combined knowledge and experience provides a formidable, and authoritative, overview of the subject.

Chapter 1 traces the history of ASR and describes in detail the nature of alkali–aggregate rections in concrete, the basic requirements for the reaction to initiate and propagate, the factors controlling the reaction, and the observed effects of the reaction. Chapter 2 presents a thorough analysis of the chemistry of alkali–aggregate reactions and relates these to the chemistry of cements, mineral admixtures, and the mineralogy of aggregates, and discusses the chemical mechanisms of ASR in considerable detail. Chapter 3 provides a critical review of the various test methods curently available for ASR. These include petrographic studies, well-established traditional methods, new rapid test methods, and non-destructive test methods. Special attention is drawn in this chapter to the limitations of the many new rapid test methods, particularly the so-called Duggan test, as well as to the caution required when artificially high alkali levels are used to initiate ASR. There is considerable controversy over the role and effectiveness of mineral admixtures such as fly ash, slag and microsilica in counteracting the effects of ASR. Unfortunately conclusions derived from tests or reviews of published papers appear to some extent to be biased depending on the interests and proclivities of the author concerned. Should all mineral admixtures mitigate the deleterious effects of ASR at the same rate or to the same extent at the same time? There is no concensus here. Chapter 4 discusses some of these factors critically and shows conclusively that, when used correctly and in the required amount, mineral and pozzolanic admixtures play a very positive role in controlling the expansive strains, cracking, loss of strength and stiffness, and structural distortions arising from ASR. But they cannot and do not always remove completely all the deleterious effects all of the time. This chapter gives a positive and realistic assessment of the role and effectiveness of mineral admixtures in concrete.

As pointed out earlier, the factors involved in ASR are complex, interdependent and interactive, and the one major parameter—the environment—which is influential in determining the occurrence and physical effects of ASR, is so unpredictable and uncontrollable that experience and knowledge gained by just one individual or even by one country is unlikely to be adequate and sufficient to get to the roots of the problem. Chapters 5 to 11 remedy this situation and bring to the reader the knowledge and understanding gained worldwide in experiencing ASR. These chapters encompass a wide range of geological and environmental conditions and provide a unique insight into ASR.

Chapter 5 discusses the UK experience of ASR and describes the diagnosis, prognosis and management of ASR, with emphasis on the steps taken to minimise the future risks of ASR. There is now considerable confidence that the existing stock of ASR-affected structures can be brought back into safe service lives, and that ASR can be made to be a much less frequent occurrence in practice. Chapter 6 describes the Danish experience, which is perhaps unique because most of Denmark's aggregate sources contain reactive components, and the first identification of ASR goes as far back as the 1910s. The extensive fundamental studies on ASR are reported, particularly on the mechanisms of reactivity, and the various preventative measures developed during the last two decades are outlined.

In chapter 7 we find that the Icelandic experience provides another view of the ASR spectrum. Unlike Denmark, geological conditions have determined that the cements will have high alkali contents, that many concrete aggregates contain crystalline glass of volcanic origin, and that the high-rainfall environment provides sufficient moisture for alkali–silica reactivity. All the ingredients for ASR are readily present in Icelandic concretes. These problems are described and the preventative and protective measures adopted as a result of research and field surveys are discussed. Significantly ASR damage has not been observed in Icelandic concrete since about 1979.

Chapter 8 discusses the Canadian experience of ASR. It is probably the only country where current extensive studies are in progress both on various aspects of ASR as well as on supplementary cementing materials as a method of counteracting the deleterious effects of ASR. Again, Canada is probably the only country where all three types of alkali–aggregate reactivity, namely, alkali–silica reaction, alkali–carbonate reaction, and the more intricate slow/

late expanding alkali–silicate/silica reaction have been observed and experienced. The chapter discusses these, the Canadian Standards, and other new test methods to determine the potential reactivity of aggregates and the various methods of counteracting ASR. Chapter 9 reports on the New Zealand experience of ASR, and provides an example of how the problem was tackled in a country where nearly 30% of the aggregate sources contain reactive materials. A combination of petrographic examination and field inspection of ASR-affected structures produced satisfactory validation of laboratory test results, and also revealed how previously unknown combinations of reactive aggregates contributed to the problem. This chapter points out that it can take a long time to develop an adequate database of information that can be of value to specifiers, designers and engineers, and that field surveys and laboratory studies can jointly pave the way to recognising and controlling the problem.

Chapter 10, on the Japanese experience of ASR, discusses three aspects of ASR-affected concrete structures. The extensive research on the mechanisms of ASR, microstructure of the reaction products, and petrographic studies, is summarised. Examples of damaged structures and the repair methodologies adopted are discussed. Standard and new rapid tests developed in Japan to monitor ASR are reported, and the preventative measures used to avoid ASR are described. There is a wealth of information presented in this chapter, and readers should derive a great measure of help and advice in the management of the ASR problem. The final chapter discusses the Indian experience of ASR. Here, the reactivity of quartzite and granitic aggregates is believed to be due to the presence of strained quartz, even when secondary siliceous minerals are absent. These are again slowly-reactive materials which necessitated modified test methods and revised threshold values for mortar bar expansion tests. This chapter describes these, and how the problem of ASR has been effectively tackled in India.

This volume conveys several messages to those countries that have either not experienced, or have ignored, the possibility of the existence of alkali–aggregate reaction in concrete. Testing, field surveys and evaluation of affected structures can take a long time to produce a large enough body of data to be of use in specification, design and construction. The nature of the alkali–aggregate reactivity is such that to control and prevent its occurrence requires a thorough knowledge of the aggregates being used, their potential reactivity as determined by laboratory study, and a clear understanding of the implication of the reactivity on the behaviour and performance of plain and reinforced concrete elements. These activities need to be supported by field surveys and *in-situ* evaluations which are needed to confirm laboratory testing, develop effective control measures, and formulate repair management techniques and methodologies. It is important to note that detailed discussion of test results and the experience of specific reactive materials will tend to be fairly unique to the region or country concerned. Attempts to simply translate such results to materials in other regions or countries is unwise, as is the mistake that idealised microstructural information provides solutions to engineering problems. Whether alkali–aggregate reactivity can only be minimised or completely eliminated even when the problem is fully recognised and understood, and all reasonable and practicable controls are exercised over the various components that contribute to the reactivity, only time will tell.

In summary, experience from many countries teaches us many lessons. ASR is an unusual and unpredictable phenomenon. The factors involved are too complex, very interdependent and highly interactive. Simplistic models belie the complex nature of the reactivity, but the problem is not intractable. External agents, environmental conditions and microclimates can have as much of a devastating effect as the basic ingredients of the reaction. The reaction can occur at a very early stage if all the conditions necessary coalesce at that time. On the other hand, the reactivity can lie dormant for many years (or continue at an unidentifiable and insignificant expansion level), only to be triggered off dramatically by simple changes in environmental conditions or failure of some other structural or construction component unconnected with ASR. Engineers should not allow themselves to be overwhelmed or blinded either by field occurrences or by laboratory findings at microstructural and engineering level. What knowledge and experience tells us is that structures need not and should not collapse because of ASR. We need to accept that detailing and designing for durability are just as important as designing for external loads, and that there is probably no single solution that will obviate the deleterious effects of ASR all the time in all situations. It is serviceability, stability and structural integrity rather than ultimate strength that is most at risk in structures affected by ASR. If this book is able to communicate some of these essential messages, its purpose will have been fulfilled.

R.N.S.

Contributors

Dr S. Chatterjee The Danish Technological Institute, Gregersensvej, DK-2630 Taastrup, Denmark

Mr Z. Fördös Research Engineer, Cement and Concrete Laboratory, Aalborg Portland A/S, PO Box 165, DK-9100, Aalborg, Denmark

Professor F. P. Glasser Department of Chemistry, University of Aberdeen, Meston Walk, Aberdeen AB9 2UE, UK

Dr P. E. Grattan-Bellew Institute for Research in Construction, Building Materials Section, M20 Ottawa, Ontario K1A OR9, Canada

Professor M. Kawamura Professor of Civil Engineering, Department of Civil Engineering, Kanazawa University, Kanazawa, Ishikawa, Japan

Professor K. Kobayashi Professor of Civil Engineering, Osaka University of Technology, Osaka, Japan

Professor T. Kojima Professor of Civil Engineering, Department of Civil Engineering, Ritsumeikan University, Kyoto, Japan

Dr T. Miyagawa Associate Professor of Civil Engineering, Department of Civil Engineering, Kyoto University, Kyoto, Japan

Dr A. K. Mullick Director, National Council for Cement and Building Materials, New Delhi 110 049, India

Dr K. Nakano Director and Head of Central Laboratory, Osaka Cement Company, Osaka, Japan

Professor S. Nishibayashi Professor of Civil Engineering, Department of Civil Engineering and Ocean Civil Engineering, Tottori University, Koyama, Japan

Professor K. Okada Dean of College of Engineering, Fukuyama University, Fukuyama, Hiroshima, Japan

Mr H. Ólafsson Director, The Icelandic Building Research Institute, Keldnaholt, 112 Reykjavik, Iceland

Dr K. Ono Director of the Department of Civil Engineering, Konoike Constructions Company, Osaka, Japan

Dr A. B. Poole Geomaterials Unit, Queen Mary and Westfield College, University of London, Mile End Road, London E1 4NS, UK

Dr I. Sims   Messrs Sandberg, Consulting, Inspecting and Testing Engineers, 40 Grosvenor Gardens, London SW1W 0LB, UK

Mr D. A. St. John   Department of Scientific and Industrial Research, Chemistry Division, Private Bag, Petone, New Zealand

Dr R. N. Swamy   Structural Integrity Research Institute, University of Sheffield, Mappin Street, Sheffield S1 3JD, UK

Dr N. Thaulow   Chief Chemical Engineer, G. M. Idorn Consult A/S, International Consulting and Research Engineers, Blokken 44, DK-3460 Birkerød, Denmark

Contents

4   Role and effectiveness of mineral admixtures in relation to alkali–silica reaction   96
R.N. Swamy

5   Alkali–silica reaction—UK experience   122
I. Sims

11 Alkali–silica reaction—Indian experience   307
A.K. Mullick

Index   335

1 Introduction to alkali–aggregate reaction in concrete

A.B. POOLE

Abstract

This chapter introduces and defines the nature of alkali–aggregate reactions in concretes. Alkali–silica reactivity in concrete is a particular variety of chemical reaction within the fabric of a concrete involving alkali hydroxides, usually derived from the alkalis present in the cement used, and reactive forms of silica present within aggregate particles. This chemical reaction also requires water for it to produce the alkali–silica gel reaction product which swells with the absorption of moisture. The amount of gel and the swelling pressures exerted are very variable depending on reaction temperature, type and proportions of reacting materials, gel composition, and other factors, but they are often sufficiently high to induce the development and propagation of microfractures in the concrete which, in turn, lead to expansion and disruption of the affected concrete structure or element. Typical deleterious features of alkali–silica reaction in concrete structures include cracking, expansion and consequent misalignment of structural elements, spalling of fragments of surface concrete as 'pop-outs', and the presence of gel in fractures or associated with aggregate particles within the concrete. The reaction typically takes between 5 and 12 years to develop, though there are many exceptions, and it is most severe where alkali concentrations in the concrete pore fluids are high. A very wide variety of aggregate rock types in structures from many parts of the world have been reported as being alkali–silica reactive. This is consequent on the reactive forms of silica often only forming a minor mineral component of the aggregate such as the cement between mineral grains. This silicious material must be amorphous or cryptocrystalline with a large surface area if it is to react sufficiently to produce deleterious effects in the concrete.

1.1 Historical background

Forms of concrete have been used as a durable building material at least since Roman times, and concrete structures of Roman origin can still be seen in many parts of Europe today. As early as the nineteenth century it was realised that, although normally a very durable material, concrete could deteriorate, with frost and seawater being considered the principal agents causing deterioration of concrete structures. Cases of concrete failure that could not be attributed to one or other of these causes were left unexplained.

During the 1920s and 1930s numbers of concrete structures in California, USA, were observed to develop severe cracking within a few years of their construction, although quite acceptable standards of construction and quality control of materials were employed. It was a major scientific achievement with far-reaching consequences when Stanton in 1940 was able to demonstrate[1] the existence of alkali–aggregate reaction as an intrinsic deleterious process between the constituents of a concrete. It soon became clear that exposure to external environmental conditions was of less importance to the type of concrete deterioration observed in California than the characteristics of the cements and aggregates used. Stanton's subsequent experimental studies showed that cracking and expansion of concrete were caused by combinations of the high-alkali cement and opaline aggregates used. In 1941, shortly after Stanton had published his work, Blanks[2] and Meissner[3] described cracking and deterioration in the concrete of the Parker Dam. They were able to show that an alkali–silica reaction product was being produced in the concrete and that the reactive components in the aggregate were altered andesite and rhyolite fragments which together only represented about 2% of the total aggregate.

During the following decades research on the alkali–aggregate reaction was carried out at many laboratories, first in the United States but later in Europe, Canada and other parts of the world. Research studies have progressed rapidly in a number of different directions, ranging from identification of the aggregate mineral components which are involved in the reaction, through the mechanisms and controls of the reactions themselves, to diagnosis, testing and assessment of reaction effects. Substantial contributions to the knowledge of this subject have been made by research workers from many parts of the world, but perhaps three of the most significant names in this field of research who have produced work now regarded by many as of particular importance are Swenson of Canada, who recognised alkali–aggregate reactions which involved carbonate aggregates,[4] Idorn of Denmark, as one of the first European scientists to investigate concrete deterioration due to alkali–silica reaction[5] and Vivian of Australia, who has contributed a great deal to understanding the mechanisms of the reaction over many years. Many other research workers may also be attributed with contributing research of great value in this field, so that it becomes difficult to select particular names for special mention. Nevertheless the reference list for this text bears witness to the efforts and dedicated research work of a large number of scientists, many of whom are involved with the alkali–aggregate problems of concrete at the present time.

Since Stanton published his first findings in 1940 an enormous volume of research papers has been published on this subject, with contributions from workers in all five continents appearing in numerous national and international journals. In 1974 the first of a series of international meetings of scientists interested in alkali–aggregate reactivity in concrete was held in Denmark; the Portland Research and Development Seminar on Alkali–Silica Reaction at Køge. Since that first meeting a series of international conferences has been held, which has attracted increasing numbers of research workers and engineers. These conferences were held in Iceland (1975)[6], the UK (1976)[7], the US (1978)[8], South Africa (1981)[9], Denmark (1983)[10] and Canada (1986)[11]. The series of published proceedings of these conferences perhaps provides the most important source of research information available, including valuable reviews of national and international experience relating to research findings, case histories and to preventive and remedial measures.

1.2 Alkali–aggregate reactions in concrete

The expansion, deterioration and perhaps even failure of concrete structural elements resulting from alkali–aggregate reaction in the concrete are due to swelling pressures developing as a result of the reactivity within the fabric of the concrete which are sufficient to produce and propagate microfractures. In general terms these reactions involve chemical interaction between alkali hydroxides which are usually (but not always) derived from the cement used in the concrete and reactive components in the aggregate particles used.

Detailed research studies have shown that there are a number of different materials which produce chemical interactions in concrete which can be described as alkali–aggregate reactivity. One of these is alkali–carbonate reactivity, which is quite distinct from the alkali–silica reaction and was first described by Swenson[4] in 1957. He described sections of concrete pavement in Kingston, Ontario, as exhibiting excessive expansion which closed the joints and produced cracking of the slabs within 6 months of placing. The cracks defined roughly hexagonal areas 50–100 mm across and extended deeply into the concrete. Since these original observations other cases of alkali–carbonate reactivity in concretes have been reported from many parts of the world. These include areas in the USA reported by Newlon and Sherwood[12], Axon and Lind[13], Welp and De Young[14], Hadley[15] and others, Iraq, reported by Alsinawi and Murad[16], Bahrain by Hussen[17] and in Canada by Rogers[18]. Various types of alkali–carbonate reaction have also been reported, though not all of them appear to be expansive or deleterious. They may be classified into the following broad groups principally according to the type of reaction rim or reaction products they produce.

(a) Carbonate reactions with calcitic limestones. Dark reaction rims develop within the margin of the limestone aggregate particles. These rims are more soluble in hydrochloric acid than the interior of the particle.

(b) Reactions involving dolomitic limestone aggregates characterised by distinct reaction rims within the aggregate. Etching with hydrochloric acid shows that both rim zones and the interior of the particles dissolve at the same rate.

(c) Reactions involving fine-grained dolomitic limestone aggregate which contains interstitial calcite and clay. Reaction with alkalis produces a distinct dedolomitised rim. Etching of reacted particles with dilute hydrochloric acid clearly shows up a reaction rim, which in the majority of cases reported is found to be enriched in silica.

The third type of reaction (c) appears to be the only type to produce significant expansion. The cause is not properly understood at present but Gillott[19] has suggested that the dedolomitisation of the crystals in the aggregate particles opens channels, allowing moisture to be absorbed on previously dry clay surfaces. The swelling caused by this absorption causes irreversible expansion of the rock and subsequent expansion and cracking of the concrete. The reaction process is essentially one of dedolomitisation together with the production of brucite [$Mg(OH)_2$] and regeneration of alkali hydroxide and is believed to proceed in two stages, which may be expressed as follows:

$$CaMg(CO_3)_2 + 2ROH = Mg(OH)_2 + CaCO_3 + R_2CO_3$$

and

$$R_2CO_3 + Ca(OH)_2 = 2ROH + CaCO_3$$

where R may represent sodium, potassium or lithium.

Detailed discussions of this reaction and the possible mechanisms causing it are given in papers by Walker[20] and by Poole[21].

A second group of reactions reported in concretes in recent years are referred to as alkali–silicate reactions. These reactions appear to occur in alkali-rich concretes which contain argillite and greywacke rock types in the aggregate. The reaction of these rock types with alkalis is generally slow and is not completely understood. Silicate mineral constituents in these rock aggregates appear to expand and cause disruption of the concrete. The expansion of individual rock particles suggests absorption of water on previously 'dry' aluminosilicate surfaces in the microcrystalline portions of these rocks. There appears to be a direct relationship between the amount of microcrystalline material, the porosity and the expansion of the concrete containing these aggregates[22]. Complications in identifying alkali–silicate reactions of this type can arise in that alkali–silica reactions are sometimes also present in the concrete. Furthermore the results of the diagnostic tests used for detecting the alkali–silica reaction may give misleading and unreliable results in such cases.

The most common reaction between alkali hydroxides and silicious material in the concrete aggregate is usually referred to as alkali–silica reactivity. It produces an expansive reaction product that can develop sufficient swelling pressure to crack and disrupt concrete. Typically the reaction progresses slowly so that it is usually some years before expansion and damage to the structure become apparent.

An example illustrating the effects of this reaction and the timescales involved is the Chambon Dam in the Romanche River in France[23]. The dam was commissioned in 1935. Cracks were first noticed in 1950. In 1985 the dam was still expanding, the level of the crest was moving upward at between 2.6 and 4 mm per year and movement towards the right bank was recorded as between 0.6 and 0.7 mm per year. In the upper part of the dam towards the left abutment compressive stresses of between 5 and 8 MPa were recorded by flat jacks.

1.3 The alkali–silica reaction in concrete

This reaction differs from the alkali–carbonate and alkali–silicate reactions in that, as a result of the reaction between the alkali pore fluids in the concrete and silicious components of the aggregate particles, an alkali–silica gel is produced which is hydrophilic. As it absorbs moisture it increases in volume, thus generating pressures sufficient to disrupt the fabric of the concrete. The reaction may be considered to progress according to the following idealised equations[24]:

$$4SiO_2 + 2NaOH = Na_2Si_4O_9 + H_2O$$
$$SiO_2 + 2NaOH = Na_2Si_3 + H_2O$$

However, the chemical composition of alkali–silica gel is variable and indefinite. Also, some research studies by Vivian[25] and others indicate that it is the OH^- concentration which is important to this reaction and that the alkali metal cation is only relevant in so far as it becomes incorporated into the gel. A more satisfactory way of representing the chemical mechanism of the reaction is perhaps a two-stage process[26]:

(1) Acid–base reaction

$$H_{0.38}SiO_{2.19} + 0.38NaOH = Na_{0.38}SiO_{2.19} + 0.38H_2O$$

(2) Attack of the siloxane bridges and disintegration of the silica

$$Na_{0.38}SiO_{2.19} + 1.62NaOH = 2Na^{2+} + H_2SiO_4^{2-}$$

In more general terms the reaction will proceed in stages, with the first stage being the hydrolysis of the reactive silica by OH⁻ to form an alkali–silica gel as indicated and a later secondary overlapping stage being the absorption of water by the gel, which will increase in volume as a result. The swelling pressures produced by the gel induce the formation of microcracks close to the reaction sites, and these propagate and coalesce to produce cracking within the fabric of the concrete and overall expansion of the structural element affected. It is commonly observed that once cracking has developed these cracks provide access to the interior of the concrete and allow other deleterious mechanisms to operate. Leaching by percolating water, often with the precipitation of calcium carbonate on surfaces, is common. The replacement of gel by ettringite ($C_3A \cdot 3CaSO_4 \cdot 3H_2O$) has also been observed in cracks, together with the development of secondary coarsely crystalline ettringite and secondary developments of portlandite [$Ca(OH)_2$] in the surrounding cement paste. In the majority of cases the sulphate levels in the concrete are normal, suggesting that an external source of sulphate attack has not caused the development of ettringite, but rather that the necessary sulphate is derived from within the concrete itself. The replacement of gel by ettringite also suggests that the alkali–silica reaction is well advanced before the formation of ettringite, which appears to develop prefentially in the gel. Transport of sulphate ions along with the water to the hydrating alkali–silica gel may be the mechanism by which the ettringite crystals develop and grow. Their growth in fine microcracks and pores in the cement paste may exert sufficient pressures within the concrete fabric to contribute to the observed expansions.

It might also be expected that the development of cracking due to alkali–silica reaction would facilitate the corrosion of any steel reinforcement within the concrete. However, these is little observational evidence supporting this at the present time, and it has been suggested that the highly alkaline gel produced from the reaction fills the pore spaces and microcracks surrounding the reinforcing bars, hence passivating them and preventing corrosion.

1.4 The alkali requirement in alkali–silica reaction

The great majority of cases of concrete structures reported as showing deterioration due to alkali–silica reaction were made using a high-alkali cement. Ordinary Portland cement will normally contain a small proportion of sodium and potassium present as sulphates and double sulphates (K, Na)SO₄, which tend to coat other clinker minerals, and also as minor constituents in the other cement minerals[27].

These alkalis are derived from the raw feed materials used in the manufacture of the cement, usually the argillaceous fraction and the coal (if this is used for firing the kiln). If the argillaceous material used as raw material contains mica or illitic clay, then the clinker produced will be enriched in potassium, while if degraded feldspar is present the clinker may contain more sodium, potassium or both depending on the composition of the feldspar in the raw feed. The final proportion of the alkalis present in a clinker will be dependent on the proportions of mica, illite or feldspar in the feed and upon the details of the cement manufacture in the particular plant concerned. The alkali phases tend to be a volatile fraction in the kiln environment, and some 50% will be volatilised during the burning process. Much of this alkali is redeposited in the chain section of the kiln and in the preheaters, dust precipitators and filters. In the interests of fuel economy and clean flue gases, recirculation of the dusts is common practice in modern plants, but this procedure has an adverse effect on the alkali composition of the cement clinker (Figure 1.1). Reduction in burning temperatures can also have a disadvantageous effect on alkali contents (Figure 1.2). In the UK, much of Europe and North America potassium is usually the principal alkali present, but the ratio of sodium to potassium can vary widely.

In order to assess the total alkalis present in a cement or concrete it has become standard practice to express the alkali content in terms of 'sodium equivalent'. This correlates the sodium and potassium oxides in terms of molecular proportions. The calculation of sodium equivalent is as follows:

$$\text{Sodium equivalent} = \text{weight percentage } Na_2O + 0.658 \text{ weight percentage } K_2O$$

The chemical determination of these alkalis in samples of cement or concrete normally follows the nitric acid extraction procedure as outlined in BS 4550:

Residual K_2O in clinker of total input

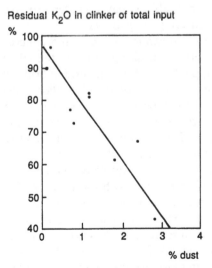

% dust

Figure 1.1 Residual K_2O of clinker versus amount of dust removed according to Goes and Keil (1960). From Worning, I. and Johansen, V. (1983) *Proceedings of the Sixth International Conference on Alkalis in Concrete, Research and Practice*, p. 44.

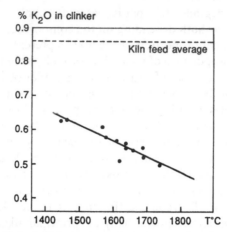

Figure 1.2 Percentage K₂O of clinker versus estimated average burning zone temperature data from a precalciner kiln. From Worning, I. and Johansen, V. (1983) *Proceedings of the Sixth International Conference on Alkalis in Concrete, Research and Practice*, p. 46.

Part 2 (1970), Clause 16.2. In the analysis of concrete samples it has been suggested that the acid extraction method is likely to give alkali concentrations higher than are normally readily available for alkali–silica reaction in the concrete because additional alkalis may be extracted by the acid from aggregate minerals such as feldspars. Although this may be correct it is generally considered that the acid extraction method is appropriate, and account is taken of the probability that it will tend to give a maximum figure for total alkalis in engineering judgements.

The alkali–silica reaction cannot proceed in a concrete if the alkali concentration is below a certain threshold value. In the UK a careful review has been made of existing published data concerning both alkali concentrations necessary for reaction and forms of reactive silica found in aggregates. A committee chaired by Hawkins has used these data to prepare guidance notes[28] concerned with minimising or avoiding the risk of damage to concrete by alkali–silica reaction. This document recommends that the alkali content of the cement used should be below 0.6% by weight and that the mass of alkalis from all sources should be kept below 3.0 kg/m³ in the concrete.

The Hawkins report is concerned with specifications for new concrete only. In existing concrete structures in which deleterious effects due to alkali–silica reaction have occurred, alkali levels are sometimes found to be lower than the figures given above. In some cases the effects of the reaction are observed to be continuing, although alkali levels are well below the 3.0 kg/m³ thought to be the minimum to initiate reaction. Several possibilities arise from these observations. The low alkali figures obtained for concretes from these structures may be the result of alkalis having been leached from the structure with time. Where the reaction is continuing (and is distinct from continuous deterioration due to other causes), the already initiated reaction is either able to generate sufficient chemical gradients to continue in spite of low alkali levels, or very localised high alkali concentrations within the concrete are able to maintain the reaction at those sites. There is certainly evidence from detailed electron probe microanalysis[29] that, although the overall alkali levels in the cement paste may be very low, reacting aggregate particles in the same concrete can be very high in alkalis.

Although the principal source of alkalis in concretes is the cement used, the addition of alkali to the concrete from other sources should not be overlooked. Normally the mix water will not contain significant alkali concentrations. However, where there is a possibility that sodium chloride will be incorporated into the mix, for example by the use of seawater, the alkali contributed by the sodium chloride should be incorporated into calculations for total alkali contents. The Hawkins Report recommends that, if alkalis from sources other than the cement exceed 0.2 kg/m^3 of the concrete, they should be taken into account when calculating the total reactive alkali available. Some mineral constituents of the aggregates, notably the micas, feldspars and illitic clays, contain alkali metal cations in their structures. Opinions are divided as to whether alkalis from these minerals become available for reaction within the concrete or are too firmly held within the crystal lattice. It seems likely that where the aggregates have suffered a degree of 'geological weathering' before being used as aggregate the partial degradation of these minerals from this cause will facilitate the leaching of at least some of the alkali into the pore fluids of the concrete.

In addition to alkalis derived from the constituents of the concrete, consideration should be given to the absorption of alkalis by hardened concrete in contact with seawater, some groundwaters and other materials, such as de-icing salts. The absorption of alkalis from such sources by a concrete will, of course, depend primarily on the porosity and permeability of the concrete, and the length of time and the nature of the exposure to the alkali-containing fluids. Consideration must also be given to the mechanisms driving the absorption process. An example of certain alkali–silica reactive concrete plinths for electrical transformers and other electrical power supply equipment in south-west England is interesting in that detailed investigations of the alkali concentrations of the affected concretes showed the top of the plinths to contain up to four times as much total alkali as the lower parts which were in contact with moist ground. Two explanations for these observations have been proposed. The first is that moisture movement from the moist lower surfaces to the wind- and sun-dried top surfaces accounts for this alkali migration. An alternative or possibily complementary mechanism involves the electrical potential difference between the two faces of the slab induced by the electrical installations. The alkalis migrate through the pore fluid electrolyte to concentrate in the upper layers of the plinth. Experimental studies carried out by Moore[30] certainly appear to confirm that the reactivity can be increased, and cation migration will take place in alkali–silica affected concretes under the influence of electrical potential difference.

The role of alkalis in the pore fluids of a concrete in propagating the alkali–silica reaction is undisputed and has led to investigations of pore fluid chemistry in hardened concrete[31]. Initially, on setting, the pore solutions in a concrete are complex, containing sulphate, calcium and alkali hydroxides, but within 48 hours there appears a pore solution which remains constant over long time periods and which contains a stable mixture of sodium and potassium hydroxides together with a very low concentrations of calcium hydroxide. The exact composition of the pore fluid will depend on the alkalis available and will have a normality of around 0.8 N, though evidence is presented by Diamond[31] which suggests that the pore solution normality may be reduced in concretes undergoing active alkali–silica reactivity.

1.5 The reactive silica component in the aggregate

The majority of rock materials used as aggregate for concrete are composed of more than one mineral component. Limestones and dolomites are mono-mineralic if pure, but many typically contain a small percentage of other minerals, with quartz and clay perhaps the most common. Sandstones and quartzites when pure are considered monomineralic but can also contain a variety of minor constituents such as feldspar and micas. Even when comprising only of quartz, the silica cementing the grains together is often different from the silica in the grains themselves. Greywacke and hornfels rocks may in the simplest terms be considered to be very impure varieties of sandstones containing a variety of other mineral, rock and clay species. Again the material cementing the grains together may be different from the silica of the quartz grains.

Some sedimentary rocks are largely formed of organic material, with perhaps coal and limestones formed from shell fragments being the most familiar. Certain organisms use silica to form their hard parts, and important among these are diatoms, which produce outer coatings (tests) of opaline silica, and certain sponges which develop microcrystalline quartz stiffening filaments called spicules. These silicious materials may form an important component of certain rock types. Alternatively the silica they contain may be redistributed within the sediment in which they are buried. Such redistribution of silicious material by groundwater or other means is generally considered to have produced cherts, flint, chalcedony, agate and related silicious materials in the sedimentary rocks in which they occur.

Hornfels and other metamorphic rocks are recrystallised versions of sedimentary and igneous rock types. The severity of the metamorphism will control which new minerals will be formed from the original rock minerals.

Common minerals which form from clay rich or sandy rocks are micas, chlorites, garnets and feldspars, but during the process the original quartz will partly or wholly recrystallise, perhaps developing strains in the crystal lattice as a result. Igneous rocks, such as granites and basalts, typically contain several different mineral species. There are usually one or more of the ferromagnesian minerals such as mica, amphibole or pyroxenes, together with feldspars, and sometimes quartz if the melt from which the rock crystallised is sufficiently silicious. There are also a number of other minor mineral constituents such as iron ores in nearly all igneous rocks. It should be remembered that since igneous rock minerals crystallised at high temperature, some of them are only metastable at normal temperatures.

Certain igneous rocks, particularly volcanic varieties, cool very rapidly on exposure to the atmosphere and in some cases do not crystallise fully but contain a glassy phase which is usually interstitial to the early-formed crystals. This glassy phase typically contains the most volatile low-melting-temperature components of the original melt. These typically include silica and alkali constituents.

A form of 'reactive silica' is an essential requirement for alkali–silica reaction in concrete to take place, though the volume needed to produce deleterious effects need only be very small. As little as 2% of reactive component has been reported in certain cases where severe distress in the concrete has been observed. There are a wide variety of rock types used as aggregates for concrete and, with the exception of particularly pure varieties of rocks such as certain limestones, any of them might possibly contain a small proportion of a form of reactive silica as either an original, primary or secondary constituent. It is therefore only on the basis of service record or careful test result data that the vast majority of aggregates can be declared non-reactive in concretes containing high alkali concentrations.

It is also clearly incorrect to consider rock type as a criterion for an aggregate's potential for reactivity. Instead attention should be focused on the mineral constituents of the rock itself. Although most rocks are capable of containing reactive forms of silica, the number of types of silica that exhibit reactivity is quite small. Perhaps the overriding requirements for a silicious material to be alkali-reactive are that it should be a form of silica that is poorly crystalline or contains many lattice defects, or alternatively it should be amorphous or glassy in character. It should also be microporous so as to present large surface areas for reaction. The natural mineral materials that meet these criteria are listed in Table 1.1.

Certain granites, granitic gneisses, hornfels and greywackes have been found to be reactive when used in concrete. It has been noticed that, although the precise reactive mineral constituent within these rocks cannot be identified, the crystal quartz grains they contain show optical strain shadows when examined using a polarising microscope. A further observation was that overall alkali–silica reactivity of the concrete can be correlated in a general

Table 1.1 Some natural materials that have been identified as alkali-reactive in concrete.

Mineral	Comments
Opaline silica	Always highly reactive, can occur as primary or secondary material in rocks
Chalcedony	Moderately reactive. A minor constituent in some cherts and flints
Volcanic glass	Reactive in some cases. A common minor constituent of fine-grained volcanic rocks such as basalts and rhyolites
Silicious cement/ cryptocrystalline quartz	Cementing or marginal materials at grain boundaries of certain greywackes and hornfels not clearly identified at the present time

way with the severity of the strain, though there are many exceptions. These observations led many petrographers to follow methods developed by Dolar-Mantuani[32] for the measurement of the angle between the onset and completion of undulatory extinction of strained quartz grains in polarised light as a means of estimating the possible reactivity of a particular aggregate type. The method used is illustrated in Figure 1.3 and requires a number of grains to be measured in each rock to allow statistical analysis.

Recently doubt has been cast upon the suggestion that optical strain in quartz relates directly to the reactivity of the quartz grains in the rock[33]. Grattan-Bellew suggests that the lattice defects in quartz grains only show up as strain shadows once they have migrated into zones as a result of applied

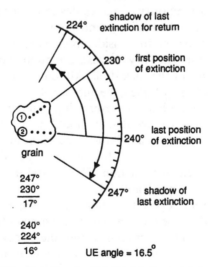

Figure 1.3 Illustration of the procedure for determining the undulatory extinction (UE) angle. From Ref. 32.

stress. Thus, provided no annealing has occurred, the extent of strain shadow development will be an indication of the level of geological stresses the rock has been subjected to. The principal causes of geological stresses on rock are tectonism and metamorphism. These effects are usually also associated with recrystallisation of the mineral species within the rock, which for quartz begins with incipient recrystallisation at grain margins to produce new crypto-crystalline quartz with a grain size which is close to the resolution limit of the optical microscope. It has been suggested that the severity of strain shadow development in quartz in favourable circumstances may be correlated with stresses applied to the rock, and this stress level in turn may be correlated with the development of cryptocrystalline quartz at grain margins. If it is assumed that the cryptocrystalline quartz is the alkali-reactive component, this provides an explanation for the observations concerning strain shadows and reactivity.

1.5.1 *The role of moisture in alkali–silica reactivity*

There are many reports of observations that those parts of a concrete structure which are exposed to the elements are more severely cracked and damaged by the effects of alkali reactivity than other parts of the structure which are protected from the weather. There are cases where there is a noticeable difference in degree of surface damage between the weather and lee sides of the same structure. A commonly used method of visually assessing surface damage to concrete elements of structures is to use an arbitrary scale of points, 1 to 5 for example, ranging from undamaged to severely deteriorated. Such a scale requires a series of 'benchmark' examples and is rather subjective; however, where alkali–silica reactivity has caused damage there is typically a difference of one to two points between the weather and lee side of a structure. It may well be that cyclic wetting and drying of an exposed surface contributes, by 'pumping', to the movement of the alkalis towards the sites of reaction though moist environmental conditions alone undoubtedly aid the progress of reaction.

As with most chemical reactions, alkali–silica reactivity requires water to proceed. However, water has a dual role in this particular case: firstly, it is essential as the 'carrier' of alkali cations and hydroxyl ions and, secondly, it is absorbed by the hydroscopic gel which swells, developing pressures sufficient to crack the concrete.

Concrete, even in dry conditions, will retain pore fluids, so that with the exception of a dried outer layer up to a few tens of millimetres thick the relative humidity (RH) within the concrete will remain at 80–90%.

Experiments and case study investigations have been able to demonstrate that the effects of alkali–silica reaction such as expansion vary directly with the percentage relative humidity of the concrete. The type of relationship is illustrated in Figure 1.4. It can be seen that below 70% RH expansion and

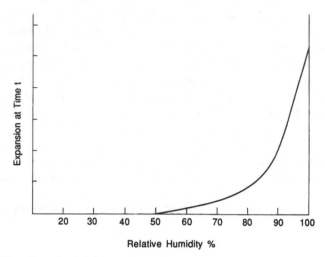

Figure 1.4 The effect of relative humidity on the expansion of concrete due to alkali–silica reaction.

expansive reaction are negligible, but above 80% RH the expansive effects are seen to increase dramatically. The two-stage process of reaction and of absorption of water by the gel reaction product can also be demonstrated in the laboratory. If an alkali–silica reactive concrete prism is prepared in the laboratory but not stored in a humid environment, expansion due to development of gel will be negligible. However, if such a specimen is removed to humid storage conditions, rapid expansion will ensue such that within a few days it will have expanded to approximately the same extent as an identical prism which had been stored in moist conditions throughout. These observations imply that, although reaction may occur at relatively low humidities, high moisture levels are required in order for the gel formed by the reaction to absorb water and increase in volume.

1.6 The concept of the 'pessimum' proportion

Provided the reactants, alkalis and appropriate forms of silica, are present in sufficient local concentrations within the concrete, reaction will occur. Evidence from numerous studies has shown that there is a certain concentration or proportion of reactive aggregate present in an otherwise inert aggregate that causes a maximum expansion in concrete made from it. This proportion of reactive material producing a maximum expansive effect is referred to as the 'pessimum proportion'[34,35]. Similarly, according to Ozol, there is also a 'pessimum proportion' for alkali content which will give maximum expansion. Variations above or below that value reduce or inhibit the expansion.

The pessimum proportion relationship can be illustrated using a graphical model in which the expansions for a series of similar concrete prisms are recorded. The prisms all contain a fixed concentration of alkali but contain a range of concentrations of a reactive aggregate constituent. The reaction will proceed and produce expansion as the reactive aggregate constituent and alkali become used up. The balance of the alkali remaining in the cement paste after a given time will be different for each prism, with highest concentrations in the prisms containing least reactive aggregate and decreasing as the percentage of reactive aggregate increases. As time proceeds reaction and expansion continue, so that for prisms with small percentages of reactive material the aggregate becomes used up with consequent changes in the ratio of alkali to reactive constituent. Similarly, with high concentrations of reactive aggregate, it is the alkalis that become used up, with the ratio changing in the opposite direction to the first case. If the reaction, and hence expansion, is considered to progress according to the balance of alkalis to reactive constituent, a model for the form of a pessimum proportion situation can be illustrated as in Figure 1.5. The diagram illustrates the simplest case where the reaction rate and expansion are being controlled by the ratio of alkali to silica or silica to alkali, both reactant concentrations being equally important for the reaction.

Pessimum proportions obviously do not and need not occur at the 50% level. If a reactive aggregate has a higher rate of removal of alkalis than the simple case illustrated, each unit increase in reactive constituent will remove a larger proportion of the starting concentration of alkali. This will effectively move the pessimum to the left, as illustrated in Figure 1.6.

Most research studies concerned with pessimum proportions deal with a range of concentrations of reactive aggregate material and with a single alkali level only in all the concrete or mortar samples considered. Variation in alkali concentration should, of course, produce similar pessimum effects to the variation in reactive aggregate component. Hobbs[36] suggests that both alkali concentration and percentage of reactive aggregate component should be considered together and that there will be a 'critical ratio' of water-soluble

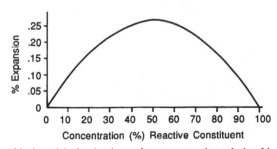

Figure 1.5 Graphical model of a simple pessimum proportion relationship. From Ref. 34.

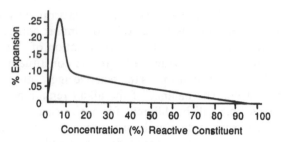

Figure 1.6 The pessimum proportion model showing the effect of a high alkali removal rate. From Ref. 34.

alkali to reactive silica in order to produce expansion and cracking of a mortar specimen in a minimum time period. Such a 'critical ratio' will depend on the nature of the reactive material in the mortar or concrete and the mobility of pore fluids within the sample. This mobility will, of course, depend on the porosity and permeability of the concrete, and this in turn will depend on a number of factors, but perhaps most important among these will be the water–cement ratio.

1.7 Gel reaction product

The interaction between alkali and reactive silica in the presence of moisture produces an alkali–silica gel which swells on the absorption of water. This gel varies considerably in composition, depending on the composition of the alkali pore fluids and probably on the nature of the particular form of reactive silica, temperature of reaction and concentration of reactants.

When first exposed by cracking open a sample of affected concrete, the gel is usually transparent and resinous in appearance, with a viscosity between that of thick motor oil and resin. There are considerable variations in viscosity between samples of gel. Some examples have clearly been sufficiently fluid to flow along cracks and fill or part-fill voids in the concrete. In some cases examples of half-filled voids indicate the original orientation of the sample with respect to gravity. Typically, gels carbonate with time and exposure to air, becoming white and hard with desiccation cracks similar to those observed in thin layers of dried mud. Often drying of the gel on the surface of a concrete renders it much less visible so that research workers are well advised to look for gel immediately upon breaking a fresh surface on a suspect concrete sample. Cores taken from structures affected by alkali–silica reaction are commonly wrapped in layers of clingfilm and sealed in polythene while still moist from the coring process. This procedure typically allows gel (if present) to be identified on the surface of the core when it is unwrapped, because the hygroscopic nature of the gel causes it to produce shiny or 'sweaty' patches on the surface which disappear as the core dries out.

Analyses of gel indicate that a wide range of compositions are possible, though care is necessary for their correct interpretation because the analysis is usually made using an electron probe microanalyser which requires a completely desiccated specimen. Also many gel samples are partially carbonated by exposure to air before analysis. A further point to note is that many modern computer-corrected microanalyses are also normalised to 100% for the elements determined, with those elements below atomic number 11 such as carbon not being detected. Table 1.2 has been compiled from a number of sources and gives an idea of the range of gel compositions reported.

Gel viscosities also exhibit a wide range of values, and can produce pressures extending up to 11 MPa. It is clear that as an alkali–silica gel first develops at a reaction site within a concrete it will at first be very viscous, but as it absorbs moisture from the pore fluids and swells it will gradually become less viscous until eventually it will become fluid enough to flow along cracks, fill voids and permeate the micropores in the cement paste.

Table 1.2 Some chemical analyses of alkali–silica gels (weight per cent).

Na_2O	K_2O	SiO_2	CaO	MgO	Difference from 100%	Reference
12.9	—	53.9	2.9	0.6	29.8	Stanton (1942)
12.9	—	53.4	2.6	0.8	30.2	
14.9	5.2	61.7	0.6	—	17.6	Idorn (1961)
13.4	5.1	65.5	0.5	0.2	15.3	
12.4	4.9	69.9	0.3	0.5	12.0	
17.9	8.2	73.7	1.1	0.1	0.0	
9.4	4.1	72.8	1.3	0.2	11.2	
14.6	6.2	61.9	—	0.1	17.2	
16.2	5.7	56.8	—	—	21.3	
8.2	4.1	56.1	17.4	0.2	14.0	
8.3	5.0	28.5	22.4	0.2	35.6	
1.2	0.4	51.4	29.9	10.0	17.1	Poole (1975)
7.4	0.7	53.0	22.1	10.0	16.7	
1.5	13.9	38.9	27.3	—	17.4	Gutteridge and Hobbs (1980)
0.4	4.7	51.1	21.5	—	22.3	Regourd (1983)
—	0.6	27.9	35.2	—	36.2	
1.0	6.9	61.5	9.2	—	21.3	Oberholster (1983)
1.0	6.2	53.8	8.2	—	30.7	
1.8	5.5	49.9	12.8	—	29.9	
1.0	5.2	50.4	12.0	—	30.7	
1.4	9.0	62.9	12.5	—	13.8	
0.8	7.4	53.2	10.0	—	28.4	
1.2	4.1	66.5	6.5	—	21.7	Baronio (1983)
3.7	12.9	43.3	21.8	0.8	14.0*	Mullick and Samuel (1986)
3.9	11.7	49.4	15.9	0.5	16.7†	

* Loss on ignition: Al_2O_3 2.8%, Fe_2O_3 0.7%.
† Loss on ignition: Al_2O_3 1.8%, Fe_2O_3 0.5%.

The relationships between the composition of the gel, its viscosity and ability to swell with absorption of water, and the pressure it will develop as it swells, are complex and not clearly understood at the present time. The relatively few studies that have been undertaken are principally concerned with simple sodium silicate gels. Using the few published vapour pressure values for the Na_2O–SiO_2–H_2O system, Moore[37] was able to plot the viscosities for various compositions, as indicated in Figure 1.7. Diamond[38] studied sodium silicate gels experimentally and found that sodic gels which swelled on exposure to water under free-swelling conditions were of two types, very high-swelling types (60–80%) and quite low-swelling types (less than 4%), but with no clear correlation with the sodium to silica ratio. Calcium-containing gels were found only to expand moderately. In measuring the swelling pressures of synthetic sodium silicate gels he also found that they fell into two groups, those with swelling pressures of between 4 and 11 MPa and others exerting pressures of less than 0.5 MPa. Again there was no clear correlation with the chemical composition or with the high and low grouping for freely swelling gels. The effects of composition on free-swelling gels and on the pressure gels developed are illustrated in Figures 1.8 and 1.9. An

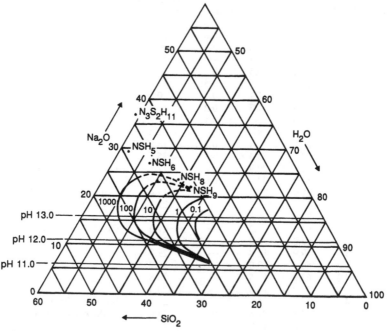

Figure 1.7 Calculated relationship between gel composition and viscosity. From Moore, A.E. (1978) *Proceedings of the Fourth International Conference on the Effects of Alkalis in Cement and Concrete*, p. 367.

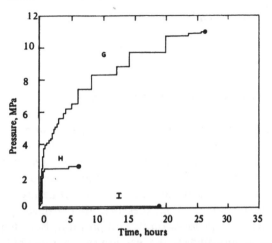

Figure 1.8 Pattern of swelling pressure development of synthetic alkali–silica gels. Gel compositions are as follows (%): (G) $Na_2O:SiO_2$, 0.34 (moderately sodic gel); (H) $Na_2O:SiO_2$, 0.30 (moderately sodic gel); (I) $Na_2O:SiO_2$, 0.42 (highly sodic gel). From Ref. 38.

unexpected effect observed by Diamond was a pronounced ageing effect. Gels tested within a few days of preparation when retested after 4 months' sealed storage at room temperature were found to have changed from high-expansion gels (63%) to practically inert gels (less than 2% expansion). Clearly the behaviour of gels within a concrete undergoing alkali–silica reaction is most

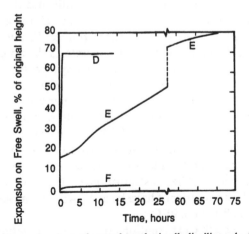

Figure 1.9 Free-swelling expansions of several synthetic alkali–silica gels. Gel compositions are as follows (%): (D) $Na_2O:SiO_2$, 0.53 (highly sodic gel); (E) $Na_2O:SiO_2$, 0.27 (moderately sodic gel); (F) $Na_2O:SiO_2$, 0.30 (moderately sodic gel). From Ref. 38.

complex and not as dependent on chemical composition as was originally thought.

1.8 Controlling factors in alkali–silica reaction

The appropriate relative proportions of the reactive components of alkali in the pore fluids and reactive silicious material in the aggregate particles, together with sufficient moisture, are all essential requirements for the initiation and progression of alkali–silica reaction in concrete. The composition of the gel is dependent in part on the exact nature and proportions of the reactants, but the swelling pressures that it develops as it absorbs water are only partly dependent on its composition. Both temperature and rate of migration of pore fluids to the reaction site are thought to be important factors which relate to swelling pressures and hence expansion and disruption of the concrete.

Most chemical reactions are speeded up by increased temperature, as can be demonstrated qualitatively in the case of alkali–silica reaction by means of the Jones and Tarleton[39] 'gel pat' test, in which particles of aggregate are exposed to strong alkali solution on the surface of a cement tablet. At ambient temperature highly alkali-reactive materials, such as opal, develop gel on their surfaces within a few days of storage in alkali. If the temperature of storage is raised to, say, 50°C, the development of gel on the surface of such particles will occur within 24 hours and certain less reactive materials, such as some types of flint, will begin to show gel development, though at more normal temperatures no sign of reaction can be observed[40].

A clear indication that temperature can also influence the second part of the reaction which involves the absorption of water by the gel, causing it to develop the swelling pressures which in turn produce the microcracking and hence the expansive disruption within the concrete, is shown by the influence of temperature on laboratory expansion tests. Many reports are now available detailing similar effects to those illustrated by Diamond[38] in Figure 1.10 shown here. At high temperature reaction and expansion are initiated early and develop rapidly, but as reaction continues both rate of reaction and rate of expansion slow down, as in Figure 1.10. By contrast, reactive concretes and mortars stored at lower temperatures react more slowly, but eventually expansion reaches the same level and then exceeds the expansions attained at the higher temperature. The illustration given is for expansions at 20°C and 40°C, but similar comparisons have been drawn for various temperatures in the range between 8°C and 50°C.

There are several possible explanations for these observations, both probably contributing to the observed effect. At lower temperature the reaction proceeds more slowly. The migration of alkalis to the reaction sites will also

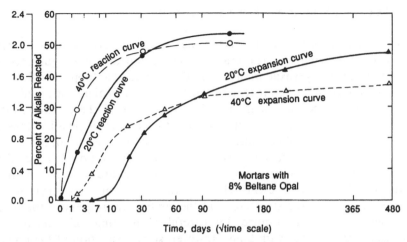

Figure 1.10 Comparison of reaction parameter curves versus expansion curves for sealed mortar specimens at 20°C and 40°C. From Ref. 38.

be slower so that the gel expands more slowly. Also there must be a stage at which the water content of the gel reaches a level which will give a maximum swelling pressure, after which the gel will become increasingly fluid and able to migrate along cracks without exerting significant pressure. At the lower temperature the period at which gel exerts its maximum pressure will be prolonged, hence providing greater measured expansions. An alternative explanation, for which there are no experimental data at the present time, is that the structure or composition of gels varies for different temperatures, with the low temperature varieties capable of producing the highest swelling pressures.

The influence of temperature variation and cycling on the reaction and expansion is still a matter for debate. Effects resulting from rapid variation in temperature, if present, will only relate to the outermost skin and exposed parts of a concrete structure, since changes of temperature within a concrete structure in normal circumstances will be limited and slow. Measurements made of concrete temperatures within large structures such as dams typically only show very small changes associated with the seasons.

The controlling influence of moisture availability has already been noted. The alkali–silica reaction appears to be capable of forming gels of low moisture content initially which will swell and exert pressures leading to expansion immediately water becomes available. There is also evidence that at least under laboratory conditions partially dehydrated silica gel can be rehydrated and will re-expand when additional water is added to the specimen. However, dried gel that has become white and carbonated cannot be reconstituted and is not readily soluble in water.

1.9 The observed effects of alkali–silica reaction

Concrete structures affected by alkali–silica reaction typically show at least one of the following features: cracking, expansion and misalignments of structural elements, gel exudations and spalls referred to as 'pop-outs'. Unfortunately, none of these features can be said to be uniquely diagnostic of alkali–silica reactivity since, particularly at field inspection level, these features may be considered to be the result of any one of a number of possible causes. Thus, although a careful field inspection of a structure is essential, the observations made can at best only give an indication of the possibility of deleterious alkali–silica reaction being present in the structure. The confirmation of the reaction being present relies on the laboratory investigation of samples taken from the structure, and the continued progress of the reaction can only be determined by careful monitoring of dimensional changes in the structure with time. However, some indication of the potential for future reactivity can be obtained from laboratory studies, in particular the continued expansion of cores taken from the structure and stored under controlled conditions of temperature and humidity.

The most common field observation relating to alkali–silica reactivity is the development of cracks in the concrete surface. If the concrete element is not subject to directional stress, the crack pattern developed forms irregular polygons with boundaries reminiscent of the political boundaries on a map, and is sometimes referred to as 'map cracking' for this reason. Typically these cracks show margins that are bleached, pinkish or brownish in colour and extend several millimetres each side of the crack. An example of this map cracking is shown in Figure 1.11.

The cracks will first form as a three- or four-pronged star pattern resulting from the expansion of gel around a reacting aggregate particle beneath the surface of the concrete, the cracks relieving the stress because of the local expansion. As other particles react more cracks develop until they join up to form the 'map cracking' shown in Figure 1.11. The bleaching or discoloration often associated with these cracks is probably due to the strong alkali solutions which permeate along them, killing and bleaching lichen and fungus growths which occur in the vicinity of the crack on the surface of the concrete.

The sequence of cracking as the reaction in the concrete progresses is first the development of three- or four-pronged star cracks (sometimes referred to as 'Isle of Man' cracks), then the coalescence into a 'map cracking' pattern. At later stages new cracks continue to develop, subdividing the original polygonal areas between cracks into smaller polygons, and in addition existing cracks widen and lengthen.

If a concrete element is affected by alkali–silica reaction sufficiently severe to cause cracking but is under pre-existing stress, then instead of developing map cracking cracks orientated with respect to the stress directions will

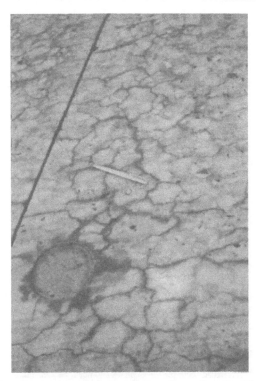

Figure 1.11 A portion of concrete road pavement, Cape Town. Length of the pencil is 125 mm.

develop. An example that illustrates this effect is that of a bridge support column (Figure 1.12), the load of the bridge deck on the column being sufficient to induce vertical cracking in the column. This particular column was eventually removed from the bridge and cut into sections. Once the loading was removed, map cracking developed on the surface of the column as the alkali–silica reaction continued.

Since deleterious effects due to alkali–silica reactivity are largely the result of expansion within concrete elements in the structure, and since these effects are usually uneven in their severity, dimensional misalignments are a typical feature of the reaction. In many case study examples certain concrete pours are more badly affected than others, and in the case of the Val de la Mare Dam expansion of one of the concrete segments has caused it to move upstream by 50 mm compared with neighbouring segments in the dam, as is illustrated in Figure 1.13.

Other features indicative of reaction are the exudation of alkali–silica gel from cracks in the faces of concrete structures affected. This gel, when fresh, will be transparent or brownish and resinous in texture. On dehydration it becomes white, sometimes with curving desiccation cracks. Such hardened

Figure 1.12 A section of a road bridge support column at the Mott MacDonald test exposure site, Devon.

Figure 1.13 Crest parapet, Val de la Mare Dam, Jersey.

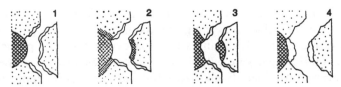

Figure 1.14 Idealised development of a 'pop-out'. 1. Bond loosened between pebble and adjacent mortar; 2. Rupture along inner boundary of reaction rim in pebble; 3. Rupture through interior of pebble without reaction rim; 4. Area of rupture hidden by extensive exudations of gel on broken face of pebble. From Ref. 5.

gel deposits are very similar to and are often mistaken for the very much more common calcium carbonate deposits resulting from the leaching of calcium hydroxide from the cement paste and its subsequent carbonation when it comes into contact with the air at the surface of the concrete.

One other feature often symptomatic of alkali–silica reactivity is the development of circular or elliptical spalls from the concrete surface. As with the other features noted above, spalling of surface concrete can also arise for a variety of reasons, and care is needed to identify the particular mechanism for a given example. The development of gel at a particular reactive aggregate particle which is close to the surface of the concrete causes excessive pressure to develop as it expands. Thus a circular area is spalled from the surface of the concrete as a shallow semicone whose apex centres on the point of reaction. Such a feature is illustrated in Figure 1.14. In good examples the presence of gel associated with the reactive aggregate may be observed at the apex of the spall or 'pop-out'.

1.10 Tests for the identification of alkali–silica reactivity in concrete

Numerous tests have been devised to ascertain whether a particular aggregate or concrete mix will develop alkali–silica reactivity. These tests centre on checks concerning either the aggregate alone or the concrete mix or a simplified variant of it.

A number of tests have been introduced concerning the direct evaluation of the aggregate materials which, if applied with care, identify alkali-reactive materials in aggregate particles satisfactorily. Petrographic examination using a polarising microscope with thin sections cut from the aggregates is becoming increasingly important as a first step in identifying components which may be potentially reactive. The method does not confirm that such materials will be reactive and deleterious in a concrete so other tests need to be applied if suspect materials are observed in the aggregate particles.

Opal and opaline materials are often difficult for the petrographer to identify when in association with other silicious materials or forming a cement between grains. In such circumstances the Jones and Tarleton[39] 'gel pat' test provides a reliable guide concerning the alkali reactivity of a particle. It is particularly useful for identifying which particular types of particles are reactive in a mixed aggregate but, like the petrographic method, if the test is used with care it will identify alkali-reactive particles but it cannot provide information as to whether the reactive particles identified would produce deleterious effects in a concrete. One of the most commonly used direct tests for potential reactivity of aggregates is the ASTM C289-66 test. In this test aggregate samples are stored at 80°C for 24 hours in 1 N sodium hydroxide. After this treatment the dissolved silica from the aggregate and the reduction in alkalinity of the solution are determined. These two results are then plotted against each other on a graph. This graph is divided into three fields: 'innocuous', 'potentially deleterious' and 'deleterious'. Thus a particular aggregate can be classified according to its plotted position. Although the test is rapid and reasonably simple to use, many results fall in the potentially deleterious field, leaving room for uncertainty concerning their use. Also, a few results reported for cherts in the UK appear to plot as deleterious, although these aggregates have a satisfactory service record.

The development of gel from a reactive aggregate stored in alkali would appear to be a sensible criterion for establishing a particular aggregate's potential for reaction. If a sample of suspect aggregate is crushed and treated with alkali, it is possible to identify the gel if it develops and to quantify the amount produced in a given time by arranging for it to absorb coloured cuprammonium ions[41]. Even so this test does not indicate whether the aggregate would prove deleterious in practice.

Direct tests on the aggregate materials do not provide information concerning their use in a concrete, and as a result a number of tests have been devised or are currently under investigation which make use of mortar or concrete mixes containing the suspect materials, or the concrete mix it is proposed to use, and casting these as prisms, cylinders or cubes. After curing, these specimens are stored under controlled conditions of temperature and humidity and are monitored for dimensional change, development of cracks or changes in other physical characteristics such as ultrasonic pulse velocity.

Until recently the American ASTM test C227-69 was widely used; this monitors length change in 25-mm^2-section, 250-mm-long, prisms of mortar made with crushed and graded aggregate and stored at 38°C for up to 6 months. More recently developed new test methods have tended towards larger, thicker prisms of a concrete mix which attempts to reflect the behaviour of actual concretes to be used more closely. Nearly all these tests monitor expansion of the mortar or concrete as a length change and, although often storage is at elevated temperature, the tests take a considerable length of time to complete and considerable care is required in interpretation of results.

Recently autoclaving methods have been used in Japan and China in order to speed up the expansion and hence shorten the test period[42,43]. These methods appear to work satisfactorily, but it has been suggested that the extreme storage methods used make correlation of their results with actual in-service performance of concrete structures unreliable.

1.11 Conclusions

The problem of concrete deterioration resulting from alkali–silica reaction between alkalis from the pore fluids and reactive silicious components within the concrete itself is of worldwide significance. Although the number of cases of serious deterioration is very small indeed compared with the number of concrete structures that have been built, no continent, and few countries of the world, have no examples of affected structures. Figure 1.15 indicates on a world map those countries that have reported problems with concrete structures arising from this cause. The effects of the reaction normally do not appear for a number of years after construction, 5–15 years is perhaps most common, but some reported examples have not exhibited distress for 25–40 years post construction. This implies that many more structures will begin to show signs of deterioration in the coming decades. Most countries are now well aware of the possibilities of problems arising from alkali–silica reactivity and take precautions concerning both selection of aggregates and alkali contents of cements. Thus it is to be hoped that deterioration due to this particular cause will slowly decline and disappear in the future.

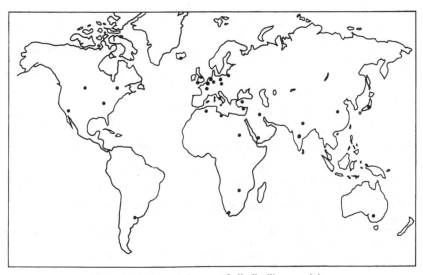

Figure 1.15 Recorded cases of alkali–silica reactivity.

References

1. Stanton, T.E. (1940) Expansion of concrete through reaction between cement and aggregate. *Proc. ASCE* **66**, 1781–1811.
2. Blanks, R.F. (1941) Concrete deterioration at Parker Dam. *Eng. News. Rec.***126**, 462–465.
3. Meissner, H.S. (1941) Cracking in concrete due to expansion reaction between aggregate and high-alkali cement as evidenced in Parker Dam. *Proc. Am. Conc. Inst.* **37**, 549–568.
4. Swenson, E.G. (1957) A reactive aggregate undetected by ASTM tests. *ASTM Bulletin* No. 226, 48–57.
5. Idorn, G.M. (1967) *Durability of Concrete Structures in Denmark*. Danish Technical Press, Copenhagen.
6. Asgeirsson, H. (Ed.) (1975) *Proceedings of a Symposium on Alkali–Aggregate Reaction, Preventive Measures*. Icelandic Building Research Institute, Reykjavik.
7. Poole, A. B. (Ed.) (1976) *Proceedings of the Third International Symposium on the Effect of Alkalis on the Properties of Concrete*. Cement and Concrete Association, Wexham Springs, Slough, UK.
8. Diamond, S. (Ed.) (1978) *Proceedings of the Fourth International Conference on the Effects of Alkalis in Cement and Concrete*. Publication No. CE-MAT-1-78. School of Engineering, Purdue University, W. Lafayette, USA.
9. Oberholster, R.E. (Ed.) (1981) *Proceedings of the Fifth International Conference on Alkali–Aggregate Reaction in Concrete*. National Building Research Institute of the Council for Scientific and Industrial Research, Pretoria, South Africa.
10. Idorn, G.M. and Rostam, S. (Eds.) (1983) *Proceedings of the Sixth International Conference on Alkalis in Concrete, Research and Practice*. Technical University of Denmark, Copenhagen.
11. Grattan-Bellew, P.E. (Ed.) (1986) *Proceedings of the Seventh International Conference on Concrete Alkali–Aggregate Reactions*. Noyes Publications, Far Ridge, NJ.
12. Newlon, H.H. and Sherwood, W.C. (1962) An occurrence of alkali-reactive carbonate rock in Virginia. *Highway Res. Board Bull.* **355**, 27–32.
13. Axon, E.O. and Lind, J. (1964) Alkali–carbonate reactivity—academic or practical problem? *Highway Res. Record*, No. 45, 114–125.
14. Welp, T.L. and De Young, C.E. (1964) Variations in performance of concrete with carbonate aggregates in Iowa. *Highway Res. Record* No. 45, 159–177.
15. Hadley, D.W. (1964) Alkali reactivity of dolomitic carbonate rocks. *Highway Res. Record* No. 45, 1–20.
16. Alsinawi, S.A. and Murad, S. (1976) On the alkali–carbonate reactivity of aggregates from Iraqi quarries. *Proceedings of the Third International Symposium on the Effect of Alkalis on the Properties of Concrete* pp. 255–265.
17. Hussen, T.K. (1975) The natural aggregate materials of Bahrain and their influence on concrete durability. *PhD Thesis*, University of London.
18. Rogers, C.A. (1965) Alkali aggregate reactions, concrete aggregate testing and problem aggregates in Ontario, a review. Ontario Ministry of Transportation and Communications. Engineering Materials Office, Canada.
19. Gillott, J.E. (1975) Alkali–aggregate reactions in concrete. *Eng. Geol.* **9**, 303–326.
20. Walker, H.N. (1978) *Chemical Reactions of Carbonate Reactions in Cement Paste*. ASTM STP 169-B, 722–743.
21. Poole, A.B. (1981) Alkali reactions in concrete. *Proceedings of the Fifth International Conference on Alkali–Aggregate Reaction in Concrete*, S252/34, pp. 1–7.
22. Grattan-Bellew, P.E. (1978) Study of expansivity of a suite of quartzwackes, argillites and quartz arenites. *Proceedings of the Fourth International Conference on the Effects of Alkalis in Cement and Concrete*, pp. 113–140.
23. Millet, J. C. (1985) Quelques problèmes actuels de conception, construction, entretien et confortement. *Journées d'Etudes* No. 256.
24. Hansen, W.C. (1944) Studies relating to the mechanism by which alkali–aggregate reaction produces expansion in concrete. *J. Am. Concr. Inst.* **15**, 213–217.
25. Vivian, H.E. (1951) Studies in cement aggregate reaction. XVI. The effect of hydroxyl ions on the reaction of opal. *Aust. J. Appl. Sci.* **2**, 108–113.
26. Dent-Glasser, L.S. and Kataoka, N. (1981) The chemistry of alkali–aggregate reactions.

   Proceedings of the Fifth International Conference on Alkali–Aggregate Reaction in Concrete, S252/23, pp. 1–7.
27. Skalny, J. and Klemm. W.A. (1981) Alkalis in clinker: origin, chemistry, effects. *Proceedings of the Fifth International Conference on Alkali–Aggregate Reaction in Concrete*, S252/1, pp. 1–7.
28. Concrete Society (1987) *Alkali–Silica Reaction, Minimising the Risk of Damage to Concrete*. Guidance notes and model specification clauses. Technical Report No. 30.
29. French, W.J. (1986) A review of some reactive aggregates from the U.K. with reference to the mechanism of reaction and deterioration. *Proceedings of the Seventh International Conference on Concrete Alkali–Aggregate Reactions*, pp. 226–230.
30. Moore, A.E. (1978) Effect of electric current on alkali–silica reaction. *Proceedings of the Fourth International Conference on the Effects of Alkalis in Cement and Concrete*, pp. 69–72.
31. Diamond, S. (1983) Alkali-reactions in concrete—pore solution effects. *Proceedings of the Sixth International Conference on Alkalis in Concrete*, pp. 155–166.
32. Dolar-Mantuani, L.M.M. (1981) Undulatory extinction in quartz used for identifying potentially alkali-reactive rocks. *Proceedings of the Fifth International Conference on Alkali–Aggregate Reaction in Concrete*, S252/36, pp. 1–12.
33. Grattan-Bellew, P.E. (1986) Is undulatory extinction in quartz indicative of alkali-expansivity of granitic aggregates? *Proceedings of the Seventh International Conference on Concrete Alkali–Aggregate Reactions*, pp. 434–439.
34. Ozol, M.A. (1975) The pessimum proportion as a reference point in modulating alkali–silica reaction. *Proceedings of a Symposium on Alkali–Aggregate Reaction, Preventive Measures*, pp. 113–130.
35. Mielenz, R.C., Green, K.T. and Benton, E.J. (1947) Chemical test for the reactivity of aggregate with cement alkalis: chemical processes in cement aggregate reaction. *Am. Conc. Inst. Pr.* **44**, 193–224.
36. Hobbs, D.W. (1981) Expansion due to alkali–silica reaction and the influence of pulverised fuel ash. *Proceedings of the Fifth International Conference on Alkali–Aggregate Reaction in Concrete*, S252/30, pp. 1–10.
37. Moore, A.E. (1978) An attempt to predict the maximum forces that could be generated by alkali–silica reaction. *Proceedings of the Fourth International Conference on the Effects of Alkalis in Cement and Concrete*, pp. 363–365.
38. Diamond, S., Barneyback, R.S. Jr and Struble, L.J. (1981) On the physics and chemistry of alkali–silica reactions. *Proceedings of the Fifth Conference on Alkali–Aggregate Reaction in Concrete*, S252/22, pp. 1–11.
39. Jones, F.E. and Tarleton, R.D. (1958) *Reactions between Aggregates and Cement*. National Building Research Paper 25, Pt VI. HMSO, London.
40. Poole, A.B. and Al-Dabbagh, I. (1983) Reactive aggregates and the products of alkali–silica reaction in concretes. *Proceedings of the Sixth International Conference on Alkalis in Concrete*, pp. 175–185.
41. Poole, A.B., McLachlan, A. and Ellis, D.J. (1988) A simple staining technique for the identification of alkali–silica gel in concrete and aggregate. *Cement Conc. Res.* **18**, 116–120.
42. Tong, M., Hon, S., Zhen, S., Yuan, M., Ye, Y. and Lu, Y. (1986) Applications of autoclave rapid test method in practical engineering projects in China. *Proceedings of the Seventh International Conference on Concrete Alkali–Aggregate Reactions*, pp. 294–298.
43. Nishibayashi, S., Yamura, K. and Matsushita, H. (1986) A rapid method for determining the alkali–aggregate reaction in concrete by autoclave. *Proceedings of the Seventh International Conference on Concrete Alkali–Aggregate Reactions*, pp. 299–303.

2 Chemistry of the alkali–aggregate reaction

F.P. GLASSER

2.1 Introduction

The knowledge that some mineral aggregates are reactive with cement is not new. Indeed, this is the reason why many natural pozzolans can be activated by cement or lime. However, while many reactions occurring internally in cemented systems have beneficial consequences, or at least have no significant adverse effects, the alkali–aggregate reaction does give cause for concern. Its harmful effects arise from the expansive nature of the products of reaction, which lead to cracking. In turn, cracking may accelerate other mechanisms of deterioration, typically those likely to lead to loss of physical integrity, e.g. freeze–thaw, or those which arise from ingress of foreign ions: sulphate ions from saline waters, salts from de-icing run-off, etc. If, however, the alkali–aggregate reaction results in undesirable *physical* consequences, why study its *chemistry*? The answer to this question is, in general terms, straightforward: the physical process of expansion is generated as a consequence of chemical reactions. This potential for reaction is conditioned by the chemicals and mineralogical nature of the components of the concrete system—cement, aggregate, water, etc.—as well as by its conditions of service—temperature, humidity, etc. The magnitude of the expansive potential is, however, complex and as yet poorly understood, arising from a mixture of chemical and physical causes: factors such as the microstructure of the paste and the sizing and distribution of the reactive aggregate influence the magnitude of the observed expansion. It is important, therefore, to recall that the chemical and physical components of the reactions are virtually inseparable. However, the basic origins of the chemical potential involved in alkali–aggregate reactions is also clear: certain siliceous materials which occur in natural mineral aggregates react with the alkaline components of cement, and the increased molar volume of the products, relative to those of the reactants, creates swelling pressures. When these pressures cannot readily be accommodated or restrained within hardened concrete, it expands with cracking. The kinetics of the alkali–silica reaction are not well understood, and quantitative characterisation of the potential for further, continuing, reaction at any particular point in time has not been achieved. As a consequence it is not yet possible to measure a single chemical parameter, or even several parameters, and correlate these with the expansive potential of the system. Thus it is not at present possible to relate chemical factors quantitatively to the prime engineering parameter—dimensional stability. But an understanding of the basic chemistry has been achieved.

2.2 Nature of concrete

Large masses of concrete used in engineered structures consist of three components: cement, aggregate and water. Each plays an essential role in forming a rigid, dimensionally stable solid. Cement, comprising Portland cement together with any pozzolanic addition, forms the essential binder. Examples of pozzolanic additions include fly ash (PFA) resulting from coal combustion, glassy blast furnace slag (BFS), silica fume (SF) and various naturally occurring pozzolanic materials. They have certain characteristic features relevant to the alkali–aggregate reaction which will be described in more detail subsequently.

Water is the activating agent for hydration; it is essential since the precursor binder materials are anhydrous, or nearly so, while the set product consists principally of a gel-like calcium silicate hydrate, of which water is an essential component. This gel-like hydrate has a variable composition; its chemical Ca/Si ratios lie in the range 1.9–0.9 approximately although for Portland cement the reported values cluster in the range 1.9–1.7: lower ratios are only achieved in the presence of blending agents. This gel-like material is often designated in shorthand notation as C–S–H, thereby indicating its principal components (C = calcium, S = silica, H = H_2O, water) but without being specific as to its precise composition. Other solid phases are also present. Modern Portland cements develop much $Ca(OH)_2$ in the course of hydration, although its quantity may be reduced by the presence of blending agents. Other phases are also present but apparently play little role in the alkali–aggregate reaction.

Water may also contain dissolved salts. Most national codes lay down strict standards for mixing-water quality, but these standards often have to be relaxed, as in offshore oil constructions, where salt water may be used, or in desert climates where only saline waters may be available.

Typical aggregates used for making concretes are graded according to size, so as to achieve good space filling and economy of cement. The aggregate has other important functions; for example, it helps control the shrinkage which would otherwise occur in cement-rich formulations. The setting of cement is strongly exothermic, i.e. it liberates heat, and large masses of neat cement would experience unacceptable thermal cycling during their initial set and strength gain; by acting as a heat sink, and also by lowering the total cement content per unit volume, the aggregate fulfils an important role during both early and late hydration stages.

Since aggregates are drawn from various sources, depending on local conditions, a wide variety of rock types are used. Examples of quarried rocks, typically used as coarse aggregates, range from carbonates, e.g. limestone, to igneous rocks such as granites. However, finer aggregate materials, e.g. 'sand', are typically obtained from surficial deposits. These may have been laid down by water, by wind transportation, or even by glacial action. Such aggregates are likely to have an uncertain geological history and complex mineralogical constitution. However, since the average chemical composition of the access-ible portions of the earth's crust is about 60% SiO_2 by weight, it is not surprising that free SiO_2 in one or more of its polymorphic variants (e.g. quartz) should be abundant in natural sands. The quartz polymorph of SiO_2 is especially abundant on account of its insolubility and hardness, which makes it resistant to abrasion and degradation. Of course other silicates occur in sands: principally, these comprise minerals formed with SiO_2, but containing in addition one or more common oxides such as those of aluminium, iron, sodium, potassium, magnesium and calcium.

2.3 Chemical and mineralogical nature of cement

The Portland cement which is commonly used in concrete is a manufactured product, made by calcination at high temperatures of naturally occurring raw materials. Limestone and clay are often used; they are intimately ground and fed into a rotary kiln. The kiln is frequently fitted with a suspension preheater to reduce the overall energy requirements of the process. A complex series of chemical reactions ensue as the feed materials are progressively heated; these reactions involve dehydroxylation, decarbonation and solid-state reactions among the oxide components. At the highest calcination temperatures, typ-ically $\sim 1450°C$, some melting of the kiln charge occurs and new mineral phases, not present in the original feed, develop in response to the high-temperature regime. Reaction continues in the solid state but is greatly facili-tated by transport through a reactive high-temperature melt. The resulting product, known as clinker, is finally cooled to ambient and ground to a high specific surface area, typically 3000–5000 cm^2/g. The microstructure and mineralogy of clinker are characteristic of a manufactured product but also show certain affinities to those observed in igneous rocks.

Although the chemical CaO content of Portland cement is very high, typically 63–69% by weight, the fired product contains little free lime, typi-cally 1–3%, and consists essentially of four phases: two calcium silicates,

Ca_3SiO_5 and Ca_2SiO_4, one aluminate, $Ca_3Al_2O_6$, and an aluminoferrite, $Ca_2(Fe, Al)_2O_5$. These formulae are somewhat idealised. Other minor components also occur in the clinker feed—MgO, Na_2O, K_2O, manganese oxide, etc. These components may occur as solid solutions in one of the four principal phases, or else may be present in discrete phases. The behaviour of the alkalis, Na_2O and K_2O, are especially important in the present context. Owing to the high calcination temperatures used, some material transport also occurs in the kiln by vapour phase transport as well as by the slow movement of the solid clinker. CaO, Al_2O_3, iron oxides and SiO_2 have low vapour pressures, and are transported with the clinker but others—such as sodium, potassium, carbon and sulphur oxides—have significant vapour pressures at clinkering temperatures. The result is that the hot vapour stream flowing through the kiln is rich not only in water and carbon dioxide, but also in sulphur, sodium and potassium species. The vapour pressures of some of the sulphates in cement clinkering have been restudied and reviewed[1], and it is apparent from these studies, as well as from mineralogical examination of the nature of the solid condensates which form on cooling clinker, that the circulations of alkali and sulphur in the kiln are linked. The vapour constitution of the kiln atmosphere is complex, but it appears that at some points in the clinkering cycle evaporation predominates, while at other points, notably where the gas stream is cooling (as in suspension preheaters), condensation occurs. When the vapour attains its dew point it becomes supersaturated with respect to alkali sulphates, which then condense onto the cooling clinker, coating the surface of clinker grains, which are removed from the kiln along with the clinker.

The alkali content of clinker may thus be differentiated broadly into two generic types: alkali which is normally present condensed on the surfaces of clinker grains, perhaps as sulphate salts, and alkali which is locked into the crystal structures of the clinker minerals. All the clinker minerals have some potential to retain alkalis in solid solution, but much of the sodium appears in C_3A, while potassium tends to be more evenly distributed, with C_2S (belite) and associated glass playing an important role as hosts.

When alkali-containing clinker is hydrated, surface and clinker alkalis behave differently with respect to their release rates. Alkalis present as water-soluble sulphates become available for solution almost immediately, whereas alkalis locked into clinker grains become available for solution rather more slowly in the course of hydration of their host minerals. The characteristic overall release rate, therefore, may be variable from one cement to another, depending on the distribution of alkali between rapid-release and slow-release sources as well as on total alkali.

However, the alkali–aggregate reaction itself typically occurs rather slowly and, over the period of time required for its action, virtually all the cement alkali, irrespective of source, is likely to become potentially available for release.

2.4 Chemical nature of blending agents

Three blending agents find extensive use in concrete technology. All three—silica fume, fly ash, and blast furnace slag—are by-products of industrial processes, and as a consequence their chemical and physical properties are somewhat variable.

Silica fume has perhaps the simplest constitution. It is derived from ferrosilicon smelting; silicon metal, which has a high vapour pressure at the temperatures used in the process, evaporates and on coming into contact with oxygen-containing gas is subsequently reoxidised to SiO_2. The resulting silica precipitate is X-ray amorphous and has a very high specific surface area, even when compared with a finely ground cement clinker. The chemical SiO_2 content of silica fume is typically $\geqslant 90\%$, the principal inert impurity being unburnt carbon together with lesser amounts of other evaporable elements: aluminium, magnesium, etc. Silica fume is, however, normally low in alkali.

The main source of slag used in blended cements derives from iron-making blast furnaces. These are typically operated at $\sim 1550°C$. When the furnace is tapped, the molten silicate slag is readily separated from denser, liquid crude iron with which it is immiscible. Once separated from metal, the slag is quenched or quickly chilled to preserve a high proportion of glass; normally, slags intended for blending with cement have at least 70% glass content, ranging upwards to $>98\%$. On account of the strongly reducing conditions obtaining in the furnace, the iron oxide content of the slag is low: its principal constituents are calcium, magnesium, aluminium and silicon oxides. However, some reduced sulphur (S^{2-}) also substitutes for oxygen in the glass 'structure'.

The chemical composition of the slag is somewhat variable and represents a metallurgical compromise. It must melt fully at $\sim 1550°C$, have good desulphurising power, be able to flux away unwanted oxide impurities in the ore, such as Al_2O_3 and SiO_2, yet give a minimum slag volume, since the energy required for its melting is essentially irrecoverable. Not surprisingly, therefore, blast furnace slags are rather variable in composition depending on plant practice, on the ore source, and on the metallurgical proportioning formulae which are employed. Alkalis are generally restricted in slags, since they may potentially damage the blast furnace refractories. Nevertheless, the alkali contents of slags are often high relative to the alkali contents of many cements; the total alkali content of slags frequently exceeds 1%.

Coal combustion products, such as fly ash, are even more variable than slags. Several symposia have been held which give accounts of their compositional and mineralogical variation[2-4]. Briefly, their composition and mineralogy reflect (i) the bulk composition of the ash fraction of the coal and (ii) the effect of a short thermal exposure, typically 4–8 seconds, during which the coal particle may attain 1600–1700°C, followed by rapid quenching. The more alumina- and silica-rich fly ashes (designated type F in the ASTM classification) tend to have a substantial glass content which is intimately mixed with crystalline components. Some of these minerals are primary, e.g. unmelted quartz, while others are secondary in the sense that they were developed either during the high-temperature excursion or during subsequent cooling of the ash. Typical secondary minerals include mullite, spinel and haematite. The pozzolanic activity of class F ashes is associated with their glass content, the crystalline minerals being inert or nearly so. The more calcium-rich ashes, designated type C, tend to contain rather less glass and develop more crystalline components. These crystalline phases often include essential calcium, e.g. melilite, dicalcium silicate, tricalcium aluminate, and these may exhibit substantial pozzolanic activity in their own right. Both types of fly ash may also contain substantial alkali. During combustion, some of the alkali in coal is evaporated, but of that which is evaporated some subsequently condenses on or near the surface of the ash particles. Some alkali may also be present in the crystalline phases, e.g. sodium in melilite. However, in aluminosilicate class F ashes much of the alkali is dissolved in the glass, and since this glass phase is eventually reactive with cement— indeed, this furnishes the basis of its pozzolanic behaviour—the bulk of the alkali content of class F fly ashes must be considered to be at least potentially available for release. In slags, where almost the whole of the alkali content is dissolved in the glass, the alkali must also be considered as potentially available for release. It should be recalled, however, that (i) the alkali may not be homogeneously distributed throughout the blending agent, as in certain fly ashes, where it may concentrate at or near the surface of particles, and (ii) the blending agent may react rather slowly with cement, as in the case of slag cement blends where considerable unreacted slag may persist for decades.

The release of alkali from slag is the simplest case to consider because its alkali content is evenly and homogeneously distributed. As hydration progresses, the alkali content of the glass is progressively available for potential release. 'Potential' release in this context signifies that release could occur during the course of slag hydration but of course, at the instant of release, alkali ions may either be bound into slag–cement hydration product or be released into the pore fluid or, more realistically, partitioned between aqueous pore fluid and solid hydration product. As slag frequently hydrates rather more slowly than cement, the process of alkali release may be spread over a longer timescale. The rate of hydration of slag in slag–cement blends is the subject of some disagreement, but most investigators agree that the initial rapid rate of hydration gives way to a regime of slow reaction, or even passivity, with the result that some unhydrated slag is likely to persist for decades[5].

The chemistry of release of alkali from fly ashes tends to be more complex; in an unwashed ash, much readily soluble alkali may be present at the surface of particles. However, in general, dissolution of the ash, which occurs over a few months or years, provides the same alternatives between release, retention in the hydrate phases or partition between the two. The alkali-binding power of the hydration products, and the influence of blending agents, will be discussed subsequently.

2.5 Some chemical and mineralogical features of aggregates

SiO_2 comprises about 65% by weight of the accessible portions of the earth's crust. Not surprisingly, therefore, SiO_2 occurs both 'free', that is as a solid crystalline oxide from which other essential elements are absent, as well as in silicates, in which silicon and oxygen are combined with other elements, of which aluminium, magnesium, calcium, potassium, sodium, iron and hydrogen are the most important. Much of the concern centres on the nature of the free SiO_2 phase. In nature, SiO_2 occurs mainly as the mineral quartz. Its atomic structure and physical properties are well known: it contains a basic building block which is a unit having silicon at the centre of a tetrahedron of four oxygens; the four Si–O bonds are semicovalent. The tetrahedra are not isolated from each other; instead, each oxygen is shared by linking equally to two other, geometrically very similar, tetrahedra, so that the tetrahedra themselves are fully cross-linked into a three-dimensional framework. It is a peculiarity of tetrahedrally linked frameworks that a number of alternative framework arrangements are possible, which may differ energetically from each other only slightly. Depending on temperature and pressure, several such frameworks occur stably for SiO_2; quartz is the most abundant and is a polymorph in which the cross-linking is strong, hence it is comparatively dense and relatively unreactive, being unaffected by the presence of most strong acids or alkalis. However, two other framework types occur, designated tridymite and cristobalite. These are topologically similar to each other and both have rather open, low-density packings. They characteristically occur in rocks with high-temperature origins, e.g. in SiO_2-rich volcanics, although cristobalite has also been observed to form metastably at low temperatures. Once formed, tridymite and cristobalite may persist for geologically long periods of time without significant reversion to the more stable quartz. Since tridymite and cristobalite are crystalline, they are readily detected by their characteristic X-ray diffractions and by their distinctive optical properties under the petrographic microscope, etc. However, because their frameworks are more open relative to the denser, more tightly bonded quartz, they exhibit substantially enhanced reactivity towards alkali.

In addition to the crystalline silica polymorphs, disordered frameworks also occur. While these are thermodynamically unstable with respect to crystalline phases (normally quartz), they are nevertheless persistent in nature. Their thermodynamic metastability and comparatively open, disordered structures also give rise to an enhanced potential for reaction with cement alkalis. Moreover, SiO_2 exhibits a unique structural relationship with that of ice, H_2O, and presumably also with that of liquid water. The water molecule consists of a central oxygen, around which are arranged in approximately tetrahedral array two hydrogens and two electron pairs. The geometric similarity between SiO_4 and $OH_2 \cdot e_2$ (where e_2 stands for two electron pairs) enables 'water' to substitute to some extent in silica. In quartz the extent of such substitution is typically slight, of the order of a few parts H_2O per million. However, in less crystalline silicas the substitution may be much higher, ranging up to several per cent or more. In effect, strong bonds, e.g. Si–O–Si, are broken by hydroxylation and are replaced by more reactive Si–OH $\cdots$ OH–Si bonds, where $\cdots$ represents a weak hydrogen bond. Such amorphous, hydrous silicas may be very reactive in the presence of alkali. Depending on their physical form and mode of geological origin, they have a variety of generic names: silica gel, opal, chert and chalcedony are examples. There is thus a wide range of poorly crystallised SiO_2 varieties, some nearly anhydrous, others relatively rich in water.

Unfortunately, the poorly ordered silicas all have one other feature in common: because of their low degree of crystallinity, they are difficult to characterise and determine by techniques such as X-ray diffraction. They can, however, be determined petrographically, but this requires skill and patience on the part of the operator. Nor is ordinary chemical analysis sufficient to determine their presence: various selective dissolution methods have been proposed, based on the enhanced solubility of amorphous silicas in alkaline solutions, but these methods are far from satisfactory for the detection of potentially deleterious aggregates.

Reactive silicious aggregates thus fall into two classes. One class comprises mainly the crystalline, low-density polymorphs of SiO_2 such as cristobalite and tridymite, while the other comprises the poorly ordered as well as the essentially non-crystalline forms, such as opals, glasses, gels, etc. The latter group may range widely in water content, and no satisfactory method, other than petrographic examination, exists for their detection.

Unfortunately, there is no reliable way of determining the reactivity of various silicious aggregates with respect to their susceptibility to alkali–aggregate reaction. Partly this arises because reactivity is a function of surface area, as well as crystallinity and hydration state. In some aggregates, e.g. crushed glass, geometric and true surface areas may be equivalent, whereas in others, such as cherts, the true surface exposed to attack is much greater than the apparent geometric surface area, owing to their porous nature. Moreover, the harmful effects of a particular aggregate may be variable, depending upon its content: this is the well-known pessimum effect which has been described by Hobbs[6].

2.6 Origin and nature of the pore fluid in concrete

2.6.1 Origin and significance of pore fluid

Since the alkali–aggregate reaction occurs in wet environments, it is not necessary to consider reactions in dry cement concretes, although the effects of periodic wetting and drying may require to be assessed. When concretes are first mixed, their water–cement (w/c) ratio is an important design parameter. Typically, this may lie in the range 0.35–0.55. However, the amount of water required for full hydration of the cement, calculated as a w/c ratio, is substantially less—about 0.24. Therefore, water is normally present in excess of that required for its hydration, even in well-cured concrete. A well-cured concrete thus consists of four phases: aggregate, some unhydrated cement together with unreacted blending agent (if present), cement hydration products (which can, in turn, be differentiated) and an excess of an aqueous phase. Figure 2.1 shows schematically how the amounts of the various phases change as a function of time for the paste fraction of a fly ash blend; the diagram is schematic as the exact proportions of the various phases will depend on the cement type, on the reactivity of the fly ash, on the w/c ratio, temperature, etc. It is, however, important to note that considerable water remains even after the cement is essentially fully hydrated. During the initial stages of hydration the aqueous phase is most abundant, and it is more or less continuous in the sense that it wets the solid-phase grains. However, as hydration progresses the space occupied by the aqueous phase becomes increasingly filled with hydration products, mainly gel-like C–S–H and $Ca(OH)_2$, and the remaining aqueous phase gradually becomes discontinuous. Cement paste

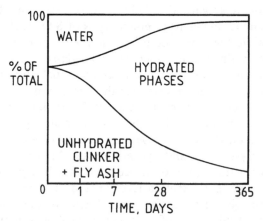

Figure 2.1 A schematic diagram, showing how the proportion of various phases changes as a function of hydration time. In this hypothetical example, blended cement is mixed to a water–solid ratio of ~ 0.3 and wet-cured.

itself is intrinsically porous, and considerable space remains to accommodate this fluid, hence the term 'pore fluid'. A paste made to very low w/c ratios may have little porosity in the range above 1–2 μm diameter (unless special steps are taken to entrain air), but much microporosity normally still remains in the minus 1–2 μm diameter range. Ordinary concretes made to higher w/c ratios thus contain macropores and, in the paste, micro- and mesopores which afford intimate contact between the pore fluid and hydration products, as well as with aggregate particles. In mortars and concretes, pore fluid may also segregate at the aggregate–cement hydration product interface, especially when mixed to w/c ratios above 0.35–0.40.

Pore fluid thus serves as a bridge between cement hydration products on the one hand and aggregate on the other. Material transport of soluble species, such as alkali ions, occurs readily through the pore fluid. In principle, transport can also occur across solid–solid interfaces—for example, between solid hydration products and aggregate—but such reactions occur at a slower rate than through an aqueous phase; hence the importance of the latter in conditioning alkali–aggregate reactions.

2.6.2 Pore fluid chemistry

The chemical composition of the aqueous phase is far removed from that of pure water. Experimentally, its composition has been studied in various ways, each method having advantages and disadvantages. Perhaps the simplest method is to prepare synthetic calcium silicates, aluminates, etc., and shake these, either individually or mixed in known proportions, with an excess of water, and periodically remove filtered aliquots of the aqueous phase for chemical analysis. This procedure enables the system to be well defined chemically, but such simulants have too simple a chemistry to reproduce the salient features of real cements. In these laboratory simulants, the aqueous solutions will be strongly alkaline, but their alkalinity will be controlled primarily by the solubility of Ca^{2+}. Nevertheless, such simulants represent a first and essential step in understanding the behaviour of real systems. Solubility data for the $CaO–SiO_2–H_2O$ system are well established and are amenable to thermodynamic treatment[7,8]; quantitative data on the synthetic alkali-containing systems are becoming available[9].

Another method of simulating pore fluid which is frequently employed enables real cements to be used: neat cement, formulated to a realistic w/c ratio, is allowed to set and mature normally but is subsequently crushed and extracted, using water as the leachant. The resulting solution is filtered and analysed. These filtrates are also strongly alkaline but, in contrast to laboratory simulants, their alkalinity is controlled primarily by soluble sodium and potassium and to a lesser extent by calcium; the common ion (OH^-) effect operates so as to reduce its solubility. A disadvantage of the procedure is that the exact composition of the filtrate is very sensitive to the ratio of leachant to cement, and if large volumes of water are used the procedure becomes increasingly unrealistic with respect to the true nature of pore fluids. A variant of this method, in which real cement clinker is simply stirred with a large volume of water, is open to similar criticism.

Perhaps the most successful and direct method of obtaining pore fluid compositions is that of pore fluid expression. Neat cement cylinders, prepared to a realistic w/c ratio, are allowed to set and cure normally and subsequently subjected to high pressure in a special device, known colloquially as a 'squeezer', to extract pore fluid. This method was pioneered by Longuet[10] and further developed by Barneyback and Diamond[11]. Very high pressures may be required to extract pore fluid; in the writer's laboratory, 100 tonnes of thrust are available for application to cement cylinders 42 mm in diameter. The expressed pore fluid is collected on a specially channelled base plate, the drain of which is shaped to accept a standard hypodermic syringe. In this way the pore fluid can be protected against contamination, e.g. by uptake of atmospheric CO_2. If further protection is needed, the entire apparatus may be enclosed in a large, nitrogen-filled box and its hydraulic mechanism operated by remote control. Only a fraction of the total pore fluid present in a cement or concrete can normally be collected in this way, although the collection efficiency may be somewhat improved by providing a small nitrogen overpressure within the high-pressure cylinder.

Numerous data are available on the composition of pore fluids of cements and cement mortars. For example, Barneyback and Diamond[11] made mortars using a 1:2:0.5 mix of Portland cement, inert sand and water respectively. The cement contained 0.30% Na_2O and 0.93% K_2O, and the alkali content of the aqueous phase was determined at intervals between 0.5 hours and 70 days. If all the alkalis present in the cement were dissolved in the pore water, it was calculated that the resulting concentrations in pore fluid would be 0.395 M (K^+) and 0.193 M (Na^+), giving a total alkali concentration of 0.588 M. In the sampling interval between 49 and 70 days, the actual alkali contents were found to be still increasing, albeit slowly, until at 70 days the pore fluid was 0.575 M in K^+ and 0.218 M in Na^+. The rise in concentration between 49 and 70 days was attributed to the balance between (i) further reduction in the amount of free water (some of which continues to be chemically combined into cement hydration products), (ii) some uptake of alkali into solid hydrate phases and, of course, (iii) the presence of a little unhydrated clinker, which may retain alkali. However, most of the alkali present in the clinker had been released into the pore fluid by 70 days. These results are broadly typical of those found in other investigations. For example, Marr and Glasser[12] used a cement containing 0.48% potassium as K_2O and 0.19% Na_2O (wt %) and a w/c ratio = 0.60. After 90 days of cure at 20°C the pore fluid was found to be 0.064 M (Na) and 0.200 M (K). These alkali contents are somewhat lower than those in ref. 11, which is probably because of the higher w/c ratio used, which somewhat dilutes the alkali, and a slightly higher retention factor in the solid hydration products. Page and Vennesland[13] have done much valuable work on pore fluid compositions. Some of their data are shown in Table 2.1. It can be seen that these data are broadly comparable with those of other investigations. In general, alkali concentrations in the range 0.1–1.0 M are sufficient to render calcium very insoluble; for example in these studies, the pore fluid calcium contents were of the order of 100 μg of calcium per litre.

Some have questioned whether the pore fluid is representative: is pore fluid expressed mainly from the larger pores and, if so, does it differ in composition . from the fluid present in smaller pores? The answers to these and other questions are at present not known, but when pore fluid analyses are included in mass balance calculations, showing the distribution of alkali between solid and aqueous phases, reasonable agreement between calculation and observation is obtained. Certainly pore fluid analyses provide the best available evidence of the internal constitution of the fluid phase and have revealed important data on the local environment which will be experienced by aggregate in concrete.

Thus numerous studies of pore fluids from real cements disclose that their alkalinity is dominated by the alkalis sodium and potassium, and not by calcium, as is the case in laboratory studies performed on synthetic calcium aluminates, silicates, etc. In real cements, the common ion effect operates to reduce calcium concentrations relative to the solubility of $Ca(OH)_2$ in alkali-free water; typically, it is decreased from 10^6 μg/l for $Ca(OH)_2$ itself to only 10^2 μg/l in pore fluids extracted from an unhydrated normal Portland cement (OPC) which, of course, still yields $Ca(OH)_2$ amongst its solid hydration products and maintains saturation, albeit at much lower calcium concentrations.

The nature of the calcium species in solution deserves comment; while it is relatively easy to measure calcium *concentrations*, its speciation and related functions, such as pH, are not so readily determined. The reasons for this lie in the high total species concentrations, and hence the high ionic strengths of the pore fluid, which result in their non-ideal behaviour. It is necessary to consider this aspect in more detail.

Table 2.1 Chemistry of the pore fluid in cement[13].

Curing time (days at 22°C)	Ion concentrations ($\times 10^{-3}$ mol l^{-1})				
	Na	K	Ca	OH	SO$_4$
7	263	613	1	788	23
28	271	629	1	834	31
56	332	695	3	839	44
84	323	639	2	743	27

2.6.3 Treatment of pore fluid data

The characteristics of an aqueous solution can be defined in various ways. One of the most useful ways is through the pH function. Historically, this function was derived by Arrhenius, who suggested that many salts in solution did not exist as molecules but instead dissociated into charged particles, which he termed ions. For example, sodium sulphate, Na_2SO_4, was viewed as dissociating according to the reaction:

$$Na_2SO_4 = 2Na^+ + SO_4^{2-}$$

Since the salt from which the ions form is itself electrically neutral, the total number of positive charges on the dissociated particles must equal the number of negative charges; furthermore, the individual ion charges must be integers. Thus, sodium is found experimentally to have one positive charge while the sulphate, SO_4^{2-}, group—which persists as an unchanged unit throughout— has two negative charges. The constitution of salt solutions can be determined experimentally by measurements of properties such as their osmotic pressure, vapour pressure lowering, depression of the freezing point, elevation of the boiling point etc.: these so-called colligative properties of solutions depend on the total number of solvent and solute particles in solution, and have been measured for numerous substances. These measurements have led to the conclusion that many salt-like substances dissociate only incompletely in solution, and that the degree of dissociation becomes complete only as infinite dilution is approached. Moreover, many molecular, covalently bonded substances do not dissociate significantly in solution: thus if a sugar, e.g. sucrose, is dissolved in water its molecules persist unchanged and no ions are formed.

If a salt such as NaOH or Na_2SO_4 were completely dissociated, the expected number of charged particles per formula unit would be two and three respectively. In practice, the actual values (technically known as the van't Hoff i factor) approach but never equal the expected values at all practical concentrations; thus dissociation is partial and increases with decreasing dilution of the salt species. Since these species interact with water, its behaviour must also be considered.

Water is an example of a substance which undergoes partial dissociation:

$$H_2O = H^+ + OH^-$$

From this relation, can be written an equilibrium constant:

$$K = \frac{[H^+][OH^-]}{[H_2O]}$$

where the square brackets indicate molar concentrations. Since the $[H^+]$ and $[OH^-]$ terms are small, $[H_2O]$ is essentially unity and may be disregarded. Thus we can define a constant, K_ω:

$$K_\omega = [H^+][OH^-]$$

At 25°C and 1 bar pressure, the numerical value of K_ω is $\sim 10^{-14}$ and the concentrations of OH^- and H^+, which must be equal to preserve electro-neutrality, are both $\sim 10^{-7}$ (molar units). At temperatures other than 25°C and pressures other than 1 bar, K_ω will change; for example, the tendency of water to dissociate is enhanced by increasing temperature, by increasing pressure, or by both. However, most reactions relevant to alkali–aggregate reactions take place in temperature–pressure regimes such that no significant corrections to K_ω are necessary, although regimes of high hydrostatic pressure, e.g. inside dams, in deep tunnels, etc., might require changes in its numerical value to be taken into account.

Since K_ω is a constant at fixed temperature and pressure, the addition of H^+ and OH^- ions from other sources displaces the equilibrium. Thus when $[H^+] > [OH^-]$, the solution is said to be *acid* and when $[OH^-] > [H^+]$, it is said to be *basic*. In cements, in which the pore fluid contains much NaOH, the $[H^+]$ and $[OH^-]$ concentrations can be determined as follows. Supposing the solution to be 0.01 M in NaOH, and assuming complete dissociation, $[OH^-] = 10^{-2}$: $[H^+]$ can be determined as follows:

$$K_\omega = 10^{-14} = [H^+][10^{-2}]$$

therefore

$$[H^+] = 10^{-12}$$

Because of the large variations in species concentration which can occur, it is convenient to use a log scale to express concentrations. This scale is known as the pH scale, p from the German 'potenz' or power, used here in its mathematical sense. Thus, by definition:

$$pH = -\log_{10}[H^+]$$

so for the NaOH solution just described, pH = 12. It is of course perfectly possible to define and use a pOH scale and, indeed, it would be sensible to do so for alkaline media such as cements. However, the use of the pH scale is so well entrenched that we continue to use it although the actual numerical values of H^+ ion concentrations may, in many instances, seem so low as to be negligible: nevertheless, the function has real physical significance as very low $[H^+]$ implies high $[OH^-]$.

The pore fluid of many real cement systems may contain relatively high concentrations of hydroxyl ions, whose charge is mainly balanced by alkali, etc.; the pH of such pore fluids may well be greater than 12.4, which is the approximate upper limit which can be achieved by the solution of $Ca(OH)_2$ at $\sim 20°C$. Measurement of 'real' pore fluids frequently gives pH values lying in the range 13–14. In these circumstances, both theoretical and practical difficulties arise in assessing the alkalinity of the system. Firstly, the problems associated with practical pH measurements in this range are substantial. Indicators—substances which exhibit colour changes at a particular pH—are not useful in the very high pH range owing to lack of suitable substances. Moreover, pH-meters, which would appear ideally suited, employ sensor elements which are only nominally selective for H^+ ions. By and large, such electrodes do work well in dilute solutions, i.e. at moderate pHs. However, in solutions of high ionic strength, such that high concentrations of alkali ions are present, the electrode response is more complex and is conditioned by the content of Na^+, K^+, etc., as well as by $[H^+]$. The electrode response may therefore show errors from the true pH, and the magnitude of these errors tends to increase as the pH increases. Special high-pH electrodes are available for use in alkaline solutions and should always be used in preference to general-purpose electrodes. But it should be realised that, while special electrodes reduce errors arising from the presence of other ions, they may not indicate the true pH.

Many investigators have therefore resorted to determining pH indirectly. The basis for the calculation rests on mass balance and charge considerations, namely:

$$\Sigma \text{ positive charges} = \Sigma \text{ negative charges}$$

The positive charge is obtained by analysis of the cation content of the pore fluid. This is readily done: sodium and potassium can be determined by flame photometry, calcium by atomic absorption; the content of other positively charged ions is generally negligible. Since anions other than OH^-, Cl^- and possibly SO_4^{2-} are essentially insoluble in the cement environment, an analysis for soluble Cl^- and SO_4^{2-} will enable OH^- to be calculated by difference:

$$\Sigma \text{ positive charges} = [Na^+] + [K^+] + [2Ca^{2+}]$$

$$= \Sigma \text{ negative charges} = [OH^-] + [Cl^-] + [2SO_4^{2-}]$$

The resulting pH is then calculated from the relationship

$$\frac{K_\omega}{[OH^-]} = pH$$

This method is certainly straightforward from the analytical standpoint, but nevertheless may not necessarily give correct values on account of the differences, discussed earlier, between the behaviour of dilute and concentrated aqueous solutions. These differences can be summed up as follows. Firstly, while K_ω is well determined for pure water and is approximately constant over a wide range of ionic strengths, in more concentrated solutions—of which cement pore fluids are an example—significant departures may occur from the nominal value for pure water. Also, the assumption that species such as calcium occur in fully ionised form may be invalid. High concentrations of salts and of OH^- tend to suppress its ionisation, with the result that neutral ion pairs, e.g. $Ca(OH)_2(aq)$, as well as partially ionised species, e.g. $Ca(OH)^+$, may occur in significant concentrations. Therefore, the assumption that every calcium ion will give rise to two OH^- in solution is not necessarily valid, and a more rigorous treatment of solution data may be required.

When soluble phases have active masses which differ from their actual masses, the differences are expressed by introduction of a correction factor, γ, known as an activity coefficient. A more rigorous definition of pH, taking these factors into account is as follows:

$$pH = -\log_{aH^+} = -\log C_{H^+} \times \gamma_{H^+} = -\log\left(\frac{K_\omega}{C_{OH^-}}\right) \times \gamma_{H^+}$$

where H^+ and C_{H^+} are the activity and concentration respectively of the hydrogen ion, etc. and γ is the activity coefficient of the species. Evaluation of γ is possible: numerous relationships have been devised. One of the most widely used is the Davies equation:

$$\log \gamma = -AZ^2\left(\frac{\sqrt{I}}{1+\sqrt{I}} - 0.20I\right)$$

where A is a constant equal to 0.512 at 25°C, Z is the charge on the ion and I is the ionic strength of the solution[14]. Given the species concentration, γ can be evaluated. Similarly, for ionic strengths other than zero, K_ω and other thermodynamic parameters have been evaluated[14]. The equation has been applied to cement pore fluids, for which after application of correction factors, it was found that agreement between observed and calculated pH values improved, normally to within ± 0.15 pH units[9,16].

2.6.4 Effect of salt addition on pH

Problems may also arise with respect to the treatment of salts which may come into contact with concrete. For example, concretes may be exposed to solutions of salts such as sodium chloride, sodium sulphate, etc. Such salts are near-neutral in aqueous solutions. This gives rise to a common misconception in which it is argued that uptake of these salts should not affect cement pH because the salt is itself neutral and hence cannot contribute to the alkali–aggregate reaction. In fact, this argument is false. In the presence of cements, the anion of a neutral salt—chloride, sulphate, etc.—may react with components of the paste. For example, chlorides are partly removed from solution by incorporation into chloroaluminates, while sulphate is rather more completely removed by incorporation into monosulphate and/or trisulphate (ettringite)-type phases. However, electroneutrality must still be maintained and, in general, additional anions must become available to the solution to balance the charges on the soluble alkali which remains. In effect, the cement phases act as anion exchangers, removing Cl^-, SO_4^{2-}, etc. from pore fluids or percolating solutions and releasing OH^- sufficient to maintain neutrality. Or, in other words, the relative insolubility of hydrates such as chloro– or sulphate–aluminates ensures that, whatever the alkali salt source and its pH, and regardless of the initial pH of the cement pore fluids, uptake of concentrated alkali salts tends to increase their pH. It needs to be emphasised that the alkali–aggregate reaction is not directly conditioned by the alkali ion, but by the hydroxide counter ions which are present: reference to the 'alkali–aggregate reaction' is strictly incorrect: what is really meant is a hydroxide ion–aggregate reaction. Adding concentrated alkali salts generates additional OH^-, and they are thus effective at raising pH levels. The process cannot continue indefinitely: eventually the anion exchange capacity of the hydrates is exhausted, so the mechanism of anion removal gradually becomes less effective and pHs are no longer affected. Other pH-controlling mechanisms, such as straightforward dilution, are also operative and become increasingly important in structures exposed to leaching or to very dilute alkalis.

The implications of these chemical factors with respect to the alkalinity of pore fluids may be summarised as follows. Wet cements contain appreciable amounts of an aqueous phase, which is normally very alkaline. Its alkalinity is expressed on a log scale, units of which are termed pH. The pH scale relates the contents of H^+ and OH^- to each other through K_ω, the ion product of water. In neutral or near-neutral solutions, the concentrations of both H^+ and OH^- are small and equal, or nearly so. But in concentrated alkali solutions $[OH^-] \gg [H^+]$. Moreover, in concentrated solutions the simple, idealised relationships between concentration and pH must also be corrected for several factors: for example, departure of the ion product, K_ω, from its value in pure water. The direct experimental determination of pH in highly alkaline solutions is handicapped by deficiencies in the performance of pH sensors. Mass balance calculations, often used to determine pH indirectly, need sophisticated treatments of the data to allow for non-ideality of the aqueous phase. The addition of concentrated sodium or potassium salts from external sources tends to increase the alkalinity of the cement system, at least initially: the pH of solutions of pure salts, as measured in the laboratory, is not necessarily relevant to their behaviour in cements. These practical and theoretical limitations do not detract from the validity of the concepts outlined here, but do indicate that care is required in making calculations from analytical data and in drawing conclusions from both calculations and measurements concerning the internal alkalinity of cement systems.

2.7 Chemical mechanism of the alkali–aggregate reaction

Figure 2.2 shows in schematic form the essentials of the alkali–aggregate reaction: mass transport is an essential feature. Aggregate particles, nominally

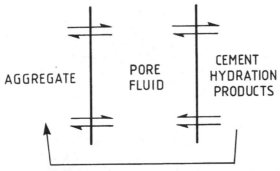

Figure 2.2 Mass transport initiates the alkali–aggregate reaction. Transfer pathways are shown between pore solution and aggregate, but secondary pathways may also occur between solids: aggregate and cement paste.

consisting of SiO_2, are thermodynamically unstable in the cement environment. Reaction commences, leading to a lowering of the free energy of the system. This reaction, or series of reactions, is accompanied by mass transport of OH^- and alkali ions: the pore fluid is in intimate contact with cement hydration products as well as with aggregate particles and it serves as the main agent of transport.

The details of reaction can be seen in more detail in Figure 2.3, showing the complex nature of the aggregate surfaces and cement paste microstructure. The figure is not drawn to an exact scale but is meant to convey the impression that many of the relevant events occur on an atomic scale. The cement microstructure is shown: it consists of solid hydration products, of which the drawing shows two features—platelets of $Ca(OH)_2$ and bundles of fibres or lath-like C–S–H structures. The hydration products are not normally space-filling, so considerable pore volume exists as meso- and micropores. The micropores are not shown, but water-filled mesopores are suggested by areas of water molecules, depicted as H–O–H.

The surface of a silicious aggregate particle is also shown. Normally, its surface oxygens are hydroxylated, even in pure water. Surface studies show that this disturbed region is normally several atoms or even tens of atoms deep. When aggregates are placed in a hydroxyl-rich medium, their potential to undergo further hydroxylation is enhanced. With well-crystallised quartz, this potential for further reaction exists, but the rate of hydroxylation is so slow as to be almost imperceptible on the normal civil engineering timescale. Temperature markedly accelerates the reaction: in hydrothermal conditions, finely ground but crystalline quartz is very reactive. However, susceptible aggregates typically react much more rapidly than crystalline quartz, even at normal ambient temperatures.

The principal consequences of the reaction at ambient have been explored in various ways, and numerous reports, symposia and conference proceedings

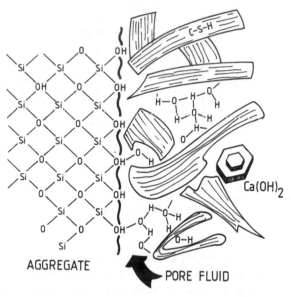

Figure 2.3 The complete microstructure and mineralogy, as well as the structure of the aggregate–paste interface, influences the course of the alkali–aggregate reaction.

describe these studies; several recent references are cited[17-21]. However, this chapter concentrates on the chemical mechanisms involved, which have been elucidated by Dent-Glasser and co-workers[22,23]. In their explanation, a gel-like layer of indefinite constitution forms at the aggregate–cement paste interface. As this gel forms it imbibes water and soluble ionic species, principally Na^+, K^+ and Ca^{2+}. The gel is not very soluble, and generally concentrates around the margins of susceptible silicious aggregate particles, but softer gels may also be exuded by mechanical swelling pressures.

Thus three reactions occur. In the first stage the high-pH pore fluid reacts with Si–O–Si bonds to form silanol bonds:

$$Si–O–Si + H_2O \rightarrow Si \overset{\displaystyle OH \ldots OH}{\diagup \hspace{3em} \diagdown} Si$$

Of course, hydrous silica aggregates may already contain substantial silanol bonding. These silanol groups are 'acidic' in the Lewis definition, and participate readily in further reaction with 'basic' cement pore fluids. Thus, a four-coordinate silicon with one coordination site already occupied by OH may be abbreviated as Si–OH, and in the second stage it reacts with further hydroxyls:

$$Si–OH + OH^- \rightarrow Si–O^- + H_2O$$

liberating more water in the process. The negatively charged $Si–O^-$ species attracts positive charges; mobile, readily abundant species such as sodium, potassium and calcium diffuse into the gel in sufficient numbers to balance the charge on the negatively charged groups. Dent-Glasser and Kataoka[22] have represented the approximate stoichiometry as:

$$H_{0.38}SiO_{2.19} + 0.38Na_2O \rightarrow Na_{0.38}SiO_{2.19} + 0.38H_2O$$

where the charge compensation is achieved by Na_2O, although, of course, the implication is that other cations may also participate.

In the third stage, more siloxane, Si–O–Si bridges are attacked:

$$Si–O–Si + 2OH^- \rightarrow Si–O^- + {}^-O–Si + H_2O$$

At the limit of the process, some silica may pass into solution: at high pHs the principal soluble silicate species is $H_2SiO_4^{2-}$.

The gel itself has a significantly greater specific volume than that of the SiO_2 which it replaces, and it is this which creates the swelling pressures and expansion which are so characteristic of the alkali–aggregate reaction. The extent of the swelling is difficult to predict; silica gels have water contents and densities which are variable over wide ranges, and the additional possibilities for imbibition of water, Na^+, K^+, etc., cause its density and hence specific volume to vary widely. Moreover, silica gels may maintain a local steady-state, quasi-equilibrium with the local pore water. If the local pore water is diluted, as may occur by percolation of fresh water, the gel may spontaneously lower its ionic content: during drying cycles, when local salt concentrations tend to increase in the pore fluid, the reverse reactions occur. The limits of wet–dry cycling are clearly important, because in structures cycling may induce high local concentrations of alkali and hence localised alkali–aggregate attack, even though the mean alkali content of the bulk concrete may remain relatively low.

The imbibition properties associated with gels are sometimes confused with osmosis. It is true that the gel product of alkali–aggregate attack is semipermeable, and that it contains substantial alkali, calcium, etc., but it is not the alkali which actually initiates reaction. Rather it is the local concentration of OH^- ions and specific susceptibility of the aggregate particle that condition the potential for reaction. Osmosis, as such, plays at most a minor part in the process: imbibition is the appropriate physicochemical process which governs the gel shrink–swell properties.

The indefinite constitution of the gel-like reaction product has a number of consequences. Gels vary in physical properties, mainly as a function of their water content, from hard rigid materials to plastic, readily extrudable substances. Examination of most real examples of alkali–aggregate reaction shows that this is probably a feature of most real reactions: hard rigid gels are most likely to cause expansive forces leading to cracking, while the more fluid, watery gels are readily extruded into cracks and may eventually emerge as exudates; soluble silicate species also migrate and subsequently reprecipitate, mainly in zones where the local pH is lowered, as occurs during carbonation.

2.8 Control of alkali levels in concrete

Various additives have been claimed to control the expansivity of susceptible aggregates in concrete. Considerable disagreement exists concerning the mechanism whereby slag and fly ash reduce expansions. Four theories have been proposed to explain their action:

(1) The blending agent, which is typically less reactive than cement, acts as a diluent for the rapid-release sources, e.g. Portland cement. Thus, if 50% of cement were replaced by slag, and only half the slag were to react within, say, 1 year, not only would some alkali remain bound in the slag, but the effective w/c ratio of the system would remain higher than in the corresponding mix with cement only, with concomitant dilution of the alkali.
(2) The products of hydration of blended cement systems differ significantly from those of OPC systems and, as a consequence, they have a higher binding power for alkalis: removal of alkali from the pore fluids into cement solids, it is argued, results in less immediate potential for reaction with aggregate.
(3) Migration of alkali towards alkali-sensitive aggregate particles is inhibited by the lower permeability of blended cement matrices.
(4) Blending agents lower the $Ca(OH)_2$ content of the cement paste, and thereby reduce its pH.

Of course, some combination of these mechanisms may be invoked.

The four theories may be dealt with individually. The first hypothesis can certainly be checked, at least approximately, by mass balance calculations. Data on the kinetics of hydration of slag and fly ash (especially the latter) are somewhat imprecise, but calculations show that the effect of the additive is apparently substantial. Indeed, the high alkali content of some fly ashes might seem to suggest that alkali contents should rise, even though the fly ash may have only partially reacted. The second theory has rather more weight of evidence behind it. One of the principal hydration products of cement and its blends is C–S–H: C–S–H, it will be recalled, has a variable composition. In Portland cements, its typical Ca/Si ratio is about 1.8, but some blending agents, such as slag, fly ash and silica fume, tend to decrease its Ca/Si ratio. Slags and class C fly ashes, on account of their high calcium contents, lead in general to only a slight reduction in Ca/Si ratios, which may decrease to 1.4–1.5 (the exact ratio depends of course on the composition and proportioning of the blending agent, as well as the time allowed for homogenisation: the figures given are only intended as a guide). However, class F fly ashes, and especially SiO_2 fume, can lead to lower ratios still, perhaps as low as 1.0 in blends containing much silicious ash. The surface charge on C–S–H is dependent on its Ca/Si ratio: at high ratios its charge is positive, with the result that C–S–H tends to sorb anions but not cations. Hence incorporation of the alkalis Na^+ and K^+ into the C–S–H of Portland cements is rather limited; alkalis preferentially concentrate in pore fluid. But as the Ca/Si ratio decreases, the surface charge on C–S–H drops to zero, eventually becoming negative below about Ca/Si $\sim$ 1.2–1.3. Negatively charged C–S–H shows enhanced binding power for cations, especially alkalis. Thus the alkali-binding power of C–S–H is enhanced by inclusion of reactive silicious blending agents even though the blending agent is itself consumed. Pore fluid analyses of well-matured fly ash blends may thus show some lowering of alkali content, provided of course that the fly ash was itself not too high in alkali. Examples of mass balance calculations of the alkali levels in blended cements have been given by Taylor[24], who concluded that the enhanced alkali content of fly ashes could be approximately compensated by increases in sorption capacity of the hydration products: other factors were examined but had a lesser impact on the calculation. After 1 year, the calculations showed that a slight lowering of pore fluid alkalis could be expected to occur despite the higher alkali content of the PFA relative to the cement. He also distinguished between the behaviour of Na^+ and K^+, which differed. Thus this study, as well as others, shows that the concept of 'equivalent alkali' or 'equivalent Na_2O content' is at best a very crude approximation: if data permit, the two alkalis should always be treated separately.

Silica fume, however, has the most dramatic effect on pore fluid pHs, and is the subject of a recent review[25]. Not only does it sorb alkali in its own right, but if well dispersed it rapidly and efficiently lowers the Ca/Si ratio of C–S–H, leading to a substantial reduction in pore fluid pH. Thus, in summary, the second theory has some supporting evidence to show that slags and fly ash can have markedly higher alkali contents than cement without notably raising pore fluid pHs. Silica fume, which is virtually alkali-free and is more reactive than either fly ash or slag, is known to lower pore fluid alkalinities in well-dispersed, well-cured blends; because of its reactivity, this lowering is achieved rapidly.

A note of caution concerning the effectiveness of silica fume is required. In order to apply silica fume effectively, fairly high replacement levels (10–20%) may be necessary, and in order to maintain workability a plasticiser is often used. Perry and Gillott[26], in their study of such mixtures, found that superplasticiser very markedly increased the rate of expansion relative to identical mixes made without superplasticiser. The chemical influence of superplasticisers on the reaction has not been explored, but it is potentially significant that the lower w/c ratios attained with superplasticisers also tend to raise pore fluid pHs, other factors being equal.

The theory of hindered alkali migration finds some indirect support, especially for slag blends, in which exceptionally low diffusivities have been observed. Some reservations must, however, be expressed concerning the significance of the numerical values of permeability and diffusivity which have had to be obtained using high driving pressures.

The reduction of $Ca(OH)_2$ theory has little to commend it. The pH of alkali-free cement blends is essentially controlled by the equilibrium between $Ca(OH)_2$, C–S–H and an aqueous phase. This equilibrium is independent of the *amounts* of the individual phases. Thus a reduction in $Ca(OH)_2$ content should not affect pH, other factors being equal, although its total elimination could, of course, affect pH. It is, however, possible to envisage that local departures from equilibrium can occur if $Ca(OH)_2$ grains become physically isolated from contact with pore fluid, but no evidence exists that this occurs, and the chances of it happening so as to affect essentially all of the $Ca(OH)_2$ in a typical paste must be remote.

In summary, therefore, silica fume clearly has a beneficial effect in removing alkali from pore fluid, thereby lowering its pH. Such mechanisms may also operate in the case of silicious fly ashes and possibly slags, but with the last two this mechanism may be partly or wholly compensated by the intrinsically higher alkali contents of the blending agent relative to a low-alkali cement or to silica fume. Other purely chemical effects appear to be relatively of less importance, and other explanations of the efficaciousness of slag and fly ash must be invoked, at least in part, for the reported amelioration of alkali–aggregate reaction in these blends.

References

1. Choi, G.S. and Glasser, F.P. (1988) The sulphur cycle in cement kilns: vapour pressure and solid phase stability of the sulphate phases. *Cement Concr. Res.* **18**, 367–384.
2. McCarthy, C.P., Glasser, F.P., Roy, D.M. and Hemmings, R.T. (1988) Fly ash and coal conversion by products: characterization, utilization and disposal IV. Materials Research Society, Pittsburgh, PA, *Proceedings* Vol. 113, p. 347.
3. McCarthy, G.J., Glasser, F.P., Roy, D.M. and Diamond, S. (1987) Fly ash and coal conversion by products: characterization, utilization and disposal III. Materials Research Society, Pittsburgh, PA, *Proceedings* Vol. 86 p. 399.
4. Malhotra, V.M. (Ed.) (1986) Fly ash, silica fume, slag and natural pozzolans in concrete. *Proceedings of the Second International Conference, Madrid* (2 vol.) American Concrete Institute, SP-91.
5. Luke, K. and Glasser, F.P. (1987) Selective dissolution of hydrated blast furnace slag cements. *Cement Concr. Res.* **17**, 273–282.
6. Hobbs, D. W. (1988) *Alkali Silica Reaction in Concrete*. Thomas Telford, London, p. 183.
7. Glasser, F.P., Lachowski, E.E. and Macphee, D.E. (1987) Compensational model for calcium silicate hydrate (C–S–H) gels, their solubilities and free energies of formation. *J. Am. Ceram. Soc.* **70**, 481–485.
8. Luke, K. and Glasser, F.P. (1988) Time and temperature dependent changes in the internal constitution of blended cements. *Il Cemento* No. 85 (Aug–Sept), 1979–191.
9. Macphee, D.E., Luke K., Glasser, F.P. and Lachowski, E.E. (1989) Solubility and ageing of calcium silicate hydrates in alkaline solution at 25°C. *J. Am. Ceram. Soc.* (in press).

10. Longuet, P., Burglen, L. and Zelwer, A. (1973) La phase liquide du ciment hydrate. *Rev. des Matériaux de Construction et de Travaux Publics* No. 676, 35–41.
11. Barneyback, R.S. Jr and Diamond, S. (1981) Expression and analysis of pore fluids from hardened cement pastes and mortars. *Cement Concr. Res.* 11, 279–285.
12. Marr, J. and Glasser, F.P. (1983) Effect of silica, PFA and slag additives on the composition of cement pore fluids. *Proceedings of the Sixth International Conference on Alkalis in Concrete, Research and Practice*, pp. 239–242.
13. Page, C.L. and Vennesland, O. (1983) Pore solution composition and chloride binding capacity of silica fume cement pastes. *Matériaux et Constructions* 16, No. 919, 19–25.
14. Pommersheim, J.M. and Clifton, J.R. (1982) Mathematical modelling of tricalcium silicate hydration. II. Hydration sub-models and the effect of model parameters. *Cem. Concr. Res.* 12, 765–772.
15. de Robertis, A., de Stefano, C., Rigano, C. and Sammaranto, S. (1986) Ionic strength dependence of complex formation enthalpies: a literature data analysis. *J. Chem. Res. (S)* 194–195.
16. Macphee, D.E., Lachowski, E.E. and Glasser, F.P. (1988) Polymerization effects in C–S–H: implications for Portland cement hydration. *Advanc. Cement Res.* 1, 127–133.
17. Oberholster, R.E. (Ed) (1981) *Proceedings of the Fifth International Conference on Alkali–Aggregate Reaction in Concrete*. National Building Research Institute of the Council for Scientific and Industrial Research, Pretoria, South Africa.
18. Poole, B. (Ed.) (1976) *The Effect of Alkalis on the Properties of Concrete*. Cement and Concrete Association, Wexham Springs, Slough, UK.
19. Idorn, G.M. and Rostam, S. (Eds.) (1983) *Proceedings of the Sixth International Conference on Alkalis in Concrete, Research and Practice*. Technical University of Denmark, Copenhagen.
20. Grattan-Bellow, P.E. (Ed.) (1986) *Proceedings of the Seventh International Conference on Concrete Alkali–Aggregate Reactions*. Noyes Publications, Far Ridge, NJ.
21. Stuble, L.J. (1987) *The Influence of Cement Pore Solution on Alkali–Silica Reaction*. US Department of Commerce, Washington DC., Report NBSIR 87-3632.
22. Dent-Glasser, L.S. and Kataoka, N. (1981) The chemistry of alkali-aggregate reactions. *Proceedings of the Fifth International conference on Alkali–Aggregate Reactions*, S 252/23, p. 66.
23. Dent-Glasser, L.S. (1979) Osmotic pressure and the swelling of gels. *Cement Concr. Res.* 9, 515–517.
24. Taylor, H.F.W. (1988) A method for predicting alkali ion concentrations in cement pore solutions. *Advanc. Cement Res.* 1, 5–17.
25. Sellevold, F.J. et al. (1988) *Condensed Silica Fume in Concrete*. Thomas Telford, London.
26. Perry, C. and Gillott, J.E. (1985) The feasibility of using silica fume to control expansion due to alkali–aggregate reaction. *Durability of Building Materials* 3, 133–146.

3 Testing for alkali–silica reaction

R.N. SWAMY

Abstract

This chapter is devoted to a review of the various laboratory tests that are available to identify alkali-reactive aggregates and their potential for deleterious expansion in concrete. Laboratory testing can often be difficult, complex, confusing, time-consuming and even inconclusive, but nevertheless it is the surest way to assess and evaluate the potential reactivity of aggregates prior to their use in concrete construction. Petrography studies, well-established traditional tests and the newer, rapid test methods relevant and appropriate to the alkali–silica reaction (ASR) are discussed and their advantages and limitations evaluated in the light of published literature. Modifications to the test methodologies are suggested to improve their reliability and precision. The limitations of core testing, without or with accelerated heat-curing regimes, when applied to ASR-affected structures are emphasised. Special attention is given to the role of non-destructive testing in identifying the initiation and progress of ASR, and the advantages and limitations of the pulse velocity and dynamic modulus tests are described. A combination of petrographic studies, field performance and a concrete prism test which can realistically model all the major influences on ASR, including environmental effects, is recommended. It is emphasised that testing for ASR requires care, skill and a good understanding of the reaction process, and the results need to be interpreted with engineering judgement. When specifying acceptable limits of expansion for practical design, it should be borne in mind that expansive strains, *per se*, are not the only ASR effects to be considered in design, since different reactive aggregates affect the engineering properties of ASR-affected concrete at different rates.

3.1 Introduction

Previous chapters in this book have shown that alkali–aggregate reactions occur in Portland cement concrete between cement alkalis, such as sodium and potassium hydroxides, and certain reactive aggregates. These reactions have also been shown to lead to expansive strains and consequent cracking of the concrete. The progress of cracking in turn reduces both the strength and elasticity properties of the material. The cracking also allows the penetration of moisture and other aggressive agents such as atmospheric pollutants and de-icing chemicals which lead to other deteriorating processes in the concrete, often superimposed on the effects of alkali–silica reactivity.

Although there are several methods of containing or even avoiding the effects of ASR, laboratory testing of aggregates, and of mortar and concrete containing these suspect aggregates, is one sure way of assessing and evaluating the potential reactivity of these aggregates prior to their use in concrete construction. Laboratory testing can often be difficult and time-consuming, but is nevertheless an essential part of the process of limiting damage to concrete construction arising from ASR.

There are several standard test methods established over many years to test the potential reactivity of aggregates and of cement–aggregate combinations. In recent years a large number of new and rapid test methods have been proposed and developed[1,2]. Most of these test methods rely on temperature or pressure to speed up the expansion, and in the process tend to give expansions greater than those obtained with traditional test methods. Another major problem with several of these new and rapid test methods for ASR is that they tend to show not only poor precision when subjected to multilaboratory tests but also poor correlation with field performance. Nevertheless, laboratory testing is the starting point in the evaluation of aggregates and concretes for their susceptibility to ASR.

Laboratory testing can sometimes be confusing and apparently inconclusive, particularly when evaluating the source of deterioration in structures showing the characteristic signs of ASR, such as crack patterns without or with gel exudations. Such testing may then lead to wrong diagnosis. Structures suffering from ASR can thus be diagnosed to be not subjected to ASR[3], whilst structures showing apparent signs of expansion and crack patterns can be wrongly diagnosed to be suffering from ASR. In-situ monitoring is the best way to assess and estimate the additional potential for further expansion and deterioration of structures already affected by ASR. Such testing can be long-term, but very valuable and useful information can be obtained from testing of concrete cores drilled from such structures[4].

Laboratory testing of concrete cores taken from ASR-affected concrete structures is, however, not easy to carry out, nor is it easy to interpret the test results, therefore both should be performed with extreme care and engineering judgement. Core testing to determine in-situ strength, concrete quality and soundness of even ordinary structures unaffected by ASR is influenced by a large number of parameters[5,6]; and these factors become much more complicated when the strength of the concrete in the structure is low and the concrete is showing signs of deterioration due to ASR or other environmental conditions[7]. When testing cores for ASR, additional factors such as the temperature of the test, the nature of the solution in which the cores are immersed (water or alkaline solutions), the alkali content of the concrete and the repeatability and reproducibility of the results all have significant influence on the test results[8]. It is therefore strongly recommended that laboratory testing of cores taken from ASR-affected structures should only be carried out by personnel with a good understanding of ASR backed by considerable experience, expertise and professional judgement.

The aim of this chapter is to present a critical review of the well-established standard test methods and of the rapid test methods which are considered to be relevant and appropriate to ASR testing, and which have received some assessment through interlaboratory studies. Reference is also made to non-destructive test methods relevant to ASR which do not appear to have received the attention they deserve, and some recent results using these methods are also presented.

3.2 Evaluation of alkali-reactive aggregates

The very first step to be taken when evaluating the suitability of aggregates for concrete is a thorough petrographic examination together with other instrument techniques such as X-ray diffraction and infra-red spectroscopy. It is also very valuable to investigate the known field performance of these aggregates, if such records are available. These tests will enable one to classify the aggregates, recognise the deleterious components present in them, and determine what further tests need to be carried out[9].

Types of expansively alkali-reactive aggregates and how and whether such aggregates or their constituents can be identified petrographically, and the difficulties involved in their identification, are well documented in literature. However, it is important to note that, whilst aggregates may be reactive without expansion, expansion does not occur without alkali-induced reaction.

Coarse and fine aggregates often contain components that have the potential to liberate silica. Most national codes, such as BS 812: Part 102 (1984)[10] and BS 6100: Section 5.2 (1984)[11] and the *ASTM Book of Standards*[12] provide information on the petrological description of natural aggregates and descriptive nomenclature of their constituents including such siliceous phases as chert, flint, quartz and quartzite. Many specifications, such as the *Specifications for Highway Works*[13], also contain information on undesirable phases that need to be avoided to control alkali–silica reaction. Coarse and fine aggregates that each contain at least 95% by weight of one or more specified rock or artificial types are generally considered non-reactive provided they are not contaminated by silica minerals such as opal, tridymite or cristobalite or do not contain a total of more than 2% by weight of chert, flint or chalcedony taken together. When the proportion of chert or flint is greater than 60% by weight of the total aggregate, the aggregates are again considered to be non-reactive provided the silica minerals mentioned above are not present. Quartz should not generally contain quartzite or more than 30% by weight of highly strained quartz.

The structure and texture of siliceous fractions are generally described in terms such as amorphous, cryptocrystalline, microcrystalline and crystalline. Silica minerals occur in a wide range of forms, and silica (SiO_2) itself occurs in a number of polymorphic modifications or forms. The most common form of silica minerals includes quartz, tridymite, cristobalite, opal and the chalcedony group, which covers a number of varieties of silica composed of minute crystals of quartz with submicroscopic pores. These in turn include the minerals chalcedony itself, agate, chert, flint and jasper.

In general, there are two analytical techniques available to evaluate aggregates and identify ASR, and a combination of these two techniques may be used to confirm or reject its presence. The first and simplest method is to identify the expansively reactive aggregates by petrographic examination of thin sections of the ASR-affected concrete. The petrological microscope has several functions: it enables one to view the sample in plain and cross-polarised light, locate the various deleterious components, indicate if alteration has taken place, and, further, identify the presence of fractures due to secondary mineral migration. The microscopic examination also shows alteration zones or halos around siliceous grains as well as the deposition of the gel or crystal growth.

Petrographic examination of aggregates should include a full point-count analysis of the various fractions in the material to ascertain its composition. Examination of thin sections of an aggregate will also give a good indication of the presence of any potentially reactive constituents. The limitations of petrographic evaluation of aggregates should, however, be recognised, as the number of particles examined even in large thin sections (50×75 mm) is statistically too small. Thin-section examinations should therefore be supplemented by visual examination and simple physical tests. Visual categorisation of aggregates is an important part of aggregate evaluation.

Complementary techniques to petrographic studies include X-ray diffraction analysis (XRD), infra-red spectroscopy and other instrumental techniques. The resulting analytical charts can give useful information on the minerals present and the complex array of secondary minerals produced as a result of the chemical reactions.

Both these techniques will provide a valuable list of mineralogical contents and their proportions, but it is important to recognise that neither technique will guarantee information on the development or absence of ASR. In interpreting results of petrographic examination and related studies, it should be borne in mind that other characteristics of aggregates such as bulk density, porosity, particle size distribution, amount of reactive particles, and the environment, all have profound effects on observed expansions. These factors are discussed in other chapters. It is therefore important to recognise that petrographic examination and related thin-section studies are only the first step in evaluating the suitability of concrete aggregates.

3.3 Conventional mortar/concrete tests

There are three tests which have traditionally been used to evaluate the potential reactivity of aggregates—the ASTM C227 mortar bar test[14], the Canadian concrete prism test, CSA A23.2-14A[15], and the chemical method, ASTM C289[16]. Recent research has shown that all these test methods have serious drawbacks, which are briefly discussed below.

3.3.1 *Mortar bar tests*

Mortar bar expansion tests for evaluating alkali–silica reactive aggregates have been shown to be not always reliable[17,18]. As mandated in ASTM C227, mortar bars are stored in containers with wicks. Rogers and Hooton[17], testing mortar bars with a wide range of containers with and without wicks, found that containers with efficient wick systems may cause excessive leaching of alkalis out of the mortar bars and may thus reduce expansion significantly. On the other hand, mortar bars stored in containers without wicks or sealed in plastic bags showed significant expansion (Figure 3.1). After 1 year the

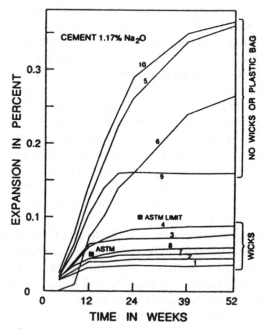

Figure 3.1 Influence of the presence or absence of wicks on mortar bar expansion tests. (Numbers refer to type of container.) From Ref. 17.

amount of expansion was found to correlate well with the amount of alkalis remaining in the mortar bars. Similar studies also show that Pyrex glass aggregates are not good aggregates for alkali reactivity tests[19]. It was found that using the same cement, with Pyrex glass aggregates, bars stored in containers 1, 2, 3, 4, 7 and 8 (Figure 3.1) showed expansion in excess of 0.3% whilst much less expansion was found in containers 5, 6, and 10. Pyrex contains significant amounts of alkalis, and these are readily released in the alkaline environment of the mortar bar test. In efficiently wicked containers, alkalis find an easy path into the water at the bottom of the container, but the Pyrex contributes more alkalis to continue the chemical reaction. Therefore, mortar bar tests, to be successful, require the removal of wicks from containers or sealing the specimens in plastic bags.

Mortar bar test results are also very much influenced by the test conditions[18]. A large number of factors are known to affect ASR expansion, and these include the proportion of reactive aggregate in the mortar, the particle size distribution of the aggregates, the alkali content of the cement, storage temperature and humidity. The size of the mortar bars also has an important influence on the expansion measured: the larger the cross-sectional area of the mortar bar the greater the expansion observed.

3.3.2 Concrete prism test

The concrete prism expansion test is supposed to work for all types of aggregates, but experience again has shown that the test can produce misleading results[18]. The cement content, water–cement ratio, temperature and humidity of the storage containers, the fineness of cement and particle size of the aggregate all seem to influence the test results, to a greater or lesser extent. High cement contents can be particularly harmful, leading to the possibility of satisfactory aggregates being classed as deleteriously expansive. The draft BS812 method[20] is particularly relevant here because it recommends the use of an unnaturally high cement content of 710 kg/m^3 for the concrete prism test. For example, a well-documented non-reactive limestone from Ireland was diagnosed by the draft BS test as being expansive[18]. Further, when slow expanding siliceous aggregates are tested for their potential reactivity, the prisms need to be stored at 38°C. It has been reported that, unlike in the mortar bar test, the removal of the wicks from the storage containers in the concrete prism test with such aggregates reduced the expansion considerably[18]. Examples of this kind indicate the extreme care required in developing tests of this nature to identify potentially reactive aggregates.

The dependence of the concrete prism expansion test, carried out with a known alkali–carbonate reactive aggregate, on storage conditions has also been reported by Rogers and Hooton[17]. The storage conditions had a clear influence on the amount of alkalis leached out, and the results shown in Figure 3.2 highlight the interrelationship between concrete prism expansion,

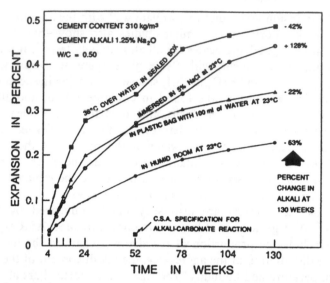

Figure 3.2 Interrelationship between concrete prism expansion, type of storage condition and water-soluble alkalis remaining in the concrete after $2\frac{1}{2}$ years. From Ref. 17.

type of storage condition and the water-soluble alkalis remaining in the concrete after $2\frac{1}{2}$ years. These results show that, depending on storage conditions, between 63% and 22% of alkalis had been lost after $2\frac{1}{2}$ years, but there appears to be no relationship between the change in alkali content and expansion. The significantly different expansions measured in the concrete prism expansion test when exposed to different storage conditions due to the variable leaching of alkalis from the concrete is also confirmed by a multilaboratory study on alkali–carbonate reaction reported by Rogers[21]. The coefficient of variation in these tests was 23.4%. Prevention of leaching of alkalis can be achieved by various methods such as increasing the size of the specimen, modifying the storage condition or by using high alkali levels with or without high cement contents.

Additional problems arise when slow/late-expanding alkali–silicate/silica reactive aggregates have to be evaluated for their potential reactivity. Unlike alkali–silica reactions, this slow/late type of reaction is associated with the expansion of coarse aggregate particles in addition to gel formation. Concretes containing such aggregates do not generally show cracking until many years after construction, sometimes as long as 10 years, and, further, they do not respond readily to accelerated tests. Rogers[22] has reported the use of the concrete prism expansion to evaluate the potential reactivity of these slow/late-expanding reactive aggregates. If was found that such aggregates did not cause significant expansion when tested in concrete with 5.1 kg/m^3 Na$_2$O eq., and stored in moist condition at 23°C or 38°C. Known alkali–silica reactive aggregates, on the other hand, expanded significantly under these conditions. When the concrete alkali levels were increased to about 9 kg/m^3 Na$_2$O eq., the slow/late-expanding aggregates showed significant expansion at 38°C. From these tests, it was suggested that a maximum expansion of 0.04% at 1 year can be considered to be a reasonable value to separate potentially deleteriously reactive aggregates from those which are non-reactive.

3.3.3 Chemical test methods

The chemical tests are generally rapid tests designed to give quick results when conventional mortar bar and concrete prism tests cannot be carried out. There are several tests which can be broadly classified as chemical tests[18]:

ASTM C289-86 chemical method
Weight loss method (Germany)
Gel pat test (UK)
Osmotic cell test (USA)
Chemical shrinkage method (Denmark).

The standard ASTM C289 test is the most widely known of all these tests, but it is not suitable for use with all types of aggregates, and it is unable to give the expansion potential of an aggregate. In general each of these methods is only suitable for use with certain types of aggregates and none of the tests is universally applicable.

In particular it has been shown that the chemical method is unreliable to determine the potential reactivity of carbonate aggregates. However, Berard and Roux[23] have proposed a modified version in which the chemical test is performed on the insoluble residues of the siliceous limestone or dolostone aggregates. More recently Fournier and Berube[24] have carried out an extensive investigation in which they studied the influence of different parameters such as the concentration of the acid used for carbonate dissolution and the particle size distribution of the insoluble residues under test. They also evaluated the precision of the modified test procedure through an inter-laboratory test programme; and, further, they applied the test method to more than 70 carbonate aggregates from Quebec and eastern Ontario in Canada. The results showed that there was no good correlation between the amount of dissolved silica (corrected or not) and the amount of expansion measured in the concrete prism expansion test. It was concluded that parameters other than the nature and amount of insoluble residue within the carbonate rocks governed their potential for deleterious expansion in concrete, and that factors like the permeability and porosity of the carbonate rocks had a major influence on their expansion behaviour in concrete.

3.4 Evaluation of rapid test methods

It is now universally recognised that there is an urgent need to develop and standardise the use of rapid test methods to detect alkali-reactive aggregates that cause deleterious expansion. It has been shown previously that the traditional test methods, developed early on with the recognition of the ASR problem and widely used for several decades to evaluate the potential reactivity of aggregates, are no longer considered adequate or satisfactory as they often take too long to complete and are particularly unreliable for detecting slowly reactive aggregates. Numerous rapid test methods have recently been proposed, but there are only limited data available to evaluate their validity and precision with respect to existing standard tests and known field performance of the reactive aggregates. Hooton and Rogers[25] have, however, carried out a fairly extensive study to evaluate the more promising of these test methods.

In their study, they used a single, high-alkali ($Na_2O = 0.39\%$, $K_2O = 1.18\%$, Na_2O eq. $= 1.17\%$) Portland cement. The petrographic description, reactive component and the field performance of the 12 Canadian aggregates used in the study were well known; further, the aggregates included those known to be non-reactive, deleteriously reactive and petrographically marginal aggregates (Table 3.1). Six different test methods were used:

(1) *ASTM C227* mortar bar tests[14]. From previous experience, some bars in these tests were placed in heat-sealed, polyethylene bags with approximately 10 ml of water. Expansion limits of 0.05% at 3 months and 0.10% at 6 months are recommended by ASTM C33. This was considered inadequate, and more stringent limits of 0.05% at 6 months and 0.10% at 12 months were considered. For late-expanding, alkali–silicate reactive aggregates, 0.1% should not be exceeded in 18–24 months.
(2) *NBRI method.* In this method developed by Oberholster and Davies[26], ASTM C227 mortar bars are exposed to a 1 M NaOH solution at 80°C for 14 days. The expansion is measured in the hot condition. Expansion limits suggested are: 0.10% for innocuous aggregates, 0.10–0.25% for slowly expansive aggregates and greater than 0.25% for rapidly expansive aggregates[27].
(3) *Duncan method*[28]. Expansion of four ASTM C227 bars is accelerated in an environment of 100% relative humidity (RH) and 64°C. A limit of 0.05% at 16 weeks was suggested.
(4) *Danish salt method*[29]. In this test, ASTM mortar bars are cast and three are exposed to a saturated NaCl solution at 50°C. Deleterious expansions in excess of 0.10% in the salt solution, corrected for the expansion in water, are reported within 8 weeks for some reactive aggregates and in 20 weeks for others[30].
(5) *Japanese rapid test*[31]. Mortar bars are made in this test with added NaOH

Table 3.1 Description and field performance of aggregates used in the tests. From Ref. 25.

Aggregate name	Description	Reactive component	Field performance
Nelson stone	Quarried dolostone	None	Excellent, > 30 years
Ottawa sand (Illinois)	Pure quartz sand	None	None
Fleming stone	Quarried granite	Strained quartz	None*
Guelph sand	Carbonate sand	0.2% chert	Excellent, > 30 years
Paris sand	Carbonate sand	0.3% chert	Excellent, > 30 years
Neyes sand	Silicious sand	2.5% chert	Reactive†
Paris stone	Carbonate gravel	1.6% chert	Excellent, > 30 years
Grant stone	Silicious gravel	2% argillite, greywacke, sandstone	Reactive†
Natural sand (Michigan)	Silicious sand	7.4% chert	Reactive‡
Sudbury stone	Silicious gravel	75% argillite, greywacke, sandstone	Reactive‡
Lower Notch stone	Quarried metasediments	98% argillite, greywacke, sandstone	Reactive§
Spratt stone	Quarried limestone	< 5% chalcedony	Reactive‡

*Petrographically similar to alkali–silica reactive granite in 70-year-old structure.
†Alkali reaction identified in part of one 30-year-old structure.
‡Alkali reaction identified in many structures.
§Alkali reaction identified in small part of one 20-year-old dam (but the use of fly ash has prevented reaction in most of this structure).

to raise the Na_2O eq. to 3%; then, at 24 hours they are exposed to saturated steam at 125°C (0.15 MPa pressure) for 4 hours. Based on their preliminary tests Hooton and Rogers[25] modified this test method using ASTM C227 bars with the Na_2O eq. raised to 4.0%.

(6) *Chinese autoclave test*[32]. Mortar bars are cured at 125°C in steam prior to autoclaving in a 10% KOH solution for 6 hours at 150°C. Hooton and Roberts[25] had used this test with ASTM 25 × 25 mm mortar bars and ASTM mix proportions. None of the aggregates expanded to 0.10%, and the test was abandoned. However, they report that the test worked well with Canadian aggregates when the originally proposed proportions of bars and mix were used.

The results of this study, shown in Tables 3.2 and 3.3, led to the following conclusions. It was found that to obtain expansion > 0.10%, the ASTM C227 method needed more than 18 months to indicate potentially deleterious late-expanding alkali–silicate aggregates. Bars sealed in polyethylene bags showed accelerated exansion due to reduced alkali leaching. Of the five rapid

Table 3.2 Perecentage expansion of mortar bars: ASTM and NBRI tests. From Ref. 25.

		ASTM C227						NBRI			
		MTO container			Bagged			NBRI			
No.	Aggregate name	6 months	12 months	18 months	6 months	12 months	18 months	12 days	14 days	28 days	56 days
1	Nelson	0.019	0.034	0.036	0.021	0.036	0.040	0.009	0.012	0.020	0.039
2	Ottawa	0.023	0.022	0.026	0.019	0.021	0.023	0.038	0.038	0.053	0.139
3	Fleming	0.005	0.021	0.019	0.021	0.039	0.040	0.038	0.046	0.086	0.205
4	Guelph	0.029	0.039	0.041	0.029	0.042	0.049	0.077	0.093	0.177	0.297
5	Paris sand	0.023	0.039	0.049	0.029	0.047	0.050	0.122	0.153	0.306	0.399
6	Neyes	0.024	0.024	0.030	0.020	0.020	0.028	0.118	0.164	0.369	0.513
7	Paris	0.012	0.025	0.020	0.025	0.045	0.044	0.144	0.175	0.293	0.336
8	Grant	0.011	0.024	0.014	0.030	0.054	0.057	0.143	0.185	0.423	0.533
9	Natural	0.023	0.023	0.031	0.041	0.048	0.052	0.172	0.213	0.494	0.525
10	Sudbury	0.022	0.031	0.032	0.036	0.062	0.080	0.221	0.266	0.466	0.736
11	Lower Notch	0.017	0.032	0.035	0.036	0.077	0.094	0.282	0.326	0.540	0.746
12	Spratt	0.046	0.055	0.058	0.166	0.222	0.234	0.309	0.356	0.626	0.956

mortar bar test methods, it was found that the NBRI 14-day test method was reasonably precise and able to identify all of the reactive aggregates. The other rapid test methods were unable to do this either because of the test modifications or because of the different types of reactive aggregates used in the study. Hooton and Rogers also recommended the expansion limits of the NBRI test method as appropriate, at least as an interim measure, until further test data are obtained.

The conclusions of Hooton and Rogers in relation to the NBRI test are also confirmed by Grattan-Bellew[33], who evaluated a number of Canadian

Table 3.3 Percentage expansion of mortar bars: Duncan, Danish and Japanese tests. From Ref. 25.

		Duncan test			Danish salt test			Japanese rapid test 4% NaOH
					In sat. NaCl		Corrected	
No.	Aggregate name	12 weeks	16 weeks	26 weeks	8 weeks	20 weeks	20 weeks	
1	Nelson	0.033	0.023	0.025	0.003	0.013	0.001	0.034
2	Ottawa	0.027	0.028	0.019	—	—	—	0.033
3	Fleming	0.040	0.044	0.050	0.022	0.054	0.008	0.047
4	Guelph	0.032	0.034	0.036	0.003	0.007	0.007	0.060
5	Paris Sand	0.036	0.035	0.037	0.008	0.019	0.005	—
6	Neyes	0.035	0.042	0.061	0.022	0.027	0.003	0.057
7	Paris	0.035	0.034	0.037	0.017	0.026	0.007	0.060
8	Grant	0.029	0.033	0.041	—	—	—	0.043
9	Natural	0.025	0.026	0.025	—	—	—	0.054
10	Sudbury	0.057	0.055	0.063	0.016	0.049	0.002	0.044
11	Lower Notch	—	—	—	0.016	0.023	0.008	0.069
12	Spratt	0.068	0.071	0.081	0.021	0.054	0.013	0.225

Table 3.4 Influence of test variables on expansion measured in NBRI test. From Ref. 25.

No.	Aggregate name	14-day expansions (%)			
		Constant flow	w/c = 0.50	Unwashed aggregate constant flow	w/c for constant flow
5	Paris Sand	0.153	0.197	0.147	0.46
7	Paris	0.175	0.163	0.151	0.54
10	Sudbury	0.266	0.255	0.275	0.53
11	Lower Notch	0.326	0.403	—	0.54
12	Spratts	0.356	0.375	0.441	0.54

aggregates by the mortar bar accelerated test (MBAT). It was found that this test was quite satisfactory as an aggregate screening test except for alkali–carbonate reactive aggregates. A potential problem with this was that some aggregates which might perform satisfactorily in concrete under mild exposure conditions may be classified as being deleteriously expansive.

The NBRI test has been subjected to an interlaboratory study to evaluate its precision and the influence of cement alkali content on the measured expansions[34]. Nine laboratories were involved, and three cements with various alkali contents and three aggregates exhibiting different performances (innocuous, marginal and reactive) were used. The study showed good reproducibility of results when the test procedure was used with experience. The alkali content of the cement had only a minor effect on expansions. Thus this test can only be used to evaluate aggregates and not cement–aggregate combinations.

Hooton and Rogers[25] also investigated the influence of various factors on the expansions measured in the NBRI test. The results are shown in Table 3.4. It was found that higher expansions were obtained in some cases when the water–cement (w/c) ratio was kept constant. Expansion was also not reduced significantly when aggregates remained unwashed, and in one case expansion was increased.

3.4.1 *The Duggan test*

The Duggan test is a simple and rapid concrete test which aims to distinguish between deleterious and non-deleterious expansions in concrete[35]. The advantage of the test is that the test samples are drilled cores from laboratory specimens or field structures so that various effects such as those of the mix constituents, admixtures and curing are automatically included in the expansion test. The cores used in the tests are 22 mm in diameter and 50 ± 5

mm in length. The cores are immersed in distilled water at room temperature (21°C) for 3 days, and then subjected to a dry heat at 82°C–distilled water at 21°C regime (dry heat–water immersion cycle) as shown in Figure 3.3. The initial measurements are taken at 10 days and then subsequently at 3- to 5-day intervals when the specimens are kept immersed in water. The test can generally be completed in about 6 weeks.

Scott and Duggan applied their test regime to cores taken from about 20 structures of various ages and conditions, including cores from 12- to 78-year-old concrete. They found that the core expansions correlated well with the observed concrete conditions. To examine further the scatter and statistical validity of the test, they carried out tests on 34 cores taken from each of three different concretes containing cements of varying Na_2O eq. The results are shown in Figure 3.4, in which the solid lines represent the average expansion of 34 cores and the dotted lines represent the 95% confidence limits using five cores. The results as shown confirm low scatter and good repeatability.

Scott and Duggan also carried out laboratory tests with a known reactive aggregate and five different high early-strength cements. The results shown in Figure 3.5 followed the known trends of alkali–aggregate reactions with cement alkali contents and siliceous mineral admixtures. Thus, the overall evidence was that the heating cycle developed by Scott and Duggan could be used in such a way that concretes that may take years to expand in the field could be shocked to yield very rapid expansions. However, although these tests showed the ability to differentiate between reactive and non-reactive aggregates, the authors clearly recognised that the measured expansion might be due to physical causes, such as microcracking arising from chemical

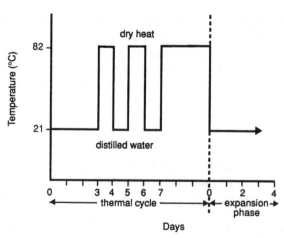

Figure 3.3 Dry heat–water immersion cycle in the Duggan test. From Ref. 35.

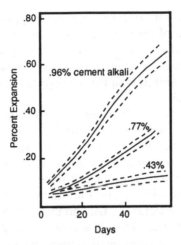

Figure 3.4 Statistical validity of the Duggan test in relation to results from field structures. —— average of 34 cores; – – – 95% confidence limits using 5 cores. From Ref. 35.

reactions due to the dry hot–cool wet immersion treatment cycle, and absorption of water by the test sample rather than by alkali–aggregate reactivity.

These concerns and suspicions of the authors were subsequently confirmed by the extensive studies reported from the University of Calgary[36], which investigated the validity and the mechanism involved in the Duggan test. Cores taken from 10 concrete mixes made with combinations of four cements

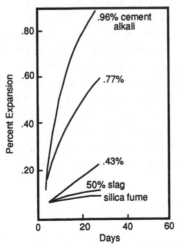

Figure 3.5 Core expansion obtained using the Duggan test from laboratory trial mixes. From Ref. 35.

with different alkali contents and three aggregate types were subjected to the Duggan test cycle. Two of the aggregates used were alkali expansive, while the third was non-expansive. The results from the Duggan test were compared with those obtained from the standard concrete prism test (CSA A 23.2–14A). The test samples were subsequently subjected to detailed petrographic examination and analysis.

The conclusions from this study prove that alkali aggregate reaction is not the major cause of expansion measured in the first 21 days of the Duggan test. The CSA concrete prism test, on the other hand, confirmed the known behaviour of the expansive and non-expansive aggregates. The expansion measured in the Duggan test was found to depend mainly on the properties of the cement. It was concluded that the amount and form of SO_3 in the unhydrated cement may affect the initial and final frequency of microcracking induced by the Duggan test cycle. It was also found that the amount of expansion in the Duggan test measured at ages up to 90 days correlated with the relative proportion of ettringite and frequency of microcrack found in the test samples.

3.5 Tentative conclusions

The overriding conclusion to be derived from all these tests, both the established conventional and the relatively new rapid tests, is that whilst they are useful the results should be interpreted with care and engineering judgement. None of the tests can be completely relied upon, whether it be the petrographic analysis, a conventional test or a rapid test. It is very clear that it is inherent in the nature of the alkali–silica reaction, and the numerous factors that influence the reaction, that no single test is likely to enable the engineer to make positive conclusions about the presence or absence of deleterious expansion. In particular, where high expansions are obtained from laboratory tests, it is important to establish whether the observed high expansions are in fact due to alkali–aggregate reactivity or the result of the methodology of the test adopted. An equally important conclusion to be deduced from a critical examination of all the available data is that the best guide to the engineer of the suitability of a given aggregate and its potential reactivity is to use a combination of petrographic examination and a laboratory test on concrete or mortar specimens incorporating the aggregates and the concrete mix proportions and constituents to be used on site.

3.6 Role of laboratory tests

It is perhaps appropriate at this stage to examine the role and value of laboratory testing in the process of controlling the effects of ASR in concrete construction. Laboratory testing for ASR will have three main objectives:

(1) To determine whether or not a set of aggregates is potentially alkali-reactive.
(2) To determine the rate of reactivity of the cement–aggregate combination in concrete, and in particular to determine whether an aggregate is a slow/late-expanding reactive aggregate and the conditions under which such reactivity is likely to be triggered.
(3) To determine the maximum expansion in concrete that is likely to occur with a given reactive aggregate, concrete constituents, mix proportions and the environmental conditions to which the structure is likely to be exposed during its lifetime.

It will be clear straight away that information on all the above three counts is needed. Previous discussions have shown that a combination of laboratory studies and known field performance is probably the best way to obtain the required information, but that the importance of utilising these test results with engineering judgement cannot be overemphasised. It should be recognised that it is just as important to identify aggregates that are or may be potentially deleteriously reactive as to ensure that aggregates that will perform satisfactorily in moderate field exposure conditions are not rejected. There is also the difficulty and possible confusion as to which test method to use, particularly when the literature appears to show that many test methods are effective and yield some form of information. With slow/late-expanding reactive aggregates, it is important to ensure that the test method used does not cause leaching of alkali from the concrete. It may then be necessary to enhance the alkali level of the mortar or concrete used in the test to identify the potential expansivity of the aggregate. It must also be decided whether the effect of leaching on the progress of the reaction should be nullified by adding extra alkali to the concrete mixture *or* by increasing the cement content to unnaturally high levels. Both approaches may initially appear to be acceptable and without objection, but the real implications of artificially increasing the alkali level of the concrete mixture are not at the moment fully known. A non-reactive aggregate may not appear to be reactive under such high alkali levels, but marginally reactive aggregates may give misleading results which could be confusing, particularly if they have a good record of field performance.

From an engineering point of view, it is also important to know the maximum expansion that is likely to arise in the field with a known reactive aggregate. The use of artificially high alkali levels in concrete prism tests can then be very unrealistic. There is also the added question as to whether mortar bar or concrete prism expansion tests should be used to determine maximum expansion levels. Mortar bar tests use high levels of cement content, at least higher than those used in practice in concrete construction, and the results obtained from such tests may not readily be applicable to concrete.

There are many reasons why concrete prism tests are preferable to mortar bar tests[37]. Although alkalis usually come from cement, they may also be derived from other concrete constituents, including aggregates, chemical and mineral admixtures and mixing water[38]. Alkalis may also penetrate into concrete from outside sources such as de-icing salts, seawater and the marine environment[39]. As shown in previous chapters, the environment also has an overriding influence on both the alkali–silica reaction and the resulting expansion. Only in a concrete prism test can all these numerous influences be realistically modelled[40]. The ideal test to determine the maximum measurable expansion is thus a concrete prism test in which the mix constituents and proportions are realistically reproduced. To obtain information on the maximum possible expansion, temperature and humidity conditions most favourable for ASR should be used as storage conditions for the prism test. Immersing these samples in water and alkaline solutions can simulate the penetration of alkalis from external sources, and can establish the full expansion potential of the reactive aggregates[39,41–43]. Using a concrete prism incorporating the correct mix constituents and proportions will have the added advantage of simulating the porosity and permeability properties of the field concrete which determine the rate and extent of ingress of external alkalis.

3.6.1 *Expansion criteria*

Associated with the assessment of the laboratory testing for ASR is the question of acceptable limits of expansion for practical design. Until recently the results from mortar bar and concrete prism expansion tests were evaluated according to the recommendations contained in ASTM C33[44], namely 0.05% at 3 months and 0.10% at 6 months. These expansion limits have long been considered to be inadequate and not stringent enough, and many national and regional organisations such as the Ontario Hydro, Bureau of Reclamation, Canadian Standards Association and the Japanese Ministry of Construction have specified their own expansion limits for reactive and slow/late-expanding reactive aggregates[1,2,22,25,33].

It is, however, important to recognise that such expansion limits, whilst being reasonably good guides to the identification of the expansive potential of aggregates and of concrete containing such aggregates, should be viewed with care and caution. Expansive strains, *per se*, are not the only factors to be considered in design, since different reactive aggregates affect other engineering properties of the ASR-affected concrete at different rates. Table 3.5, for example, shows the number of days required to reach the ASTM C227 limits for concrete containing 5.2 kg/m^3 Na$_2$O eq., two reactive aggregates and exposed to $20 \pm 1°C$ and $96 \pm 2\%$ RH, as well as the strength and elastic properties of the deteriorating concrete[45]. The data in Table 3.5 show that the two important properties most affected by ASR are the flexural/tensile strength and elastic modulus. These are also two of the most impor-

Tables **3.5** Properties of ASR-affected concrete at ASTM C227 limits. From Ref. 45.

Property of concrete	Expansion limits							
	0.05%				0.1%			
	Opal		Fused silica		Opal		Fused silica	
	Property at 0.05% expansion	Per cent loss compared with control at the same age	Property at 0.05% expansion	Per cent loss compared with control at the same age	Property at 0.1% expansion	Per cent loss compared with control at the same age	Property at 0.1% expansion	Per cent loss compared with control at the same age
Number of days	6	—	40	—	8	—	60	—
Compressive strength (N/mm²)	43.6	9.2	53.5	12.3	44.5	11	54.5	11.4
Modulus of rupture (N/mm²)	—	—	3.7	29.9	—	—	2.75	48.3
Split tensile strength (N/mm²)	—	—	3.2	27.1	—	—	2.95	28.9
Dynamic modulus (kN/mm²)	34.0	15.5	38.0	11.1	26.0	36.7	34.5	20.3
Pulse velocity (km/s)	4.08	7.2	4.5	2.1	3.95	11.2	4.35	5.8

tant parameters in engineering design related to cracking, deformation, serviceability and durability of structural members. The losses in tensile strength and flexural stiffness do not occur at the same rate or in proportion to expansion undergone by the concrete, and since expansions can be excessive or deleterious without evidence of extensive cracking, depending on the function of the concrete element, it is difficult to see how single values of critical limits of expansion can be applicable to all situations. Thus the same values of critical harmful expansive limits cannot be specified for all types of structures without regard to either their effect on engineering properties or the type of reactive aggregate and the rate of reactivity that gives rise to the rate of loss in these properties.

3.7 Non-destructive test methods

Non-destructive test (NDT) methods have been widely used to evaluate existing structures, to detect imperfections, weaknesses and deterioration, and in post-mortem examinations of structural distress. Durability-related non-destructive testing has also gained considerable prominence in recent times.

The use of dynamic NDT methods such as pulse velocity and dynamic modulus to monitor the initiation and progress of concrete deterioration is well established. Since resonance frequency and ultrasonic pulse velocity are theoretically interrelated, it is natural to expect that changes in these two properties will follow a similar pattern of behaviour for both sound and deteriorating concrete. Resonance frequency testing leads directly to the evaluation of dynamic modulus, and such tests are used extensively in assessing damage to concrete by environmental regimes such as freezing and thawing. The pulse velocity is a measure of internal deterioration and microcracking, but its wider use has been hampered to some extent because of the fact that the measurements do not evaluate a readily recognisable aspect of strength directly, and that many factors, notably the presence of moisture and reinforcing bars, influence the readings. However, apart from being fully non-destructive, the pulse velocity technique has several inherent advantages—it penetrates the full depth of the member, the velocity measurements can be taken quickly and repeatedly, and with care and skill accurately, and it can be easily combined to advantage with other in-place test methods. The method has been used fairly widely to evaluate various aspects of concrete quality, and it has been shown that the test method can be adapted to assess *in-situ* strength of concrete in real structures and the extent of concrete deterioration as in high-alumina cement concrete[46,47].

Although ASR is a relatively new phenomenon, past experience with ASR deterioration has indicated a close link between the expansion resulting from the internal chemical reactions and the damage to the microstructure through cracking and the resulting loss in material strength and stiffness, so that attempts have been made in the past to correlate ASR expansion with properties obtained from non-destructive tests. For example, resonance frequency test methods have been used to monitor the influence of various factors on ASR involving opaline sandstone—factors such as bulk density and grain size of reactive aggregate, relative humidity and curing conditions[48,49]. Figures 3.6 and 3.7 show typical data of the variation of expansion and resonant frequency with bulk density and grain size of the reactive aggregate, and the results show good correspondence between expansion and resonant frequency. Similar results were also obtained when determining the minimum humidity required for ASR (Figure 3.8). Lenzer and Ludwig[48] observed that, when the test specimens expanded rapidly, their natural frequency decreased considerably. When the RH regime was changed for a given test specimen, the resulting expansions and resonant frequency showed good correspondence. These tests carried out over more than a decade ago show that non-destructive tests can be used to monitor the changes in expansion due to ASR occurring either because of changes in the characteristics of the reactive aggregate or because of changes in the environment.

Resonant frequency has again shown good correlation with expansion in tests in which measures to counteract expansive strains, such as the addition

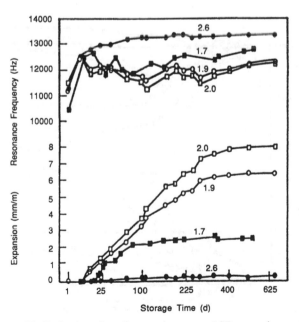

Figure 3.6 Effect of bulk density of opaline sandstone on ASR expansion and the resulting resonant frequency. From Ref. 49.

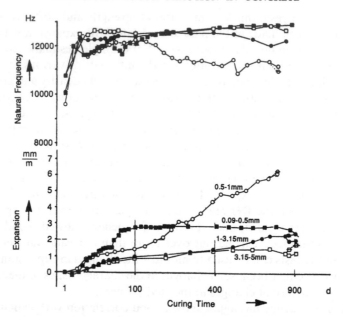

Figure 3.7 Influence of grain size of reactive aggregate on ASR expansion and resonant frequency. Specimen size: 40 × 40 × 160 mm. Mortar mix: 1 : 3 : 0.5. Cement: PZ 450F. Alkali content: 0.9% sod. ox. eq. Reactive agg.: opaline sandstone 4% by wt. of agg. From Ref. 48.

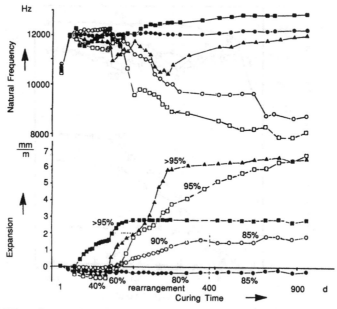

Figure 3.8 Effect of atmospheric humidity on ASR and resonant frequency. Conditions as for Figure 3.7. From Ref. 48.

or replacement of cement with fly ash (Figure 3.9) or the use of alkyl-alkoxysilane solution to impregnate the specimens (Figure 3.10), have been used. The results again confirm that the influence of mineral admixtures and surface impregnation on ASR can be clearly recognised by non-destructive tests. Although no quantitative relationships were established, these results show conclusively that non-destructive tests can be used to monitor the initiation, progress and control of ASR.

Since resonant frequency and pulse velocity are theoretically interrelated, changes in pulse velocity can also be expected to show a similar pattern of behaviour to changes in resonant frequency. However, the use of ultrasonic methods to monitor and/or estimate deterioration due to ASR has been rather limited. Akashi et al.[50] measured pulse velocity on sound and ASR-affected normal concretes; they used laboratory-made reinforced concrete specimens and cores drilled from these specimens, and found that the residual expansion of the cores could be estimated by the pulse velocity method. They also reported that the degree of deterioration of concrete due to ASR can be estimated by the spectral analysis of ultrasonic pulse waves (i.e. ultrasonic spectroscopy) passing through the concrete by applying the linear system theory.

The use of ultrasonic pulse velocity to determine the state of disruption of ASR-affected concrete in reinforced concrete bridge structures has also been reported from South Africa[51]. While the pulse velocity is highly sensitive to internal deterioration, it cannot be used directly, without special calibration, to assess the strength or changes in the strength of the concrete, because of the exponential nature of the strength–pulse velocity relationship[46]. It follows

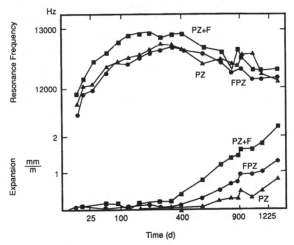

Figure 3.9 Influence of low-alkali Portland cement with fly ash and of fly ash as cement replacement on ASR expansion and resonant frequency. From Ref. 49.

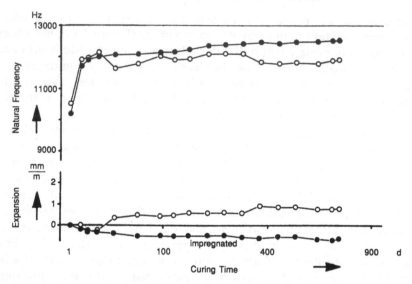

Figure 3.10 Effect of surface impregnation on ASR expansive strains and resonant frequency. Conditions as for Figure 3.7. From Ref. 48.

therefore that neither can the strength of deteriorated concrete be assessed directly from pulse velocity measurements, and this is further confirmed by work reported from South Africa[51]. Where the residual compressive strength after ASR cracking is very high, pulse velocities are unlikely to show significant reductions, and this explains why a minimum pulse velocity of 4.7 km/s was observed by Hobbs for ASR-affected cracked concrete with residual compressive strengths of 80–115 MPa[52]. At such high strength and pulse velocity levels, pulse velocity methods may appear to be too insensitive to evaluate the extent of ASR damage, although the pulse velocity may show the general trend of reduction with increasing crack width.

Pulse velocity measurements on laboratory-made reinforced concrete beams have shown that the difference in pulse velocity between the control beam and the ASR-affected beam is only of the order of 10% once the effects of reaction have stabilised, indicating that ASR can often be a surface phenomenon, which does not always extend completely into the core of the section[53]. Field measurements confirm this: measurements on the beam of a reinforced concrete portal frame showed that the reduction in pulse velocity was caused by a thin shell of deteriorated concrete surrounding an almost unaffected interior[54]. Measurements on a large badly cracked foundation block also showed similar results: core tests further confirmed that deterioration due to ASR did not extend beyond 300 mm from the surface of the block[54]. Pulse velocity techniques can thus help in establishing the extent of

ASR damage, and together with core tests or load tests can give the structural engineer an excellent method of assessment of structural distress.

Pulse velocity measurements were also used to establish the extent of ASR damage in the reinforced concrete piers of Hanshin Expressway in Japan[55]. The measured crack depth obtained from pulse velocity measurements was about 100 mm, which was almost equal to the clear cover of the reinforcing steel. Other non-destructive tests such as acoustic emission and electrical resistivity measurements have also been used to monitor ASR expansion, but these are still at a preliminary stage[2].

3.7.1 Effect of ASR on pulse velocity and dynamic modulus

In these tests two reactive aggregates were used—a naturally occurring but highly and rapidly expansive reactive aggregate, a Beltane opal, and a slow, more moderately reactive aggregate, a synthetic amorphous fused silica[56]. The concrete used in the tests had a cement content of 520 kg/m³, with an ordinary Portland cement (ASTM type 1) having a high alkali content of 1% Na_2O eq. The fine aggregate was a washed and dried natural sand, and the coarse aggregate consisted of a mixture of rounded and crushed gravel of 10 mm maximum size. The test specimens used were 100 mm cubes for compressive strength, $75 \times 75 \times 300$ mm prisms for expansion tests and $100 \times 100 \times 500$ mm prisms for pulse velocity measurements, examination of cracking and dynamic modulus test. The specimens were demoulded at 1 day and then cured under controlled conditions of $20 \pm 1°C$ and $96 \pm 2\%$ RH.

The results of the tests are shown in Figures 3.11 and 3.12, and Tables 3.6

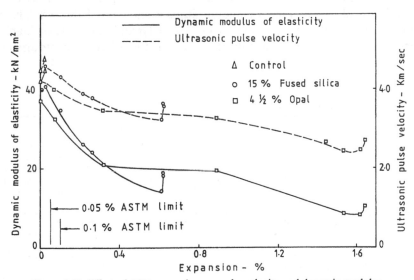

Figure 3.11 Effect of ASR expansion on pulse velocity and dynamic modulus.

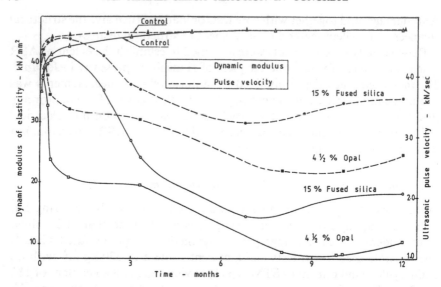

Figure 3.12 Variation of pulse velocity and dynamic modulus of ASR-affected concrete with time.

to 3.8. Table 3.6 summarises the expansion at different ages for the two reactive aggregates, opal and fused silica, and the corresponding pulse velocities and dynamic moduli. Figure 3.11 shows the variation of both these properties with expansion, while Figure 3.12 shows their variation with time. The loss in pulse velocity with expansion at different ages is summarised in Table 3.7, while the loss in dynamic modulus is shown in Table 3.8. These data show positively that both pulse velocity and dynamic modulus are highly sensitive to changes occurring in the internal structure of the concrete arising

Table 3.6 Effects of ASR expansion on pulse velocity and dynamic modulus.

Test	Mix	Age (days)							
		1	2	3	7	10	28	100	365
(1) Expansion (%)	(a) Control	0.0	0.0	0.0	0.0	0.001	0.003	0.017	0.021
	(b) 4½% opal	0.0	0.0	0.004	0.071	0.097	0.316	0.883	1.644
	(c) 15% fused silica	0.0	0.0	0.0	0.0	0.005	0.023	0.259	0.623
(2) Ultrasonic pulse velocity (km/s)	(a) Control	4.28	4.48	4.55	4.60	4.64	4.67	4.71	4.78
	(b) 4½% opal	4.12	4.27	4.32	4.02	3.70	3.48	3.29	2.70
	(c) 15% fused silica	—	4.45	—	4.57	4.59	4.61	3.80	3.64
(3) Dynamic modulus of elasticity (GPa)	(a) Control	35.6	38.1	38.8	41.0	41.1	42.5	44.2	45.4
	(b) 4½% opal	33.9	36.3	37.5	32.7	23.7	20.8	19.6	10.4
	(c) 15% fused silica	—	37.0	—	39.5	40.2	40.8	24.0	18.9

Table 3.7 Percentage loss in pulse velocity of ASR-affected concrete.

Age (days)	4½% opal		15% fused silica	
	Expansion (%)	Loss (%)	Expansion (%)	Loss (%)
2	0.0	1.1	0.0	< 1.0
7	0.071	9.5	0.0	< 1.0
10	0.097	17.0	0.005	1.0
28	0.316	23.0	0.023	1.3
100	0.883	30.1	0.259	19.3
204	1.442	44.0	0.615	32.0
300	1.618	48.9	0.625	25.3
365	1.644	43.5	0.623	23.8

from material damage due to ASR. It is also worth noting that both these properties registered measurable losses at a very early age even before any expansion could be measured, i.e. long before any change in the physical properties or visible cracking had taken place.

These data also emphasise that there is no unique relationship between expansion and loss in pulse velocity or loss in dynamic modulus. Since losses in engineering properties do not occur at the same rate or in proportion to expansion, unique relationships between expansion and pulse velocity or loss in pulse velocity or dynamic modulus cannot be expected.

Tables 3.7 and 3.8 confirm further that the magnitude of the changes in pulse velocity and dynamic modulus are different for a given reactive aggregate, and also different for different reactive aggregates. The losses are much more rapid, dramatic and high for a highly and rapidly reactive aggregate such as opal, and much less for a moderately reactive aggregate such as fused silica. The losses in dynamic modulus are again much higher than those in pulse velocity, emphasising that the effects of ASR on stiffness are more

Table 3.8 Percentage loss in dynamic modulus of ASR-affected concrete.

Age (days)	4½% opal		15% fused silica	
	Expansion (%)	Loss (%)	Expansion (%)	Loss (%)
2	0.0	4.6	0.0	2.7
7	0.071	20.3	0.0	3.8
10	0.097	42.3	0.005	2.3
28	0.316	51.1	0.023	4.0
100	0.883	55.8	0.259	45.7
204	1.442	74.7	0.615	68.0
300	1.618	81.9	0.625	59.8
365	1.644	77.1	0.623	58.4

critical than the effects on microcracking. Dynamic modulus is obviously more sensitive to changes in the concrete due to ASR than pulse velocity, but the latter is still sufficiently sensitive to enable rational engineering judgements to be made on the serviceability and integrity of a concrete structure undergoing ASR. All the data give encouraging confidence in using pulse velocity and dynamic modulus non-destructive tests to identify and assess the physical condition of concrete due to internal chemical reactions arising from ASR.

3.7.2 *Effect of expansion due to type and amount of reactive aggregate*

The data presented in Figures 3.11 and 3.12 and Tables 3.6 to 3.8 emphasise that the rate of reactivity determines the degree of deterioration due to ASR, and that the response of both the pulse velocity and dynamic modulus is very much influenced by the degree of this reactivity. The type and amount of reactive aggregate has therefore a great influence on changes in pulse velocity and dynamic modulus. This is confirmed in Figure 3.13, which shows the effect of the amount of fused silica reactive aggregate on dynamic modulus and this is compared with that due to opal. The results again confirm that dynamic tests involving dynamic modulus (and, by implication, pulse velocity) can monitor the progress of concrete deterioration due to ASR expansive strains.

In all the data presented in Figures 3.11 to 3.13, the first visible crack occurred at about 450–600 microstrains.

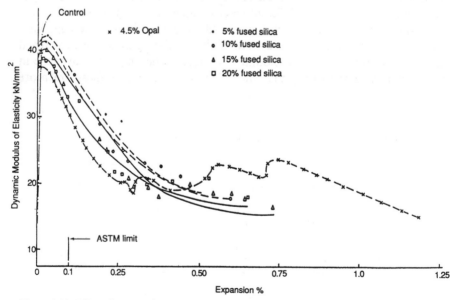

Figure 3.13 Effect of type and amount of reactive aggregate expansion on dynamic modulus.

3.7.3 *Application of NDT to concrete with mineral admixtures*

In practice controlled environmental conditions do not exist, and the ambient environment is highly variable. Further, mineral admixtures are also being increasingly used not only to improve the properties of the concrete but also to enhance its resistance to ASR. To examine the role of NDT under such conditions, tests were carried out as follows:

Series B—ASR tests with reactive aggregates
Series C—ASR tests with 50% fly ash replacement of cement
Series D—ASR tests with 50% slag replacement of cement
Series E—ASR tests with 10% microsilica added.

Only one concrete mix with a cement content of 455 kg/m^3 and mix properties of 1 : 2.2 : 1.8 : 0.5 (cement–sand–coarse aggregate–water) by weight was used throughout this series of tests. A high-alkali normal Portland cement with about 1% sodium oxide equivalent was used. The fly ash was a low-calcium type F ash, while the slag had a similar specific surface (350 m^2/kg) to that of the cement (380 m^2/kg). The cement, fly ash and slag all complied with the relevant British Standard specifications. The microsilica was added in the form of a slurry (containing 50% water and 50% solids), with appropriate corrections for the water content and sand. The fine and coarse aggregates used were the same as those reported before. The concrete mixes with fly ash and slag were proportioned by replacing the Portland cement with the mineral admixture, weight for weight, and the other constituents are then adjusted, particularly the water content[57]. The water–binder ratio was kept constant for all the concrete mixes of series B to E at 0.5. A superplasticiser, a sulphonated naphthalene formaldehyde condensate, was used in all the mixes to enhance their flow characteristics. The test specimens were initially cured for 6 days at 20 ± 1°C and 96 ± 2% RH and then subjected to a cyclic hot–wet and hot–dry regime at 38 ± 2°C.

The pulse velocity measurements for the four series of tests are shown in Figure 3.14. The fluctuation of pulse velocity with time is a reflection of the cyclic hot wet–hot dry exposure regime which alternately activated or temporarily slowed down the chemical reactivity. The results show clearly that pulse velocity measurements can follow the effects on ASR expansion arising from environmental changes, and that the role and effectiveness of the mineral admixtures are also clearly reflected on pulse velocity.

3.7.4 *Pulse velocity characteristics prior to first crack*

The pulse velocity characteristics prior to first crack occurring due to ASR are shown in Table 3.9, which provides data on the initial pulse velocity on demoulding (i.e. at 1 day), the maximum pulse velocity attained prior to the beginning of material degradation due to ASR, and the age and expansion

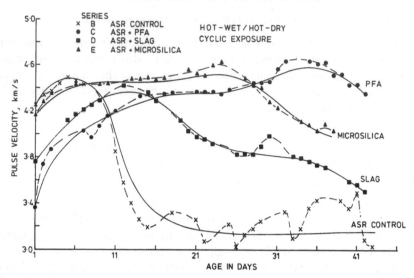

Figure 3.14 Variation of pulse velocity in ASR-affected concrete without and with mineral admixtures.

at this maximum pulse velocity. The values of initial pulse velocity soon after demoulding show the retardation in the early strength development with fly ash and slag concretes, whereas microsilica concrete achieved nearly the same pulse velocity as the concrete in series B, which had not initiated ASR reaction at this age. The initial retardation in strength development of concrete with fly ash and slag is related to mix proportioning, which is discussed elsewhere, but which can be largely counteracted by modifying the method of incorporation of the mineral admixture in concrete[57,58]. The maximum pulse velocities reached prior to ASR cracking are all of the same order for all concretes, although concrete containing both fly ash and microsilica recorded

Table 3.9 Pulse velocity (PV) characteristics prior to first crack.

| Test series | Initial PV (km/s) | Max. PV (km/s) | Properties at max. PV | |
			Age at max. PV (days)	Expansion at max. PV (%)
B—ASR	4.26	4.49	5	0.027
C—ASR+PFA	3.36	4.64	32	0.140
D—ASR+slag	3.76	4.42	12	0.039
E—SAR+MS	4.17	4.63	24	0.150

PFA, fly ash; MS, microsilica.

higher pulse velocities than concrete with the reactive aggregate alone (series B) and concrete with slag (series D). It is also to be noted that the maximum pulse velocity occurs at different ages (Figure 3.14, Table 3.9) for the different concretes B to E and at different expansions. The main point is that pulse velocity is able to follow these changes due to ASR in normal concrete and concretes with mineral admixtures.

3.7.5 *Pulse velocity properties at first crack*

Table 3.10 presents information at first cracking due to ASR—expansion, pulse velocity and age at first crack, together with the loss in pulse velocity due to first crack. A comparison with Table 3.9 shows that pulse velocity measurements can detect the formation of first crack, and that the loss in pulse velocity between the maximum value and that at first crack is less than 5%. Pulse velocity measurements are thus sensitive enough to pick up small reductions in pulse velocity due to ASR deterioration, even if such losses occur at different ages and at different expansions.

ASR cracking in series B occurred at 460 microstrains, the order of strain at which cracking generally occurs in concrete as a result of tensile strains. Suppression of the rate of chemical reaction due to ASR in series C, D and E delays the formation of cracking, and this again is picked up by pulse velocity measurements, as shown in Table 3.10. The slowing down of the rate of reaction reflected in Figure 3.14 and in Table 3.10 allows the concrete to mature, and thereby enhances the apparent tensile cracking strain of concretes containing mineral admixtures and undergoing ASR. The ASR expansive strains at first crack for concretes of series C to E are thus different, varying from 1500–1600 microstrains for fly ash and silica fume concretes to about 650 microstrains for slag concrete, emphasising again the different rates at which mineral admixtures control the alkali–silica reactions, which depend not only on the physical characteristics and mineralogical composition of the admixtures, but also on the way they are incorporated in concrete, which in

Table 3.10 Pulse velocity (PV) properties at first visible crack.

Test series	Properties at first crack			
	Expansion (%)	PV (km/s)	Age (days)	Loss in PV (%)
B—ASR	0.046	4.43	7	1.3
C—ASR+PFA	0.151	4.62	36	0.5
D—ASR+slag	0.065	4.36	15	1.4
E—ASR+MS	0.159	4.49	27	3.0

PFA, fly ash; MS, microsilica.

turn would control the development strength and tensile strain capacity[57,58]. The important point to note here again is that pulse velocity is able to monitor all these varying effects occurring at different rates and at different ages.

3.7.6 Variation of pulse velocity with expansion

The variation of pulse velocity with expansion for the concretes in series B to E is shown in Figures 3.15 to 3.18. It is again emphasised that the fluctuations in the measured pulse velocities shown in these figures are due to the cyclic environmental exposure regime of the concretes, which alternately activated and suppressed the reaction, and therefore controlled the development of the resulting expansion.

The data in Figures 3.15 to 3.18 are summarised in Tables 3.11 and 3.12. Table 3.11 shows the maximum expansions measured while the test specimens were still in the hot room and the corresponding pulse velocities and ages. Table 3.12 presents data on expansions measured after the specimens had been removed from the hot room and returned to ambient temperature and humidity, and the corresponding pulse velocities and ages. Figures 3.15 to 3.18 and the data in Tables 3.11 and 3.12 show that the progress of alkali–silica reactions and the resulting expansions can be monitored satisfactorily by pulse velocity measurements in concrete without and with mineral admixtures. The data also reveal that both the increase in expansion and its suppression are realistically reflected in these measurements.

The data presented in Figures 3.15 to 3.18 and Tables 3.11 and 3.12

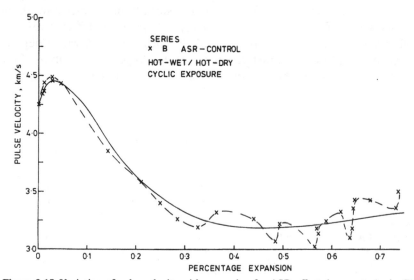

Figure 3.15 Variation of pulse velocity with expansion for ASR-affected concrete (series B).

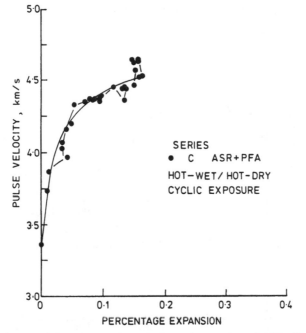

Figure 3.16 Influence of fly ash on pulse velocity and expansion of ASR-affected concrete (series C).

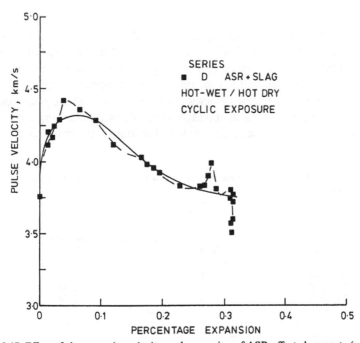

Figure 3.17 Effect of slag on pulse velocity and expansion of ASR-affected concrete (series D).

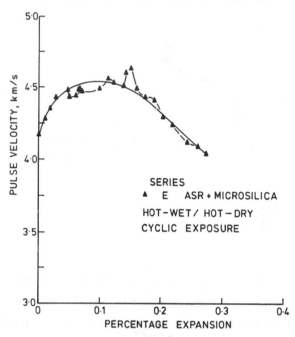

Figure 3.18 Variation of pulse velocity with expansion of ASR-affected concrete containing microsilica (series E).

emphasise two very important aspects of ASR which may not be readily recognised by researchers and practising engineers. Firstly, when the exposed environmental conditions change, resulting in the cessation of measured expansions or even their reduction, this is not immediately reflected in an increase in the measured pulse velocities. The reason is that the pulse velocity reflects the internal structural damage through microcracking and other effects, and it may take some time before healing of the internal structure

Table 3.11 Maximum measured expansions and corresponding pulse velocities.

Test series	Maximum measured values*		
	Age (days)	Expansion (%)	Pulse velocity (km/s)
B—ASR	41	0.736	3.50
C—ASR+PFA	39	0.164	4.53
D—ASR+slag	35	0.314	3.77
E—AST+MS	36	0.273	4.04

*Measured in hot room.
PFA, fly ash; MS, microsilica.

Table 3.12 Final measured pulse velocities and expansions.

Test series	Final measured values*		
	Age (days)	Expansion (%)	Pulse velocity (km/s)
B—ASR	43	0.732	3.03
C—ASR+PFA	42	0.136	4.36
D—ASR+slag	42	0.311	3.51
E—ASR+MS	38	0.256	4.04

*Measured under ambient conditions after removal from hot room.
PFA, fly ash; MS, microsilica.

begins. It can be readily seen why measured pulse velocities in real structures undergoing ASR can, if taken in isolation, result in confusion[51,52,54]. It very much depends at what point during the progress or cessation of chemical reactions the pulse velocity measurements are taken. Thus spot or isolated values of pulse velocity may not make much sense except to give an overall assessment of the extent of ASR deterioration or loss of engineering properties[45].

The second significant point is that there appears to be no unique relationship between expansion and pulse velocity, resonant frequency or dynamic modulus of ASR-affected concrete[45,48–50]. As pointed out earlier, since the rate of ASR is influenced by a host of parameters, including the nature, type and particle size distribution of the reactive aggregate as well as the unpredictable variations in the ambient environment, losses in pulse velocity do not occur at the same rate or in proportion to expansion. This is particularly true when mineral admixtures are incorporated in concrete, so that the time required to reach a given expansion (for a given set of reactive aggregate and concrete alkali conditions), and hence a given pulse velocity, will not be the same. This is illustrated by the data in Table 3.13, which

Table 3.13 Estimated pulse velocity at different expansions.

Expansion (%)	Pulse velocity (km/s)			
	B (ASR)	C (ASR+PFA)	D (ASR+slag)	E (ASR+MS)
0.05	4.43	4.20	4.32	4.50
0.10	4.23	4.40	4.27	4.53
0.15	3.90	4.50	4.08	4.47
0.20	3.64	—	3.91	4.30
0.25	3.41	—	3.83	4.13
0.30	3.33	—	3.78	—

PFA, fly ash; MS, microsilica.

summarises the pulse velocity for the concretes of series **B** to **E** to reach different rates of expansion. It should, however, be appreciated that the values of pulse velocity shown in Table 3.13 are necessarily approximate as they have been interpolated from the measured values for the expansions quoted in the same table.

3.7.7 *Pulse velocity in ASR-affected beams*

The data presented so far show that pulse velocity measurements can be used to identify both the initiation and progress of expansion due to ASR in hardened plain concrete. Recent tests have shown that these measurements can be satisfactorily extended to reinforced concrete beams undergoing alkali–silica reactions and expansive strains[59]. Figure 3.19 shows the variation of pulse velocity in two reinforced concrete beams containing opal reactive aggregate (beam B2) and fused silica (beam B5) compared with that in a control beam (B1) without reactive aggregate. These results again confirm that the response of pulse velocity measurements depends on the rate of reactivity and that this reactivity can be satisfactorily monitored.

Figures 3.20 and 3.21 show the pulse velocity variations with time when 30% and 50% fly ash were incorporated into the beams with the same two types of reactive aggregates. With opal, 50% cement replacement by fly ash was fully effective and able to restore the pulse velocity of the ASR-affected concrete to practically the same values as in the control beam without opal. This was also reflected in the behaviour of the beams, in which both the hogging deflection and cracking due to ASR were completely eliminated in beam B4 with 50% fly ash (Table 3.14).

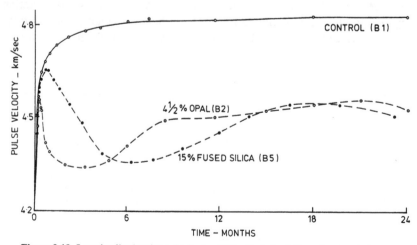

Figure 3.19 Longitudinal pulse velocity in ASR-affected reinforced concrete beams.

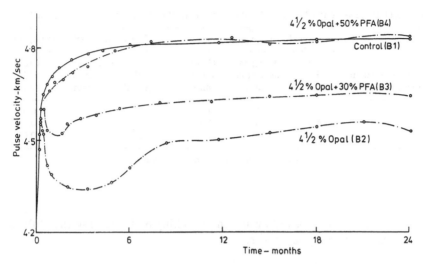

Figure 3.20 Effect of fly ash on pulse velocity values in ASR-affected beams containing opal reactive aggregate.

On the other hand, the 50% fly ash replacement was not as fully effective with fused silica reactive aggregate. The beam B7 with 50% ash retained some hogging deflection and cracking due to ASR compared with the control beam B1 (Table 3.14). These data prove conclusively that pulse velocity techniques can be confidently used in reinforced concrete beams containing fly ash and that the variations in pulse velocity truly reflect the effectiveness

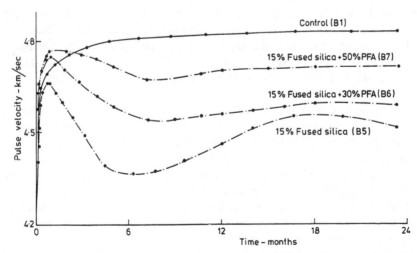

Figure 3.21 Influence of fly ash on pulse velocity variations in ASR-affected beams containing fused silica reactive aggregate.

Table 3.14 Pulse velocity (PV) and related data in ASR-affected beams containing fly ash.

Fly ash content replacement (%)	Opal reactive aggregate			Fused silica reactive aggregate		
	PV (km/s)	Hogging deflection (mm)	Crack width (mm)	PV (km/s)	Hogging deflection (mm)	Crack width (mm)
0	4.52	6	1.05	4.61	4	0.40
30	4.64	3	0.15	4.66	2	0.21
50	4.83	0	0	4.75	1	0.09

Control: PV, 4.82 km/s; Hogging deflection, 0 mm; crack width, 0 mm.

of the ash in controlling expansion and cracking and restoring the stiffness of the beam to its original values.

3.8 Conclusions

The aim of this chapter is to present a critical review of the various test methods currently available for ASR testing. These include petrographic studies, well-established traditional test methods, the new rapid test methods and non-destructive test methods. Whatever their limitations, laboratory testing is the starting point in the evaluation of aggregates and concretes for their susceptibility to ASR, supplemented by known field experience and practical performance of these aggregates in concrete.

Thorough petrographic examination together with other instrument techniques such as X-ray diffraction and infra-red spectroscopy is the first step in the evaluation of the suitability of aggregates for concrete. These test methods and the structure and texture of silica minerals are briefly discussed. The limitations of petrographic studies are highlighted, and these tests should not therefore be assumed to guarantee information on the development or absence of ASR.

The well-established traditional test methods to evaluate the potential reactivity of aggregates such as the ASTM mortar bar test, the Canadian concrete prism test and the ASTM chemical method are considered. The advantages and limitations of these test methods are fully discussed, and the modifications required if these test methods are to give reasonably reliable results are suggested. Attention is drawn to the dependence of these test methods on storage conditions and test methodologies, and the considerations to be given when evaluating slow/late-expanding alkali–silicate/silica reactive aggregates are emphasised.

Many new and rapid test methods for ASR have recently been reported but several of these show poor precision when subjected to multilaboratory tests and poor correlation with field performance. The more relevant and appropriate of these test methods are evaluated in the light of existing data, and current thinking on the most appropriate of all these tests is discussed. Special attention is given to the so-called Duggan test, and it is shown that alkali–aggregate reaction is not the major cause of expansion measured in the first 21 days of this test.

The very special care required in laboratory testing of concrete cores taken from ASR-affected structures is emphasised, and the need for a clear understanding of ASR and the engineering judgement required in interpreting these test results cannot be overstated.

The overriding conclusion to be derived from all these tests is that it is inherent in the nature of ASR, and the numerous factors that influence the reaction, that no single test, wherever its nature, can be completely relied upon. It is important to assess the role of laboratory testing, its aims and objectives, and bear these in mind when specifying acceptable limits of expansion for practical design. Expansion strains, *per se*, are not the only properties to be considered in design, since different reactive aggregates affect the engineering properties of the ASR-affected concrete at different rates.

From an engineering point of view a combination of petrographic analysis, knowledge of field performance and a mortar bar or concrete prism test is likely to give the best information required to avoid deleterious expansions in concrete construction. The concrete prism test is in general to be preferred to a mortar bar test, since only in such a test can all the numerous factors which influence ASR, such as the mix constituents, mix proportions and the effects of external environment on the porosity and permeability of the material, be realistically modelled. It should be borne in mind that, notwithstanding the effects of all the factors that influence ASR, the environment is the most critical parameter, and this is precisely the one factor that is beyond the control of human beings.

This chapter also draws special attention to the use of non-destructive test methods to identify ASR and monitor its progress and control, and their application to field testing. The pulse velocity and dynamic modulus tests are especially identified as very useful in this respect. Data are presented on the use of these two test methods to concrete containing a highly and rapidly reactive aggregate as well as a slowly and moderately reactive aggregate without and with mineral admixtures such as fly ash, slag and microsilica.

The results show conclusively that both pulse velocity and dynamic modulus can be used to quantify the deterioration of concrete due to ASR. Both these properties are highly sensitive to material and structural changes arising from ASR, and they reflect changes in their values even before any expansion and visible cracking occurs. There is, however, no unique relationship between expansion and pulse velocity or dynamic modulus, and changes in the values of these properties are governed by the rate of reactivity, i.e. by the type and quantity of reactive aggregate in a given concrete and the environment. Pulse velocity and dynamic modulus measurements also reveal that loss of elasticity and stiffness is far greater and more critical to the stability and durability of ASR-affected concrete than microcracking arising from the expansive strains.

Pulse velocity measurements can also respond to changes taking place in ASR-affected concrete prior to first crack, at first crack, and subsequently during the progress of deterioration with time. Changes in environmental conditions which activate or suppress the ASR reactions can also be satisfactorily monitored. The role and effectiveness of mineral admixtures such as fly ash, slag and microsilica can also be sensitively and satisfactorily monitored. This is confirmed by test data obtained from reinforced concrete beams containing varying amounts of fly ash and two types of reactive aggregates.

Dynamic modulus values are generally more sensitive than pulse velocity measurements, although the latter is adequately sensitive for application in real structures. Because of the exponential nature of the pulse velocity–strength relationship, isolated or spot measurements or measurements in concrete of high residual strength and with high levels of pulse velocity may appear to be too insensitive or confusing to assess changes in deterioration. However, if used with engineering judgement, and with other tests where appropriate, pulse velocity can not only reliably monitor the varying effects occurring at different rates and different ages but also provide the engineer with an excellent method of evaluating structural deterioration.

Perhaps the most important lesson to be learnt from existing literature on ASR testing is that all ASR-related tests should be carried out with considerable care and understanding of ASR, and the results need to be interpreted with engineering judgement. Valuable laboratory test data on the potential for deleterious expansion can be used to overcome structural distress in practice through design, detailing and suitable mix constituents and mix proportioning.

References

1. Grattan-Bellew, P.E. (Ed.) (1987) *Proceedings of the Seventh International Conference on Concrete Alkali–Aggregate Reactions.* Noyes Publications, Far Ridge, NJ.
2. Okada, K., Nishibayashi, S. and Kawamura, M. (Eds.) (1989) *Proceedings of the Eighth International Conference on Alkali–Aggregate Reaction.* The Society of Materials Science, Kyoto, Japan.
3. Tordoff, M.A. (1990) Assessment of prestressed concrete bridges suffering from alkali silica reaction. *Cement and Concrete Composites* Vol. 12, No. 3. Elsevier, Barking, UK, pp. 203–210.
4. British Cement Association (1988) *The Diagnosis of Alkali–Silica Reaction.*
5. Swamy, R.N. and Al-Hamed, A.H. (1984) Evaluation of small diameter core tests to determine in situ strength of concrete. ACI Publn. SP-82, *In Situ/Non-Destructive Testing of Concrete*, American Concrete Institute, Detroit, MI, pp. 411–440.
6. Munday, J.G.L. and Dhir, R.K. (1984) Assessment of in situ concrete quality by core testing. ACI Publn. SP-82, *In Situ/Non-Destructive Testing of Concrete*, American Concrete Institute, Detroit, MI, pp. 393–410.

7. Swamy, R.N. and Ali, A.M.A.H. (1984) Assessment of in situ concrete strength by various non-destructive tests. *NDT International* **17** (3), 139–146.
8. Berube, M.A. and Fournier, B. (1990) Testing field concrete for further expansion by alkali–aggregate reactions. Report EM-92, *Canadian Developments in Testing Concrete Aggregates for Alkali–Aggregate Reactivity.* Ministry of Transportation, Ontario, pp. 162–180.
9. Dolar-Mantuani, L. (1978) Practical aspects of identifying alkali-reactive aggregates by petrographic methods. *Proceedings of the Fourth International Conference on Effects of Alkalis in Cement and Concrete*, pp. 267–280.
10. British Standards Institution (1984) BS812, *Testing Aggregates*; Part 102 *Methods for Sampling*.
11. British Standards Institution (1984) BS6100: Section 5.2: *British Standard Glossary of Building and Civil Engineering Terms*, Part 5 Masonry, Section 5.2 Stone.
12. American Society for Testing and Materials, *Annual Book of Standards*, 4.02 *Concrete and Aggregates*.
13. *Specification for Highway Works*, Part 5, 1986. London, HMSO.
14. American Society for Testing and Materials (1988) ASTM C227-87, Standard test method for potential alkali reactivity of cement-aggregate combinations (Mortar Bar Method). *Annual Book of ASTM Standards*, 4.02 *Concrete and Aggregates*.
15. Canadian Standards Association (1986) CSA A23.2-14A, Supplement No. 2-1986 to *Standards Concrete Materials and Methods of Concrete Construction*, CAN 3-A 23.1-M77, and *Methods of Test for Concrete*, CAN 3-A 23.2-M77.
16. American Society for Testing and Materials (1988) ASTM C289-86, Standard test method for potential reactivity of aggregates (chemical method). *Annual Book of ASTM Standards*, 4.02 *Concrete and Aggregates*.
17. Rogers, C.A. and Hooton, R.D. (1989) Leaching of alkalis in alkali–aggregate reaction testing. *Proceedings of the Eight International Conference on Alkali–Aggregate Reaction*, pp. 327–332.
18. Grattan-Bellew, P.E. (1989) Test methods and criteria for evaluating the potential reactivity of aggregates. *Proceedings of the Eighth International Conference on Alkali–Aggregate Reaction*, pp. 279–294.
19. Hooton, R. D. (1986) Effect of containers on ASTM C 441—Pyrex mortar bar expansions. *Proceedings of the Seventh International Conference on Alkali–Aggregate Reaction in Concrete*, pp. 351–357.
20. British Standards Institution (1988) Draft BS812: Part 123, *Method, Concrete Prism Method, Testing Aggregates, Methods for the Assessment of Alkali-Reactivity Potential.*
21. Rogers, C.A. (1990) Interlaboratory study of the concrete prism expansion test for the alkali–carbonate reaction. Report EM-92, *Canadian Developments in Testing Concrete Aggregates for Alkali–Aggregate Reactivity.* Ministry of Transportation, Ontario, pp. 136–149.
22. Rogers, C.A. (1990) Concrete prism expansion testing to evaluate slow/late expanding alkali–silicate/silica reactive aggregates. Report EM-92, *Canadian Developments in Testing Concrete Aggregates for Alkali–Aggregate Reactivity.* Ministry of Transportation, Ontario, pp. 96–110.
23. Berard, J. and Roux, R. (1986) La viabilité des bétons du Québec: le rôle des granulats. *Can. J. Civ. Engng* **13**, 12–24.
24. Fournier, B. and Berube, M.A. (1990) Evaluation of a modified chemical method to determine the alkali-reactivity potential of siliceous carbonate aggregates. Report EM-92, *Canadian Developments in Testing Concrete Aggregates for Alkali–Aggregate Reactivity.* Ministry of Transportation, Ontario, pp. 118–135.
25. Hooton, R.D. and Rogers, C.A. (1988) Evaluation of rapid test methods for detecting alkali reactive aggregates, *Proceedings of the Eighth International Conference on Alkali–Aggregate Reaction*, pp. 439–444.
26. Oberholster, R.E. and Davies, G. (1986) An accelerated method for testing the potential alkali reactivity of siliceous aggregates. *Cement Conc. Res.* **16**, 181–189.
27. Davies G. and Oberholster, R.E. (1987) An interlaboratory test programme on the NBRI accelerated test to determine the alkali-reactivity of aggregates, *CSIRO Special Report BOU 92-1987*. National Building Research Institute.
28. Duncan, M.A.G., Swenson, E.G., Gillott, J. E. and Foran, M.R. (1973) Alkali–aggregate reaction in Nova Scotia. I. Summary of a five year study. *Cement Conc. Res.* **3**, 55–69.

29. Chatterji, S. (1978) An accelerated method for the detection of alkali–aggregate reactivities of aggregates. *Cement Conc. Res.* **8**, 647–650.
30. Thaulow, N. and Olafsson, H. (1983) Alkali–silica reactivity of sands, comparison of various test methods. *Proceedings of the Sixth International Conference on Alkalis in Concrete, Copenhagen*, pp. 359–365.
31. Nishibayashi, S., Yamura, K. and Matsushita, H. (1986) A rapid method of determining the alkali aggregate reaction in concrete by autoclave. *Proceedings of the Seventh International Conference on Concrete Alkali–Aggregate Reactions*, pp. 299–303.
32. Tang, Ming-shu, Han, Su-fen and Zhen, Shi-hua (1983) A rapid method for identification of alkali reactivity of aggregate. *Cement Conc. Res.* **13**, 417–422.
33. Grattan-Bellew, P.E. (1990) Canadian experience with the mortar bar accelerated test for alkali–aggregate reactivity. Report EM-92, *Canadian Developments in Testing Concrete Aggregates for Alkali–Aggregate Reactivity*. Ministry of Transportation, Ontario, pp. 17–34.
34. Hooton, R.D. (1990) Interlaboratory study of the NBRI rapid test method and CSA standardization status. Report EM-92, *Canadian Development in Testing Concrete Aggregates for Alkali–Aggregate Reactivity*. Ministry of Transportation, Ontario, pp. 225–240.
35. Scott, J.F. and Duggan, C.R. (1986) Potential new test for alkali–aggregate reactivity. *Proceedings of the Seventh International Conference on Concrete Alkali–Aggregate Reactions*, pp. 319–323.
36. Jones, T.N., Grabowski, E., Quinn, T., Gillott, J.E., Duggan, C.R. and Scott, J.F. (1990) Mechanism of expansion in Duggan test for alkali–aggregate reaction. Report EM-92, *Canadian Developments in Testing Concrete Aggregates for Alkali–Aggregate Reactivity*. Ministry of Transportation, Ontario, pp. 70–82.
37. Swamy, R.N. and Al-Asali, M.M. (1988) Expansion of concrete due to alkali–silica reaction. *ACI Materials Journal* **85**, 33–40.
38. Stark, D. and Bhatty, M.Y.S. (1986) Alkali silica reactivity: effect of alkali in aggregate on expansion. *Alkalis in Concrete*, ASTM STP 930, pp. 16–30.
39. Swamy, R. N. and Al-Asali, M.M. (1988) Alkali-silica reaction—sources of damage. *Highways and Transportation* **35**(12), 24–29.
40. Swamy, R. N. and Al-Asali, M.M. (1986) New test methods for alkali–silica reaction. *Proceedings of the Seventh International Conference on Concrete Alkali–Aggregate Reactions*, pp. 324–329.
41. Albert, P. and Raphael, S. (1987) Alkali–silica reactivity in the Beauharnois powerhouses, Beauharnois. *Proceedings of the Seventh International Conference on Concrete Alkali–Aggregate Reactions*, pp. 10–16.
42. Visvesuaraya, H.C., Rajkumar, C. and Mullick, A.K. (1981) Analysis of distress due to alkali–aggregate reaction in gallery structures of a concrete dam. *Proceedings of the Seventh International Conference on Concrete Alkali–Aggregate Reactions*, pp. 188–193.
43. Stark, D. and Depuy, G. (1987) Alkali–silica reaction in five dams in Southwestern United States. *Proceedings of the Katherine and Bryant Mather International Conference on Concrete Durability, Atlanta*. ACI SP-100, pp. 1759–1786.
44. American Society for Testing and Materials (1988) ASTM C33-86, Standard specification for concrete aggregates, *Annual Book of Standards*.
45. Swamy, R.N. and Al-Asali, M.M. (1988) Engineering properties of concrete affected by alkali–silica reaction. *ACI Materials Journal* **85**, 367–374.
46. Swamy, R.N. and Al-Hamed, A.H. (1984) The use of pulse velocity measurements to estimate strength of air-dried cubes and hence in situ strength of concrete. *In Situ/Nondestructive Testing of Concrete*, ACI Publ. SP-82, pp. 247–276.
47. Bungey, J.H. (1974) Ultrasonic pulse testing of high alumina cement concrete. *Concrete* **8**(9), 39–41.
48. Lenzner, D. and Ludwig, U. (1978) The alkali aggregate reaction with opaline sandstone from Schleswig-Holstein. *Proceedings of the Fourth International Conference on the Effects of Alkalis in Cement and Concrete*, pp. 11–34.
49. Ludwig, U. (1981) Theoretical and practical research on the alkali silica reaction. *Proceedings of the Fifth International Conference on Alkali–Aggregate Reaction in Concrete*, S252-24, pp. 1–7.
50. Akashi, T., Amasaki, S. and Takagi, N. (1986) The estimate for deterioration due to

alkali–aggregate reaction by ultrasonic methods. *Proceedings of the Seventh International Conference on Concrete Alkali-Aggregate Reactions, Ottawa*, pp. 183–187.

51. Blight, G.E., McIver, J.R., Schutte, W. R. and Rimmer, R. (1981) The effects of alkali aggregate reaction on reinforced concrete structures made with Witwatersrand quartzite aggregate. *Proceedings of the Fifth International Conference on Alkali–Aggreate Reaction in Concrete*, S252/15, pp. 1–13.

52. Hobbs, D.W. (1988) Some tests on fourteen year old concrete affected by the alkali silica reaction. *Proceedings of the Seventh International Conference on Concrete Alkali-Aggregate Reactions, Ottawa*, pp. 342–346.

53. Swamy, R.N. and Al-Asali, M.M. (1989) Effect of alkali silica reaction on the structural behaviour of reinforced concrete beams. *ACI Structural Journal*, **86**, 451–459.

54. Blight, G.E. and Alexander, M.G. (1988) Assessment of AAR damage to concrete structures. *Proceedings of the Seventh International Conference on Concrete Alkali-Aggregate Reactions, Ottawa*, pp. 121–125.

55. Imai, H., Yamasaki, T., Maehara, H. and Miyagawa, T. (1988) The deterioration by alkali silica reaction of Hanshin Expressway concrete structures—investigation and repair. *Proceedings of the Seventh International Conference on Concrete Alkali-Aggregate Reactions, Ottawa*, pp. 131–135.

56. Swamy, R.N. and Al-Asali, M.M. (1988) Expansion of concrete due to alkali silica reaction. *ACI Materials Journal* **85**, 33–40.

57. Swamy, R.N. (1988) Engineering properties of fly ash concrete. *Proceedings of the Concrete Workshop 88*. National Building Technology Centre, Australia.

58. Swamy, R.N. (1989) *Superplasticizers and Durability of Concrete*, ACI Publication SP-119, pp. 361–382.

59. Swamy, R.N. and Al-Asali, M.M. (1990) Control of alkali silica reactions in reinforced concrete beams. *ACI Materials Journal* **87**, 38–46.

4 Role and effectiveness of mineral admixtures in relation to alkali–silica reaction

R.N. SWAMY

Abstract

This chapter examines critically the role and effectiveness of mineral admixtures in counteracting the effects of alkali–silica reaction. Tests are reported on plain concrete prisms, reinforced concrete slabs and beams incorporating a slowly reactive but moderately expansive reactive aggregate and containing either fly ash, slag or microsilica. It is shown that control of expansive strains and consequent cracking is an acceptable and satisfactory solution in many instances, particularly in unreinforced concrete. However, there are many situations where additional factors such as preserving the strength and stiffness of the damaged structure and control of structural distortions are equally important if the strength, serviceability and stability of ASR-affected structures are to be maintained. Judged on these five significant criteria, data are presented to show that mineral admixtures, when used correctly and at the required amount, can control material damage and structural deterioration effectively and substantially, although they may not be able to eliminate all deleterious effects completely and at the same time. Mineral admixtures should not be expected to fulfil such a global, universal and overprotective role, but they have an unequalled, positive and promising function in contributing to preserve the safety, stability and durability of concrete materials and concrete structures affected by ASR.

4.1 Introduction

It is now universally recognised that, of all the possible methods of controlling the deleterious expansion caused by alkali–silica reaction (ASR) in hardened concrete, the use of pozzolanic or mineral admixtures can impart the most realistic advantages to the properties of the concrete to enhance its durability characteristics, and hence its resistance to the effects of ASR[1-6]. Apart from cost benefits, there are sound technical, environmental and energy-related reasons as to why such mineral admixtures could bring considerable practical benefits to concrete construction when they are used as a basic constituent of the concrete matrix[7].

However, in spite of the considerable research carried out on the role of pozzolans and of mineral admixtures in controlling ASR expansion, it is fair to say that there are still several aspects of the mechanism of pozzolanic reactions and their control of ASR expansion that are not yet clearly established[8-13]. Published literature shows that there are still many differences of opinion and conflicting results about the role and effectiveness of mineral admixtures in controlling the effects of ASR. There is a strong school of thought that the alkalis in fly ash and slag contribute to the reaction; the effects of silica fume appear to be highly variable, whilst some materials, such as fly ash, natural pozzolan and slag, are sometimes considered to be not only ineffective but even hazardous[14-27]. The situation is confusing, to say the least, and designers and users of concrete could be forgiven if they decide to stay clear of materials whose function is reported to be so conflicting and contradictory.

There are other factors influencing this current feeling of uncertainty on the universality of the effectiveness of mineral admixtures in concrete. The majority of data available to the engineer are based on cement pastes and mortars, and whilst such data contribute immensely to our understanding of the role and effectivensss of mineral admixtures in controlling the effects of ASR, there are several sound reasons as to why data obtained from such tests cannot readily and unquestioningly be translated to evaluate the behaviour of concrete in real structures in real environments. Almost all assessments of the effectiveness of mineral admixtures are based on the ASTM C441 test which has severe limitations, with regard to test methodology but also the unreliability of the criteria applied to evaluate expansion limits[28,29]. Test results also appear to suggest that the 0.6% limit defined in ASTM C150 for the alkali content of cement and the 1.5% maximum available alkalis for fly ashes in ASTM C618 are inadequate guidelines[30]. Further, mortar bar tests (ASTM C227) and tests using Pyrex glass as reactive aggregate and data obtained to define pore solution chemistry also have severe limitations, and care should be exercised in interpreting these data and in extrapolating them to situations and materials beyond what are strictly only applicable to situations similar to the methodology adopted in the tests.

There are also other engineering aspects to be considered, such as the rate at which mineral admixtures are able to control the effects of ASR. For example, the rate of pozzolanicity of fly ash and slag, which is one of the critical factors determining the effectiveness of mineral admixtures, is highly variable, and influenced by many factors. Most fly ashes and slags are also known to retard setting, and have higher rates and volumes of bleed unless the mixes are carefully proportioned. It can readily be seen that it is inevitable that the effectiveness of mineral admixtures cannot but be seen to be highly variable, and very much controlled by the test methodology and test conditions.

Further, there is the more fundamental question which does not seem to have been addressed, namely: What exactly should be the role of the mineral admixture in counteracting the effects of ASR? What do we expect the mineral admixtures to do when incorporated in concrete that is likely to undergo ASR at some stage in its life? Most published literature, and most planned research, seem to assume tacitly that control of expansion and the consequent cracking is the sole function of mineral admixtures, and their effectiveness is judged by their ability to control ASR expansion and/or cracking. This may be so in unreinforced concrete, but what happens to the structure as a whole if there is embedded steel that is inadequately proportioned or badly detailed? Or what happens when other deteriorating processes are superimposed on alkali–silica reactivity? Considering expansion alone, even if it were the sole criterion, the test conditions used by almost all investigators are so highly different from each other that it seems to be very unfair to mineral admixtures to be judged in this way, since none of the test results are really comparable. The reasons for the prevailing confusion and uncertainty are thus not difficult to see.

It will be readily seen that there are extensive data available on what mineral admixtures can and cannot do in concrete when it is affected by ASR. Equally, for reasons described earlier, and discussed in other chapters of this book, not all existing information can readily be applied, blindly and unquestioningly, to concrete construction exposed to real environments. The real danger at the moment seems to be to consider mineral admixtures as a 'universal panacea' and a 'global solution' to control the problems of ASR. The aim of this chapter is to throw additional light on existing data in order to assess critically the role and effectiveness of mineral admixtures in counteracting the effects of ASR. In order to obtain an overall picture of the interactive effect of ASR and mineral admixtures on concrete and steel, additional tests are reported here. These tests relate to:

(a) Plain concrete without and with reactive aggregate.
(b) Plain concrete containing acceptable proportions of a mineral admixture with the same reactive aggregate.
(c) Concrete slabs with reinforcement containing the same proportions of mineral admixtures and the same reactive aggregate as in (b).
(d) Reinforced concrete beams designed according to British Standard codes and containing two different types of reactive aggregates and two different proportions of fly ash.

A slowly reactive but moderately expansive reactive aggregate was used, with expansions in excess of 2000 microstrains. The test conditions were kept the same for all the mineral admixtures. In the reinforced concrete beams, an additional highly reactive aggregate, a Beltane opal, was also used.

4.2 Experimental programme

The test programme reported here consisted of three sets of tests. In the first set, two series of tests were carried out on plain concrete as follows:

Series A—control tests without reactive aggregates
Series B—ASR tests with reactive aggregate.

In the second set, three series of tests on plain concrete were carried out as follows:

Series C—ASR tests with 50% fly ash replacement of cement
Series D—ASR tests with 50% slag replacement of cement
Series E—ASR tests with 10% microsilica added.

In the third set, tests were carried out on $\frac{1}{4}$-scale models of reinforced concrete flat slabs, all made with the same reactive aggregate and containing the same type and quantity of mineral admixtures as in the second set, namely:

Series A1—control slabs without reactive aggregates
Series B1—ASR tests with reactive aggregate
Series C1—ASR tests with 50% fly ash replacement
Series D1—ASR tests with 50% slag replacement
Series E1—ASR tests with 10% microscilica added.

4.2.1 Concrete mix details and materials

Only one concrete mix 1:2.2:1.8:0.5 (cement–sand–coarse aggregate–water) by weight was used throughout the tests. A high-alkali normal Portland cement (ASTM type 1) with about 1% Na_2O eq. was used. The fly ash was a low-calcium type F ash, while the slag had a similar specific surface (350 m^2/kg) to the cement (380 m^2/kg). The cement, fly ash and slag all complied with the relevant British Standard specifications. The microsilica was added in the form of a slurry (50% water and 50% solids), with appropriate corrections for the water content and sand. The quantity of microsilica added was restricted to 10%, the amount now generally accepted as the maximum allowable in practical applications[31,32]. The fine aggregate used in the concrete mixes was a washed and dried natural sand, and the coarse aggregate consisted of a mixture of rounded and crushed gravel with 10 mm maximum size. Both concrete aggregates were considered to be completely unreactive so far as ASR is concerned.

The reactive aggregate used in these tests was a slow but moderately reactive synthetic aggregate, an amorphous fused silica identified by the author's group as a suitable reactive aggregate for laboratory studies[33]. The fused silica had a particle size distribution of 150–600 μm and contained 99.7% silica with practically no alkali in it. The fused silica was used to replace sand by an amount equal to 15% by weight of the total aggregate.

The concrete mixes with fly ash and slag were proportioned by methods developed by the author's group in which the Portland cement is replaced by the mineral admixture, weight for weight, and other constituents are adjusted, particularly the water content[34,35]. The water–binder ratio was kept constant for all the concrete mixes at 0.5. A sulphonated napthalene formaldehyde condensate superplasticiser was used in all the mixes to enhance the flow characteristics of the mixes. The amount of superplasticiser was varied to suit the type of mineral admixture to give consistent flow properties, and varied from 1.0% to 1.7% by weight of the total cementitious content. In order to reduce bleeding, a special mixing procedure was used in which the aggregates were premixed with a proportion of mixing water before adding the binder[36]. All the concrete mixes showed excellent workability with slumps in excess of 100–150 mm; the microsilica concrete had less slump, 60–80 mm.

4.2.2 Test details

The test specimens consisted of the following: $75 \times 75 \times 300$ mm prisms for expansion tests, 100-mm cubes for compression tests, and $100 \times 100 \times 500$ mm prisms for pulse velocity measurement, examination of cracking, dynamic modulus and modulus of rupture tests. All these properties were measured on three specimens so that the test data presented here are the average of measurements on all three.

The model slabs were cast in pairs, which were identical in all respects. The slabs used in these tests were $\frac{1}{4}$-scale models of a prototype flat plate structure loaded through a stub column; the slabs were specially modelled to simulate cracking, deformation and strength[36]. The size of the slabs and details of the reinforcement are shown in Figure 4.1. The main reinforcement consisted of black-annealed, plain, mild, steel wire of 3.25 mm diameter, while the column ties were of mild steel of 1.6 mm diameter. To heighten the effects of ASR, the slabs were reinforced unsymmetrically—10 bars in each direction at the

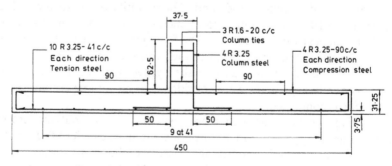

Note : All dimensions are in mm
All covers to steel are 3·75mm

Figure 4.1 Geometry and reinforcement details of slabs.

bottom with four bars in each direction at the top. The slab reinforcement and the column reinforcement were tied together to form a rigid reinforcing cage.

All the test specimens were cast in steel moulds, compacted on a vibrating table and kept covered in the laboratory for 24 hours. The specimens were then demoulded and transferred to a fog room with controlled temperature and relative humidity (RH) of $20 \pm 1°C$ and $96 \pm 2\%$.

In order to simulate realistic environmental conditions, all the test specimens were subjected to a cyclic hot–wet and hot–dry exposure regime. The specimens were initially kept in the fog room at ambient temperature for about 6 days, by which time the concrete had achieved about 70% of its 28-day strength. They were then placed in a hot room at $38 \pm 2°C$ and covered with wet hessian. The specimens were kept moist by wetting once a day but not continuously, so that in effect the specimens were exposed to a cyclic wet–dry regime at $38 \pm 2°C$. The synthetic reactive aggregate used in all these tests was chosen such that it was sufficiently slow-reacting to enable the concrete to develop its strength at early ages while being cured in the fog room before the expansive strains started to occur so that ASR chemical reactions did not interfere with cement hydration, as would happen, if, for example, highly disordered reactive aggregates such as opal were used. This was achieved: while in the fog room, the expansive strains varied from zero to 270 microstrains so that it can be safely assumed that cement hydration in the critical early stages was not adversely influenced by ASR chemical reactions.

In addition to being slowly reactive, the reactive aggregate was also chosen to ensure that it was capable of developing expansive strains sufficiently high to be deleterious to concrete and concrete structures. The tests were therefore carried out for a sufficiently long period for the uncontrolled expansive strains to reach values of about 0.70–0.75%, when the slabs were tested.

4.3 Test results and discussion

The main aim of this chapter is to throw additional light on existing data and critically examine the role and effectiveness of mineral admixtures in controlling expansive strains and other effects arising from ASR. In evaluating the data, it must be remembered that the specimens in this study were deliberately subjected to a cyclic hot wet–dry regime such that the development of expansive strains was non-uniform, and followed a start-stop pattern. From the large amount of data obtained from these tests, only those relevant to the main objectives of this chapter are discussed here.

4.3.1 *Expansive strains*

The mean expansive strains measured on the plain concrete prisms of series B to E are shown in Figure 4.2. Two points need to be emphasised here.

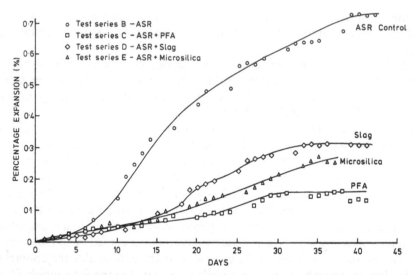

Figure 4.2 Expansive strains of concrete without and with mineral admixtures.

Firstly, each value plotted in this figure is the average of three test specimens. Secondly, the non-uniform development of strains shown in the figure is the direct result of the cyclic hot–wet/hot–dry exposure regime adopted in these tests which alternately activated or temporarily slowed down the ASR chemical reactivity.

The data clearly show the controlling effect of the mineral admixtures on concrete expansive strains. Fly ash, slag and microsilica are all effective in substantially reducing the expansive strains, but they also do this at different rates. Initially, up to about 10 days or so after the specimens were moved into the hot wet–dry cyclic exposure when the expansive strains developed rapidly (series B), all the three mineral admixtures suppressed the expansive strains almost equally well. Thereafter, the rate of reduction in the expansive strains varied, depending upon the type of admixture, its pozzolanicity and its rate of reactivity with Portland cement.

The mean expansive strains of the concretes in series B to E at various ages are shown in Table 4.1. The data show that up to about 15 days, when the ASR expansive strain averaged about 0.30–0.35%, all the mineral admixtures were able to suppress the expansion to values of about 0.05–0.07%, i.e. to about 500–700 microstrains. This is particularly significant in practice where at early ages the concrete will have only negligible or nominal tensile strain capacity. The mineral admixtures thus play a very positive role at the critical early stages of concrete strength development by suppressing through their chemical reactions both undesirable tensile strains and the resulting possible microcracking.

Table 4.1 Expansion at various ages

Age (days)	Percentage expansion*			
	Series B (ASR control)	Series C (ASR+PFA)	Series D (ASR+slag)	Series E (ASR+microsilica)
5†	0.027	0.028	0.013	0.024
10	0.120	0.042	0.036	0.060
15	0.326	0.065	0.065	0.067
20	0.390	0.079	0.166	0.113
25	0.492	0.098	0.217	0.153
30	0.608	0.151	0.278	0.203
35	0.643	0.148	0.314	0.260
40‡	0.732	0.164	0.313	0.273
42	0.730	0.136	0.311	0.254

* Expansions extrapolated to the nearest age.
† Specimens moved into hot room soon after this age.
‡ Specimens removed from hot room soon after this measurement.
PFA, fly ash.

Figure 4.2 also emphasises another important practical aspect of ASR expansive strains, namely their non-uniform and non-homogeneous development, depending upon the availability of moisture. The data show clearly that, even within a typical 24-hour period, expansion proceeds when the specimens are wet and it stops when the specimens are dried out. Further, in the tests reported here, the specimens were deliberately kept dry at the weekends, and during these periods the expansions either remained practically the same or, occasionally, dropped, even though the ambient temperature was $38 \pm 2°C$. Thus, in real structures exposed to environments which are subject to daily, seasonal or other time-dependent variations, expansive strains occur in a non-uniform disproportionate manner. On the other hand, the data also emphasise that in an environment that is basically favourable to ASR, in which the only variable is the availability of moisture expansive strains can progress in a cumulative manner, so that in time the expansive strains can reach a magnitude sufficient to develop external and visible cracking. The data also imply that ASR is a time-dependent phenomenon and that it can be dormant for a long time before being triggered off again when moisture and other favourable environmental conditions become available.

4.3.2 Control of expansive strains

Figure 4.2 and Table 4.1 also emphasise the controlling effect of mineral admixtures on expansive strains. To elucidate this effect further, the age to reach a particular expansion for concretes without and with mineral admixtures is shown in Table 4.2. The data shown in Tables 4.1 and 4.2 thus provide

Table 4.2 Average age to reach given expansions

Expansion (%)	Age* (days)			
	Series B (ASR control)	Series C (ASR + PFA)	Series D (ASR + slag)	Series E (ASR + microsilica)
0.05	7	11	12	13
0.10	8	25	16	17
0.15	11	30	20	24
0.20	12	—	23	30
0.25	13	—	27	34
0.30	14	—	34	—
0.40	18	—	—	—
0.50	25	—	—	—
0.60	29	—	—	—
0.70	39	—	—	—

* Age shown corresponding to the nearest expansion.
PFA, fly ash.

comparable quantitative information on the rate and effectiveness of fly ash, slag and microsilica mineral admixtures in controlling expansive strains developed through control of ASR for the particular set of conditions used in these tests.

Obviously, the data in Tables 4.1 and 4.2 can be neither generalised nor treated as typical, not at this stage at any rate. The effectiveness of mineral admixtures will depend on several factors: the rate of reactivity of the aggregate, the total available reactive silica and concrete alkali (internal and external) and the rate of pozzolanic/cementitious activity of the mineral admixture. This last factor will depend on the type of mineral admixture, the amount of mineral admixture and the method of its incorporation. These aspects are discussed later.

The data in Table 4.2 emphasise that the reduction in expansion achieved through control of chemical reactions followed different rates for the three mineral admixtures. Fly ash was most effective, and in concretes containing 50% fly ash replacement it took about 30 days for expansive strains to reach about 0.15%, compared with about 20 and 24 days respectively for slag and microsilica to reach the same strain. In the concrete without mineral admixtures, the 0.15% expansive strain was reached in about 11 days. In general terms, fly ash was most effective, followed by microsilica and then slag, both in the rate at which they reduced the ASR expansive strains occurring in series B tests as well as in their final control of expansive strains. In the tests reported here, the expansive strains in the series B to E had more or less stabilised by the end of about 6 weeks. At this time, none of the mineral admixtures had been able to eliminate completely the expansive strains, as shown by the concrete in series B, although in all cases substantial reductions

Table 4.3 Final expansion of all test series

Test series	Age (days)	Final expansion (%)	Pulse velocity (km/s)
ASR control —series B	41	0.736	3.50
ASR + PFA —series C	39	0.164	4.53
ASR + slag —series D	35	0.314	3.72
ASR + MS —series E	36	0.273	4.04

PFA, fly ash; MS, Microsilica.

in expansion had occurred. At the end of 6 weeks, the mean residual expansive strains in concretes containing fly ash, microsilica and slag were about 0.164%, 0.273% and 0.313% respectively compared with the uncontrolled ASR strains of about 0.732%.

The final measured expansion at the end of about 6 weeks in the concretes of series B to E, together with their corresponding pulse velocities, are shown in Table 4.3. It is interesting to note that ultrasonic pulse velocities realistically reflect the effectiveness of the mineral admixtures. The fly ash concrete of series C which was most effective had the highest pulse velocity of about 4.53 km/s; the microsilica concrete which followed behind registered a pulse velocity of 4.04 km/s. The slag concrete was less effective compared with fly ash and microsilica, and registered a pulse velocity of 3.72 km/s. The uncontrolled ASR concrete had a pulse velocity of 3.50 km/s.

These results reinforce the information presented in Chapter 3 that pulse velocity techniques may be used with confidence to monitor the deterioration of concrete due to ASR, and the changes that occur due to this deterioration. The data in Table 4.3 show that pulse velocity measurements can not only assess the degree of ASR deterioration, but also evaluate quite correctly the effectiveness of the mineral admixtures in their ability to control expansive strains[37].

4.3.3 Control of cracking

In practical applications, it is not only the control of expansive strains that is important, but also the control of cracking. In the tests reported here, the formation of the first visible external crack and the development of subsequent cracking were carefully monitored. It was found that it was generally easier to identify the formation of the first crack and trace it on the concrete surface in concrete undergoing uncontrolled expansion (series B) and in concrete containing slag (series D). In concrete containing fly ash (series C) and microsilica (series E), the first external cracks were very fine and scarce, and much more difficult to identify.

Figure 4.3 shows typical crack development in concrete undergoing uncontrolled ASR expansion (series B). In this concrete, the first visible crack appeared at about 450–500 microstrains, which is generally the limiting tensile strain capacity of concrete, and this occurred at an early age of about 6–7 days (after transferring the test specimens to the hot room). However, once the first crack became visible, both expansion and the subsequent cracking developed rapidly as the environmental conditions for the progression of alkali–silica chemical reactions were favourable. In general terms, the density of cracking and crack distribution increased up to an expansion of about 0.5%; by this stage almost all the ASR cracking had occurred and only very few new cracks could be observed beyond this state (Figure 4.3b, c and d). Beyond this expansion of about 0.5%, the cracks began to show clear signs of increasing width. Nevertheless, even at an expansive strain of about 0.732% (Figure 4.3d), the cracks were of sufficiently small size as to enable the test specimens to remain intact and be able to preserve their integrity.

The fly ash was most effective in controlling cracking, as shown in Figure 4.4, confirming the data given in Figure 4.2 and Table 4.3. At the end of 6 weeks, there were only barely identifiable surface cracks. The concrete with

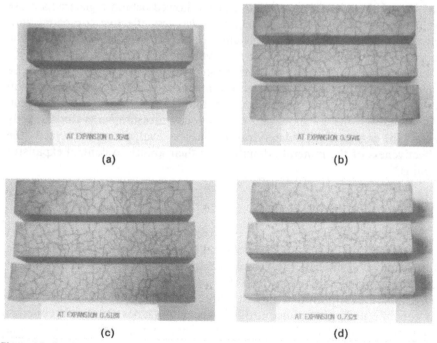

(a)   (b)

(c)   (d)

Figure 4.3 Crack pattern due to ASR at expansive strains of (a) 0.364%, (b) 0.564%, (c) 0.618% and (d) 0.732%.

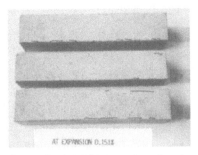

Figure 4.4 ASR cracking in fly ash concrete at an expansive strain of 0.151%.

slag (series D) showed signs of cracking at an early age but this was very fine. Figure 4.5 shows the crack configuration in slag concrete at expansive strains of 0.166% and 0.310% respectively. These cracks were clearly only surface cracks, but sufficiently widespread to cause loss in pulse velocity, as shown in Table 4.3. A comparison of Figure 4.3(b) and Figure 4.5(b) shows that slag concrete had almost the same intensity of cracking as the concrete undergoing ASR at similar expansion levels, although this occurred at a much later age, i.e. about 34 days in slag concrete compared with about 15 days in ASR concrete, and the cracks were much less penetrating.

Figure 4.6 shows the crack patterns in concrete containing microsilica (series E) at two stages of about 0.159% and 0.260% expansive strains. The first cracks were very fine, and difficult to identify, and at an expansive strain of 0.26% the cracks were fairly widespread, as shown in Figure 4.6(b).

From the data presented so far, it is clear that mineral admixtures have a beneficial and positive role in suppressing both expansive strains and the resulting cracking, particularly at early ages. But they cannot completely eliminate either high levels of expansive strains or cracking. However, it also needs to be emphasised that this cracking in concretes containing these

(a)   (b)

Figure 4.5 Cracking due to ASR in slag concrete at expansive strains of (a) 0.166% and (b) 0.310%.

(a)   (b)

Figure 4.6 ASR cracking in concrete containing microsilica at expansive strains of (a) 0.159% and (b) 0.260%.

mineral admixtures was largely surface cracking, penetrating possibly only slightly into the interior of the concrete as evaluated by pulse velocity measurements[37]. On the other hand, the density of surface cracking was sufficiently large to affect the overall pulse velocity of concrete, as shown in Table 4.3.

There are always inherent difficulties in identifying the stage at which the first microcracking occurs and becomes visible in concretes undergoing expansive strains, whether they contain mineral admixtures or not. This difficulty is further compounded when there is more than one test specimen; this is necessary in research studies, but the specimens are never identical even if all precautions are taken. In spite of these difficulties, by combining visual examination with pulse velocity measurements, a pattern of the initial cracking behaviour has been developed, and these details are shown in Table 4.4. The data again reflect the rate of effectiveness of fly ash, slag and microsilica and confirm the results shown in Figure 4.2 and Table 4.1.

The tensile strains at first crack shown in Table 4.4 are as expected. Plain concrete shows visible cracks at a tensile strain capacity of about 400–500

Table 4.4 Structural characteristics at first crack

Test series	Age at internal structural damage (days)	At first visible crack		
		Age (days)	Expansion (%)	Pulse velocity (km/s)
ASR control —series B	6	7	0.046	4.40
ASR+PFA —series C	35	36	0.150	4.60
ASR+slag —series D	15	15	0.065	4.35
ASR+MS —series E	26	27	0.160	4.50

PFA, fly ash; MS, Microsilica.

microstrains. Slag concrete is seen to crack at a strain of about 650 micro-strains, while concretes containing fly ash and microsilica exhibit first cracking at tensile strains of about 1500–1600 microstrains. These last-mentioned strains are high but should not cause surprise as at the high temperature and with moisture available, although only at times, the mineral admixtures were able to develop their pozzolanic/cementitious activity to delay development of disruptive strains, enhance strength and strain capabilities. The net result is beneficial in controlling both expansion and cracking.

4.3.4 *Influence of reinforcement*

The data presented in Figures 4.2 to 4.6 and Tables 4.1 to 4.4 relate to expansive strains and cracking in plain concrete prisms undergoing ASR, and show the effectiveness of fly ash, slag and microsilica in controlling these reactions and the resulting tensile strains and cracking in such concretes. The question can now be raised as to what happens in concrete containing reinforcement, i.e. in reinforced concrete elements.

Here the effectiveness of mineral admixtures in concrete containing reinforcement and affected by ASR cannot be assessed without a clear under-standing of how reinforced concrete elements behave when affected by ASR. In other words, how does the presence of reinforcement alter the behaviour of plain concrete affected by ASR and, further, how do mineral admixtures, seen to be so effective in plain concrete, influence the behaviour of reinforced concrete elements? That is, does ASR create any deformations in reinforced concrete elements other than tensile strains and cracking, generally observed in plain concrete, and, if so, to what extent are mineral admixtures able to control these structural deformations, which could be critical to the stability of a reinforced concrete element?

To elucidate this, test series A1 to E1 were carried out on model concrete slabs containing identical concretes, reactive aggregates and amounts of min-eral admixtures as the plain concrete prisms of series A to E. The behaviour of these slabs when subjected to ASR is discussed below.

Most reinforced concrete slabs are designed to have reinforcement at both the bottom (tension face) and top (compression face) of the slab. The effects of ASR are most pronounced when the reinforcement volume is not the same at the top as at the bottom. The slabs of this study were thus designed to have less compression steel (Figure 4.1) so that the capacity to restrain expansion was much lower at the top face of the slab than at the bottom of the slab.

The effects of ASR on the slabs of series A1 to E1 are summarised in Table 4.5. Since cracking is the first visible evidence of ASR, this is discussed first. The information presented in Table 4.5 shows that slabs of series B1, undergoing uncontrolled ASR expansion, showed extensive cracking on the top face and vertical sides of the slab as well as on the stub column. Slabs of

Table 4.5 Physical characteristics of slabs

Slab details	Hogging deflection (mm)	Cracking behaviour	
		Top face	Bottom face
Control —series A1	Nil	None	None
ASR —series B1	3.5	Extensive	None
ASR+PFA—series C1	Nil	Practically nil	None
ASR+slag —series D1	1.5	Widespread*	None
ASR+MS —series E1	Nil	Practically nil	none

* Cracks are largely surface cracks, unlike cracking in series B1.
PFA, fly ash; MS, Microsilica.

series D1 with slag also showed similar extensive cracking (Figure 4.7), but these cracks generally penetrated less into the concrete. Slabs containing fly ash and microsilica showed very little visible external cracking, and none of the slabs showed any cracks at the bottom (tension) face.

Cracking due to ASR in concrete containing steel rebars is generally very different from that in plain unrestrained concrete[38,39]. Nevertheless, the cracking observed in the reinforced concrete slabs followed a similar pattern to

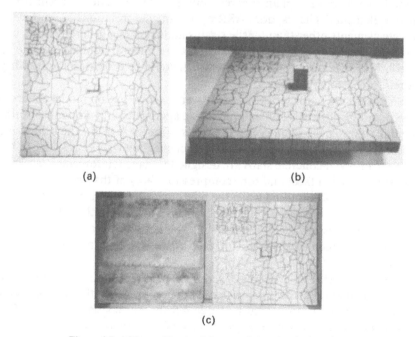

(a)   (b)

(c)

Figure 4.7 ASR cracking in slabs containing slag (a, b and c).

that of the unreinforced prisms described earlier, so far as the effectiveness of the mineral admixtures in controlling cracking is concerned. Thus the ability of the mineral admixtures to control cracking arising from ASR is confirmed in both plain (Figures 4.3 to 4.6) and reinforced (Figure 4.7, Table 4.5) concretes, and the rates at which they do this are different in reinforced concrete and in plain concrete.

Table 4.5 also shows an important difference between plain and reinforced concrete elements when affected by ASR. This relates to the hogging deflection due to differential restraint at the top and bottom of the slab. Slabs of series B1 exhibited a hogging deflection of about 3.5 mm. The effectiveness of the mineral admixtures in controlling these negative deflections varied and was similar to their effectiveness in controlling tensile strains due to ASR (Table 4.1) and cracking in plain and reinforced concrete members (Figures 4.3 to 4.7). Thus slabs containing slag (series D1) had a residual hogging deflection of 1.5 mm, whilst slabs with fly ash (series C1) and microsilica (series E1) almost completely counteracted these deformations and showed, like the control slab (series A1), no hogging deflection due to ASR.

The data presented in Table 4.5 and Figure 4.7 confirm that control of expansive strains and cracking is essential in both structural elements and plain concrete affected by ASR. All these data also show that mineral admixtures perform these duties at different rates and to different degrees, and that to a great extent what happens in plain concrete prisms is also reflected, in an engineering sense, in reinforced concrete elements. However, one critical effect of ASR, not generally observed in plain concrete specimens (unless they are subjected to severe external restraints, as in retaining walls[38,39]), was readily observed in the slabs, namely the hogging deflection due to the combined effects of expansive strains, cracking and distribution of reinforcement (i.e. internal restraint). The hogging deflections in the different slabs generally followed the pattern of the residual tensile strains and cracking observed in the concrete prisms, but they will also have significant effects on the performance of the slabs when external loads are superimposed on the effects of ASR. This is particularly important in flexural members such as slabs and beams[40]. The data presented here show that even different structural elements, such as slabs and beams, so long as they act in flexure, show similar behavioural characteristics under ASR[40].

These data also reveal another important aspect. Control of expansive strains and cracking is, by itself, not an adequate parameter to judge the effectiveness of mineral admixtures when considering ASR as a whole, and as a problem affecting concrete structures. Control of expansive strains and cracking may generally be adequate in plain concrete, but even this may not be strictly valid for all situations, since elements like plain concrete pavements and retaining walls may show structural distortions if sufficiently high external restraints occur on one side. Thus these data show that the effects of external and/or internal restraint, resulting in differential deformations and distortions, should also be considered when judging the effects of mineral admixtures in their ability to counteract the engineering effects of ASR.

4.3.5 Effectiveness of mineral admixtures in controlling the consequences of alkali–silica reactivity

The information on expansive strains, tensile strain capacity and cracking due to ASR given in Figures 4.2 to 4.6 and Tables 4.1 to 4.4 emphasise two very important aspects of the role of mineral admixtures when they are used to mitigate the effects of ASR. Firstly, the rate at which mineral admixtures control expansion and cracking is not the same; secondly, while they substantially suppress ASR chemical reactions and the consequent expansive strains and the resulting cracking, they may not always eliminate completely either the expansive strains or the resulting cracking, particularly if large expansive strains occur. For a given aggregate reactivity, concrete alkali content and environmental conditions, these are thus the key characteristics of the effectiveness of mineral admixtures in relation to their control of the effects of alkali–silica reactions.

These two fundamental properties should not come as a surprise. The critical factor which determines the controlling effect of mineral admixtures so far as ASR is concerned is their rate of pozzolanic/cementitious activity[12]. This reactivity is, in turn, influenced by three important parameters, namely (1) the particle size, specific surface and mineralogical composition of the admixture, (2) the volume per cent used, and (3) the method of its incorporation in concrete.

Fly ash has generally mean particle sizes similar to those of unhydrated ordinary Portland cement (OPC), of 10–25 μm, and a specific surface of 300–350 m^2/kg. Slag is generally ground to similar specific surfaces of 300–350 m^2/kg. Microsilica has mean particle sizes of 0.1–0.2 μm, and a very high specific surface of 20–23 m^2/g, consisting almost entirely of amorphous silica. Microsilica is thus a very highly reactive pozzolan, possessing greater pozzolanic action than most other natural and synthetic pozzolans.

Further, with siliceous admixtures such as fly ash and slag that contribute to the engineering properties of concrete, the rate of pozzolanicity depends on both the volume content and the method of incorporation of the admixture[7,34,41,42]. Early development of strength and tensile strain capacity is critical to control ASR effects if reactive aggregates are highly disordered and react very fast like Beltane Opal; with mineral admixtures, early strength development is dependent on the hydration of the cement, whereas long-term strength development is influenced more by the reactivity of the ash or slag. Thus, with fly ash and slag, the hydration mechanism of the Portland cement– siliceous admixtures is a two-stage process; this explains why in many published reports, most OPC–fly ash and OPC–slag concrete mixes exhibit lower early-age strengths than the corresponding OPC concrete. Sometimes even the long-term strength of the blended cement concrete has been lower than that of the corresponding OPC concrete[43-45]. In these circumstances, it is not surprising that fly ash and slag are not always seen to be capable of fully controlling the expansive strains and cracking effects of ASR[21-27].

Similar contradictory results also exist in literature on the effectiveness of microsilica in controlling the effects of ASR[15,16,18,28,46,47]. Small amounts of microsilica have sometimes been found to be very effective[46,47], whereas others have reported early suppression of expansive strains but long-term ineffectiveness in counteracting the effects of ASR[15,16,18,28]. At large levels of microsilica incorporation, there are fears of microsilica itself becoming a source of silica to react with the alkalis in the cement[30], and this is confirmed by the detection of silica gel in field concretes containing 15% silica fume[48]. In any case there are also other engineering implications to be considered if high levels of microsilica are incorporated in concrete.

It may readily be seen that a large number of factors contribute to the observed phenomena reported in literature on the expansion behaviour of concretes containing fly ash, slag and microsilica. Apart from the physical characteristics and mineralogical composition of the mineral admixture, its volume content and mix proportioning in the blended cement concrete, factors such as the efficiency of the admixture in refining the pore structure of the concrete and its sorptivity, the freedom of ionic diffusion and the mobility of the alkalis in the ash and slag, and of the silica in microsilica, are all critical factors, not necessarily occurring all simultaneously, in influencing the ability of siliceous admixtures in controlling the effects of ASR. The source of alkali is also another important factor influencing the ability of mineral admixtures in controlling the effects of ASR[49,50].

4.3.6 Criteria for control of ASR effects

The data presented so far emphasise that control of expansion should not be considered to be the sole criterion to determine the effectiveness of a mineral admixture in counteracting the ASR effects. In most of the published literature control of expansive strains has been assumed to be adequate and sufficient to qualify as an effective antidote to alkali–silica chemical reactions. The information presented so far shows that this assumption can be misleading and cannot always guarantee a trouble-free service life to concrete and concrete structures.

There are also other reasons why control of expansive strains cannot be assumed to be the sole criterion to judge the effectiveness of mineral admixtures in relation to ASR. Many investigators have used the ASTM C441 test method to evaluate the effectiveness of pozzolans and other mineral admixtures in preventing excessive expansion in concrete due to ASR. The test method has several drawbacks, and many test results show that the test does not represent an adequate and sufficient criterion—the criterion is unreliable and short-term data can be misleading[28,29].

So, from an engineering performance point of view, what criteria should be adopted which can be considered to be adequate and satisfactory to prevent the deleterious effects of ASR and to ensure the durable performance, stability and integrity of concrete materials and concrete structures? Two criteria are readily identified from previous data on concrete prisms:

Control of expansive strains
Control of cracking.

For load-bearing elements and structural members, these alone are not adequate and cannot ensure satisfactory performance in practice. So the following criteria must be added:

Restoration to original concrete strength
Restoration to original elastic modulus.

Both these criteria are necessary as loss of strength and elastic modulus can affect the serviceability, strength and safety of concrete structures[38,40,51]. Tests on structural elements show that there is still one further criterion, critical to sensitive structures (and this could include dams, bridges and even pavements), which should be satisfied. Tests show that the loss in structural stiffness is not the same when considering deflection or strain deformations— thus both laboratory and field studies confirm that the presence of ASR leads to acceptable loss in elastic modulus only when this is deduced from deflection measurements of load tests[38–40,52,53]. On the other hand, when structural distortions are considered, the loss in stiffness of the structure as a whole is too substantial to be ignored[38,39]. This is again confirmed by many field investigations, which show unacceptable structural distortions even though expansion levels due to ASR are not high[54,55]. Thus another criterion is required, namely:

Control of structural distortions.

To summarise, the five criteria that need to be considered to establish the effectiveness and ability of a particular material or concept to control all the effects of ASR are:

Control of expansive strains
Control of cracking
Preservation of concrete strength
Preservation of concrete elastic modulus
Control of structural deformations/distortions.

Obviously not all these criteria are either critical or need to be satisfied in a given case of ASR affliction. However, control of structural distortions cannot be ignored and is important in all types of structures. Unacceptable distortions often occur at local levels and can put parts of a structure, and sometimes the entire structure, out of service[54,55]. Control of such distortions is thus critical for structural stability and can only be ignored at the expense of the loss of function of the structure[38,39].

4.3.7 How effective are mineral admixtures in controlling ASR?

In the light of the above discussions, it is necessary to ask—how effective are mineral admixtures in controlling all the material and structural deterioration arising from ASR? Many published papers, as pointed out earlier, deal only with expansive strains; nevertheless, they show that mineral admixtures can and do control effectively, but not entirely, these expansive strains arising from alkali–silica chemical reactions. The data presented in this chapter reaffirm this—mineral admixtures can control expansive strains, cracking and hogging deflection (which could lead to structural distortions), but their effectiveness is variable and they differ in both the rate and the amount of such control. Further data are presented below to illustrate how the incorporation of mineral admixtures can not only control expansive strains and cracking but also restore the strength and elastic modulus of concrete if and when it is affected by alkali–silica reactions.

4.3.7.1 *Flat slabs.* It was shown in Table 4.5 and Figure 4.7 how mineral admixtures control expansive strains, cracking and structural deformations in slabs undergoing ASR. Further data on the effectiveness of these mineral admixtures in restoring the strength of the concrete in the slabs and of the slabs themselves are shown in Table 4.6. The ultimate strength of the slabs is shown here for information, but a detailed examination of these values and the associated structural behaviour will not be discussed here.

The data in Table 4.6 show that all the mineral admixtures were able to restore the strength of the ASR-affected concrete and of the slabs to values similar to those of the control ASR-free concrete and slabs. The concrete

Table 4.6 Strength characteristics of ASR-affected slabs with mineral admixtures

Test series	Expansion (%)	Concrete strength (MPa)		Slab ultimate load (kN)
		Compression	Flexure	
Control —series A1	Nil	52.5	4.6	10.9
ASR —series B1	0.732	41.2	1.7	9.5
ASR+PFA —series C1	0.136	51.3	4.5	10.4
ASR+slag —series D1	0.311	53.8	3.0	10.6
ASR+MS —series E1	0.256	63.9	5.9	11.5

PFA, fly ash; MS, Microsilica.

containing silica fume was able, as would be expected, to achieve compressive strengths higher than the control cube strength, and, as a result, the slabs were also able to carry higher ultimate loads than the control. The concretes containing fly ash and slag were also able to restore the compressive strength of the ASR-affected concrete to values similar to that of the control concrete.

In Figure 4.7 it was shown that slabs containing slag showed quite widespread surface cracking. The data on modulus of rupture in Table 4.6 readily explain why this happened. In spite of achieving an equivalent compressive strength to that of the control concrete, concrete containing slag and then exposed to ASR did not regain its original flexural strength completely: there was a loss in flexural strength of about 35%, and this of course resulted in widespread surface cracking, as shown in Figure 4.7.

Table 4.6 also shows that mineral admixtures were able to restore the strength of the ASR-affected slabs to a level similar to that of the control ASR-unaffected slab, even though none of the mineral admixtures was able to eliminate ASR expansive strains completely. The ultimate strength achieved by the slabs followed closely, as would be expected, the compressive strength of the concrete. Thus the concrete slab with microsilica achieved the maximum ultimate strength, about 6% higher than that of the control slab. In spite of the fact that slag was unable to eliminate the hogging deflection of the slabs and showed a residual upward deflection of 1.5 mm and widespread surface cracking, the slag concrete slab carried nearly the same ultimate strength as the control slab, and a marginally higher load than the slab containing fly ash. These data thus confirm that mineral admixtures are very effective in counteracting the material and structural effects of ASR; however, they do this at different rates and to varying degrees. They do not eliminate completely all the material and structural losses arising from ASR, but they do control them very effectively and very substantially.

4.3.7.2 *Reinforced concrete beams.* The effectiveness of mineral admixtures in controlling the effects of ASR in reinforced concrete beams is shown in Table 4.7. In these tests only one mineral admixture, namely a quality-controlled low-calcium class F fly ash, was used. Two different types of reactive aggregate were used in the beams—a very rapidly as well as highly reactive aggregate, opal, capable of inducing expansive strains of 1.6–1.7%, and a slowly but moderately reactive aggregate, a synthetic fused silica, with ability to cause expansive strains of 0.60–0.65%. The fly ash was used at two replacement levels of 30% and 50%. The results given in Table 4.7 are again conclusive proof that mineral admixtures effectively and efficiently control the various effects of ASR, but they do this at different rates and not completely.

The data show that with both types of reactive aggregate, 30% fly ash replacement was inadequate to reduce the ASR expansive strains to acceptable engineering levels. With 50% replacement, the concrete and steel strains due to ASR were reduced substantially to values at which they were unlikely

Table 4.7 Effectiveness of mineral admixtures in reinforced concrete beams.

Fly ash replacement (%)	Opal reactive aggregate					Fused silica reactive aggregate				
	Expansive strains*		Upward deflection (mm)	Crack width (mm)	Ultimate strength (kN)	Expansive strains*		Upward deflection (mm)	Crack width (mm)	Ultimate strength (kN)
	Concrete	Steel				Concrete	Steel			
Control	50	30	Nil	Nil	40.5	50	30	Nil	Nil	40.5
ASR–0	12900	2340	6	1.05	30.1	5200	1100	4	0.40	34.4
30	4400	840	3	0.15	39.2	2200	880	2	0.12	38.1
50	230	280	0	Nil	39.9	990	440	1	0.09	38.9

*Microstrains.

to lead to unacceptable structural deformations. The results of Table 4.7 in terms of upward deflection, cracking and ultimate strength confirm these expectations.

The information given in Table 4.7 reveals some other important results. For a given aggregate reactivity, the fly ash does not control the ASR strains induced in the concrete and steel reinforcement in the same beam to the same extent. With different reactive aggregates, the same amount of fly ash replacement in concrete of similar composition produces different rates of control of concrete expansive strains, steel expansive strains, hogging deflection and maximum crack width. There is of course no reason why any mineral admixture should control the expansive strains arising from ASR at the same rate with *all* reactive aggregates. The critical factor here is the interrelationship between the rate of reactivity of the aggregate and the rate of pozzolanic activity of the mineral admixture. The latter is influenced by many factors, including temperature, in the same way as there are numerous factors influencing the rate and ultimate reactivity of the aggregate. Further, siliceous admixtures such as fly ash and slag are highly variable in both their physical characteristics and mineralogical composition, even in the same country. Thus one should not really be surprised to see different rates of effectiveness of a mineral admixture such as fly ash or slag with a particular type of reactive silica. The most important thing here is to realise that mineral admixtures inherently possess the ability to suppress the alkali–silica chemical reactions and their effects on plain and reinforced concrete members. However, apart from the conditions creating the chemical reactions such as the aggregate reactivity, concrete alkalinity and environment, the way the mineral admixture is utilised in concrete is crucial to its medicinal role[34,42].

The data in Tables 4.6 and 4.7 thus show conclusively that mineral admixtures, used in the correct way and in the required quantity, can suppress effectively and substantially much of the material and structural damage caused by alkali–silica reactions. But they cannot remove all the deleterious effects completely. Indeed there is no reason why they should do so, nor should one expect them to do so. However, they go a very very long way in helping concrete materials and concrete structural elements in maintaining their strength, stiffness and stability.

4.4 Conclusions

The aim of this chapter is to examine critically the role and effectiveness of mineral admixtures in counteracting the effects of ASR. Tests are reported on plain concrete, reinforced concrete slabs, and reinforced concrete beams incorporating a slowly reactive but moderately expansive reactive aggregate and containing either fly ash, slag or microsilica. To be able to compare the effects of admixtures in plain and reinforced concrete elements, the test conditions in terms of concrete mixes, concrete alkalinity, method of incorporation of mineral admixtures and exposure regimes were all identical in the plain concrete and the slabs. The data on the beams are presented to supplement the concepts advanced by the results of the tests on prisms and slabs.

The results show that mineral admixtures control expansive strains and the resulting cracking at different rates and to varying degrees. The control of these strains and cracking in concrete with reinforcement reflects very well the behaviour observed in plain concrete, and follows the pattern shown by the concrete prisms. However, tests on structural elements show that the control of expansive strains and cracking alone is not an adequate criterion to judge the effectiveness of mineral admixtures in overcoming all the effects of ASR. It is shown that other factors, such as preserving the strength and stiffness of ASR-affected concrete at values of those of the undamaged concrete and control of structural distortions, also need to be considered if the strength and stability of structural elements are to be maintained when affected by ASR.

The data presented here show conclusively that mineral admixtures do not all control the material damage and structural deterioration due to ASR at the same rate or to the same extent and at the same time. Indeed, there is no reason for them to do so or for us to expect so. Many factors influence the effectiveness of mineral admixtures when used to counteract the effects of ASR. What is conclusive is that mineral admixtures, when used correctly and at the required amount, have a very positive role in controlling effectively and substantially the expansive strains, cracking, loss of strength and stiffness and structural distortions arising from ASR. But they cannot and do not always remove these deleterious effects completely.

References

1. Stanton, T.E. (1940) Expansion of concrete through reaction between cement and aggregate. *Proc. Am. Soc. Civ. Engrs*, **66**, 1781–1811.
2. Stanton, T.E. (1959) Studies of use of pozzolanas for counteracting excessive concrete expansion resulting from reaction between aggregate and the alkalis in cement. *Symposium on Pozzolanic Materials in Mortars and Concretes*, ASTM STP 99, pp. 178–203.
3. Pepper, L. and Mather, B. (1959) Effectiveness of mineral admixtures in preventing excessive expansion of concrete due to alkali silica reaction. *Proc. ASTM* **59**, 1178–1203.
4. Mehta, P.K. (1984) In: Ramachandran, V.S. (Ed.) *Mineral Admixtures, Concrete Admixtures Handbook—Properties, Science and Techology*. Noyes Publications, Far Ridge, NJ.
5. Aitcin, P.C. and Regourd, M. (1985) The use of condensed silica fume to control alkali silica reaction—a field case study. *Cement Concr. Res.* **15**, 711–719.
6. Olafsson, H. (1989) AAR problems in Iceland—present state. *Proceedings of the Eighth International Conference on Alkali–Aggregate Reaction*, pp. 65–70.
7. Swamy, R.N. (Ed.) (1986) *Cement Replacement Materials*, Vol. 3, *Concrete Technology and Design*. Surrey University Press, London.
8. Idorn, G.M. (1983) Thirty years with alkalis in concrete. *Proceedings of the Sixth International Conference on Alkalis in Concrete*, pp. 21–38.

9. Idorn, G.M. (1989) Alkali–silica reactions in retrospect and prospect. *Proceedings of the Eighth International Conference on Alkali–Aggregate Reaction*, pp. 1–8.
10. Roy, D.M. and Idorn, G. M. (1982) Hydration, structure and properties of blast furnace slag cements, mortars and concrete. *ACI Journal* 79, 446–457.
11. Bhatty, M.S.Y. (1983) Mechanisms of pozzolanic reactions and control of alkali aggregate reactions. *Cement, Concrete and Aggregates* 7(2), 69–77.
12. Chatterji, S. (1989) Mechanisms of alkali–silica reaction and expansion. *Proceedings of the Eighth International Conference on Alkali–Aggregate Reaction, Kyoto*, pp. 101–105.
13. Diamond, S. (1989) ASR—another look at mechanisms. *Proceedings of the Eighth International Conference on Alkali–Aggregate Reaction, Kyoto*, pp. 83–94.
14. Lenzner, D. and Ludwig, V. (1978) The alkali aggregate reaction with opaline sandstone from Schleswig-Holstein. *Proceedings of the Fourth International Conference on Effects of Alkalies in Cement and Concrete*, pp. 11–34.
15. Oberholster, R.E. and Westra, W.B. (1981) The effectiveness of mineral admixtures in reducing expansion due to the alkali–aggregate reaction with Malmesbury group aggregates. *Proceedings of the Fifth International Conference on Alkali–Aggregate Reaction in Concrete*.
16. Kawamura, M., Takemoto, K. and Hasaba S. (1984) Effect of various pozzolanic additives on alkali–silica expansion in mortars made with two types of opaline reactive aggregates. *Concrete Association of Japan Reviewer*, pp. 91–95.
17. Hogan, F.J. (1985) The effect of blast furnace slag on alkali aggregate reactivity: a literature review. *Cement, Concrete and Aggregates* 7(2), 100—107.
18. Kawamura, M. and Takemoto, K. (1985) Effects of silica fume on alkali silica expansion and its mechanisms. *Cement Association of Japan, Review*, pp. 258–261.
19. Soles, J.A., Malhotra, V.M. and Suderman, R.W. (1986) The role of supplementary cementing materials in reducing the effects of alkali–aggregate reactivity: CANMET investigations. *Proceedings of the Seventh International Conference on Concrete Alkali–Aggregate Reactions, Ottawa*, pp. 79–84.
20. National Materials Advisory Road Committee (1987) *Concrete Durability: a Multibillion Dollar Problem*. Report NMAB-437. National Academy Press, Washington DC.
21. Swamy, R.N. and Al-Asali, M.M. (1988) Engineering implications of ASR expansion in concrete and the effectiveness of mineral admixtures. *Durability of Concrete: Aspects of Admixtures and Industrial By-Products*. Document D1. Swedish Council for Building Research, pp. 133–155.
22. Kollek, J.J. (1989) Effective alkalis from pulverised fuel ash, granulated blast furnace slag and natural pozzolana deduced from mortar bar expansion results. *Proceedings of the Third International Conference on Fly Ash, Silica Fume, Slag and Natural Pozzolans in Concrete, Trondheim, Norway*, Vol. 1, pp. 373–401.
23. Mullick, A.F., Wason, R.C. and Rajkumar, C. (1989) Performance of commercial blended cements in alleviating ASR. *Proceedings of the Eighth International Conference on Alkali–Aggregate Reaction*, pp. 217–222.
24. Hobbs, D.W. (1989) Effect of mineral and chemical admixtures on alkali–aggregate reaction. *Proceedings of the Eighth International Conference on Alkali–Aggregate Reaction*, pp. 173–186.
25. Swamy, R.N. and Al-Asali, M.M., Effectiveness of mineral admixtures in controlling ASR expansion. *Proceedings of the Eighth International Conference on Alkali–Aggregate Reaction*, pp. 205–210.
26. Al-Asali, M.M. and Malhotra, V.M. (1989) *Role of High Volume Fly Ash Concrete in Controlling Exansion due to Alkali Aggregate Reaction*. CANMET, Canada.
27. Carrasquillo, R.L. and Farbriaz, J. (1989) *Alkali–Aggregate Reaction in Concrete Containing Fly Ash*—final report. The University of Texas, Austin, TX.
28. Perry, C. and Gillott, J.E. (1985) The feasibility of using silica fume to control concrete expansion due to alkali aggregate reactions. *Durability of Building Materials* 3, 133–146.
29. Swamy R.N. (1989) Silica fume and alkali silica reaction. *Durability of Concrete: Aspects of Admixtures and Industrial Byproducts*. Document D9. Swedish Council for Building Research, pp. 236–251.
30. Farbiarz, J., Schuman, D.C., Carrasquillo, R.L. and Snow, P.G. (1989) Alkali–aggregate reaction in fly ash concrete. *Proceedings of the Eighth International Conference on Alkali–Aggregate Reaction*, pp. 241–246.

31. Norwegian National Standards (1978) NS 3420 and NS 3474 (with amendments in 01-06-78).
32. Isabelle, H.L. (1986) *Development of a Canadian Specification for Silica Fume.* ACI SP 91, Vol. 2, pp. 1577–1587.
33. Swamy, R.N. and Al-Asali, M.M. (1986) Influence of alkali–silica reaction on the engineering properties of concrete, In: Dodson, V.H. (Ed.) *Alkalies in Concrete.* ASTM STP 930, pp. 69–86.
34. Swamy, R.N. (1990) Fly ash concrete—potential without misuse. *RILEM materials and Structures* 23(138), 397–411.
35. Swamy, R.N. (1989) *Superplasticizers and Concrete Durability.* ACI Publn SP-119, pp. 361–382.
36. Swamy, R.N. and Falih, F.M. (1985) Development of a small aggregate concrete for structural similitude for slab-column connections. In: Clarke, J.L., Garas, F.K. and Armer, G.S.T. (Eds.) *Design of Concrete Structures—The Use of Model Analysis*, pp. 25–37.
37. Swamy, R.N. and Wan, W.M.R. *Use of Dynamic Non-destructive Test Methods to Monitor Concrete Deterioration due to Alkali Silica Reaction*, unpublished report.
38. Swamy, R.N. (1989) Structural implications of alkali silica reaction. *Proceedings of the Eighth International Conference on Alkali–Aggregate Reaction*, pp. 683–689.
39. Swamy, R.N. (1990) Alkali silica reaction and concrete structures. *Structural Engineering Review* 2, pp. 89–103.
40. Swamy, R.N. and Al-Asali, M.M. (1989) Effect of alkali silica reaction on the structural behaviour of reinforced concrete beams. *ACI Structural Journal* 86, 451–459.
41. Malhotra, V. M. (Ed.) (1987). *Supplementary Cementing Materials for Concrete*, CANMET, Canada.
42. Swamy, R. N. and Bouikni, A. (1990) Some engineering properties of slag concrete as influenced by mix proportioning and curing, *ACI Materials Journal* 87(3) (May/June), 210–220.
43. Lamond, J.F. (1983) *Twenty Five Years' Experience using Fly Ash in Concrete.* ACI Publication SP-79, Vol. 1, pp. 47–69.
44. Hogan, F.J. and Meusel, J.W. (1981) The evaluation of durability and strength development of ground granulated blast furnace slag, *Cement, Concrete and Aggregates* 3, 40–52.
45. Malhotra, V.M. (1983) *Strength and Durability Characteristics of Concrete Incorporating a Pelletized Blast Furnace Slag.* ACI Publication SP-79, Vol. 2, pp. 891–921.
46. Gudmundsson, G. and Asgeirsson, H. (1983) Parameters affecting alkali expansion in Icelandic concretes, *Proceedings of the Sixth International Conference on Alkalis in Concrete*, pp. 217–221.
47. Hooton, R.D. (1987) Some aspects of durability with condensed silica fume in pastes, mortars and concretes. *International Workshop on Condensed Silica Fumes in Concrete.*
48. Aitcin, P.C., Regoud, M. and Fortin, R. (1986) The use of condensed silica fume to control alkali silica reaction—a field case study. *American Ceramics Society Cements Divisions Annual Meeting, Abstracts*, Vol. 63 No. 3, p. 470.
49. Swamy, R.N. and Al-Asali, M.M. (1989) The effectiveness of mineral admixtures in controlling ASR expansion. *Proceedings of the Eighth International Conference on Alkali–Aggregate Reaction*, pp. 205–210.
50. Swamy, R.N. and Al-Asali, M.M. (1988) Alkali silica reaction—sources of damage. *Highways and Transportation*, 35(2), 24–29.
51. Swamy, R.N. and Al-Asali, M.M. (1987) Influence of alkali silica reaction on the behaviour of reinforced concrete columns. *Proceedings of the Fourth International Conference on Durability of Building Materials and Components*, Vol. 2, pp. 647–653.
52. Ono, K. (1988) Damaged concrete structures in Japan due to alkali silica reaction. *International Journal of Cement Composites and Lightweight Concrete* 10, 247–257.
53. Blight, G.E., Alexander, M.G., Schutte, W.K. and Ralph, T.K. (1983) The effect of alkali–aggregate reaction on the strength and deformation of a reinforced concrete structure. *Proceedings of the Sixth International Conference on Alkalis in Concrete*, pp. 401–410.
54. Mullick, A.K. (1988) Distress in a concrete gravity dam due to alkali silica reaction. *International Journal of Cement Composites and Lightweight Concrete* 10, 225–232.
55. Albert P. and Raphael, S. (1986) Alkali–silica reactivity in the Beauharnois Powerhouses, Beauharnois. *Proceedings of the Seventh International Conference on Alkali–Aggregate Reaction*, pp. 10–16.

5 Alkali–silica reaction—UK experience

I. SIMS

Abstract

The developing awareness and acceptance of alkali–silica reaction (ASR) in the UK is traced from the late 1940s to the present day. Seemingly extensive studies at the Building Research Establishment during the 1950s were optimistically interpreted by the construction industry as reassurance, and this led to more than a decade of near-complacency when ASR was widely regarded as 'a foreign problem'. In the early 1970s, however, major structures in Jersey and on the UK mainland, particularly in the south-west and Midlands of England, were found to be damaged as the result of ASR. Some of the earlier examples of structures recognised as being damaged by ASR are described, and the particular types of aggregate combinations involved are examined. An attempt is made to assess the current extent of the occurrence of ASR in the UK and to analyse the scale of damage typically found; some unusual examples are included. Guidance and specifications developed in the UK to minimise the risk of ASR in future concrete construction are explained in detail, and the prospects of developing an adequately predictive test are fully discussed. Finally, critical consideration is given to the most recent advice on diagnosis, the appraisal of affected structures in the UK and the possibilities of management or repair. The outlook is concluded to be generally encouraging, but some newly discovered aspects, including the important role of externally derived alkalis, are briefly considered.

5.1 Introduction

When alkali–silica reactivity was discovered by Stanton[1a,b] in America in 1940, Britain was otherwise engaged, fighting the Battle of Britain and suffering the Blitz. During the war years and to the end of the 1940s, American researchers established most of the basic parameters of ASR and its consequences[2]. The first ASR tests on British aggregates were carried out in the USA in 1947 (see 5.3.2). In their 13th Annual Report in 1948[3], the Cement and Concrete Association in the UK announced an initial research programme on 'tests to determine whether the expansive reaction between aggregates and cement so often reported from the USA is a possible cause of disintegration in Britain'. It is not clear to what extent that programme was ever pursued at that time, but certainly a major programme of research had commenced at the Building Research Station (BRS) in 1946, under the supervision of F.E. Jones, and continued through most of the 1950s. The preliminary findings, as National Building Studies (NBS) Research Papers (RP), were published in 1952[4-6], and the final conclusions and recommendations, also as NBS Research Papers, were issued in 1958[7,8].

The reassurance provided by the BRS/NBS Research Papers, which was arguably more apparent than real (see 5.3.3), led to a decade of near complacency, when the UK concrete industry was generally content to regard ASR as 'a foreign problem'. In January 1971, however, a mass concrete dam in Jersey, Channel Islands, was found to exhibit displacements and cracking which were in due course attributed to ASR[9], and this finding reactivated British awareness of ASR. The reawakening was rapid, with the Cement and Concrete Association and Queen Mary College (University of London) jointly hosting the Third International Conference on ASR in September 1976[10] (the first and second conferences had been held in Copenhagen and Reykjavik respectively). In December 1976 unexplained cracking was found to be affecting concrete bases at an electricity substation in Plymouth, Devon, and by early 1977 this was known to be the result of ASR[11]. Over more than 10 years since that discovery, several hundred concrete structures have been identified on the UK mainland which are affected, to a greater or lesser extent, by ASR.

Like a disease for which the cure is uncertain, the alarm and concern generated by ASR has been generally out of proportion to the low incidence of occurrence and the typically limited scale of structural damage involved. However, the degree of interest in ASR has enabled good progress to be made in the UK in formulating measures to minimise the risk of ASR in new concrete work[12,13], in establishing agreed procedures for the diagnosis of ASR[14] and, most recently, in providing guidance for the structural assessment of affected structures[15]. Although the 50th anniversary of the discovery of ASR has been reached, it seems less likely that modern UK concrete will ever suffer ASR and even already affected structures will mostly prove to be repairable.

5.2 Reassurance: studies at the BRS

5.2.1 1952—review and initial findings[4-6]

The BRS research commenced with a detailed review of the experiences to date in the USA, including references to specific structures, an assessment of test methods and an appreciation of 'corrective' measures to avoid ASR in new works[4]. In his prefatory note to RP 14, F.M. Lea, then Director of Building Research, warned:

In the USA . . . many serious and extensive failures of concrete have since (1940) been attributed to alkali–aggregate reaction. . . . Fortunately, no evidence has yet been obtained of large-scale failures in this country arising from this cause; nevertheless, some aggregate deposits which may be reactive in some degree do occur, and unfamiliar aggregates may need examination before use.

The second NBS Research Paper, No. 15, provided a general preliminary consideration of British Portland cements and British aggregates[5]. The total alkali content of British cements, as analysed at the BRS since 1928, was found to range from 0.5 to 1.0% Na_2O eq. and to be typically higher than the 0.6% maximum level suggested by American experience for preventing ASR. Furthermore, extraction tests indicated that all of the acid-soluble Na_2O and K_2O in British cements would become available for reaction in concrete. In British aggregates, flint in the south-east England gravels was recognised to be 'slightly reactive', probably as a result of microstructure rather than the presence of opaline silica[16], and 'the acknowledged freedom from trouble in practice' was thought to be 'associated with the use of whole flint aggregate'. Thus, even in 1952, the potential importance of the flint 'pessimum' was understood and led to the cautionary statement in RP 15: 'There is a possibility that if flints are used in admixtures with inert aggregates, sufficient expansion to cause trouble might occur'. Opaline 'Malmstone' and some glassy acid to intermediate volcanic rocks were also identified as potentially reactive materials, but these were not considered to be important sources of aggregates in the UK.

The expansion bar test (a mortar bar test), which had been identified in the review as seeming to be 'the most generally useful' procedure currently available, was then applied to a limited range of British aggregates using a cement of 'medium' alkali content (0.7% Na_2O eq.) and employing storage at 20°C. The results up to 4 years were given in NBS Research Paper No. 17[6], where the interpretation was based upon comparison with the results obtained in the same test series for two opal-bearing aggregates imported from the USA (Californian siliceous magnesian limestone—10% in Thames Valley flint sand—and Nebraska sand). The American 'control' samples gave results of 0.38% and 0.03% respectively at 6 months, with 0.72% and 0.18% at 4 years, whereas the highest value obtained for any of the British natural aggregates was not more than 0.03% up to 4 years (Table 5.1 and Figure 5.1). This seemingly hopeful outcome was confirmed by the statement that 'the conclusions regarding the rock types concerned are reassuring for aggregates in this country', although this view was conditioned by the need to be mindful of the presence of any opaline silica.

With the benefit of hindsight, it is perhaps significant that low-power microscopical examination of broken mortar bar surfaces after testing identified small amounts of alkali–silica gel deposits associated with the Thames

Table 5.1 Selected mortar bar results (percentage expansion) from BRS/NBS Research Papers 17 and 20[6,7]. All results are for 'low-quality' mortar-bars (higher water–cement ratios and greater porosities) stored at 20°C.

| | Cement type (alkali content) | | | | | | | |
| | 'Medium' alkali (0.7% Na_2O eq.) Test age (years) | | | | 'High' alkali (1.2% Na_2O eq.) Test age (years) | | | |
Aggregate	1	3	4	7	1	3	4	7
American 'control' aggregates								
Siliceous magnesian limestone (10% in Thames Valley sand)	0.51	0.68	0.72	—	0.71	0.72	0.74	0.74
Nebraska Sand	0.10	0.18	0.18	—	0.16	0.25	0.27	0.30
British natural/crushed rock aggregates								
Clee Hill basalt	0.02	0.02	0.02	—	0.02	0.03	0.03	0.03
Quartzite	0.02	0.02	0.02	—	0.02	0.03	0.03	0.03
Carboniferous limestone	0.01	0.02	0.02	—	0.02	0.02	0.02	0.03
Darley Dale sandstone	0.02	0.03	0.03	—	0.03	0.03	0.04	0.04
Thames Valley flint sand	0.01	0.01	0.01	—	0.02	0.03	0.03	0.03
Cambridge flint gravel	—	—	—	—	0.02	—	0.02	0.02
Cambridge flint sand	—	—	—	—	0.02	—	0.05	0.06
Rhyolite (50% in Clee Hill basalt)	—	—	—	—	0.02	0.02	—	—
Andesite (50% in Clee Hill basalt)	—	—	—	—	0.02	0.03	—	—

Valley flint sand sample, similar to those found occurring within the 'control' specimens made with American aggregates. This would seem to indicate that ASR had occurred with Thames Valley flint sand, but that expansion had not resulted in these tests. However, this observation was not emphasised and the impression left was that British aggregates were not expansively reactive, in contrast to some of those found overseas.

5.2.2 *1958—conclusions and recommendations*[7,8]

The series of experiments described in RP 17 (see 5.2.1) were repeated using 'high-alkali' cement, actually the 0.7% Na_2O eq. cement with added NaOH to reach an equivalent 1.2%, and the findings reported in Research Paper No. 20[7]. Also, further British aggregates were tested, including representatives of those materials initially identified in RP 14 as being potentially reactive: flint, 'Malmstone' and acid to intermediate volcanic rocks. Some limited studies were carried out into aggregate mixtures and 'dilutions'.

In the 'high-alkali' cement repeat expansion bar tests, the American control aggregate exhibited higher expansions than were recorded for the earlier 'medium alkali' cement series, especially at 6 months (Table 5.1). After 4 years, none of the British natural aggregates exhibited any substantial expan-

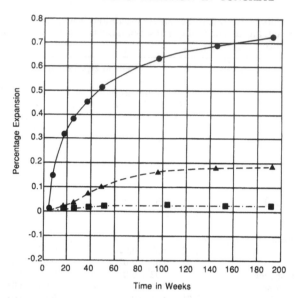

Figure 5.1 Comparison of mortar bar expansions for reactive American 'control' aggregates and Thames Valley flint sand: ■ Thames Valley sand; ▲ Nebraska sand; ● siliceous magnesian limestone 10% replacement in graded river sand. All results are for 'medium' alkali cement (0.7% Na₂O eq.), 'low-quality' mortar bars (higher water–cement ratios and greater porosities) stored at 20°C. Redrawn and adapted from Ref. 6.

sion, the highest value obtained being only 0.035%. An extension of testing to 7 years did not significantly alter the conclusion, with the highest value moving up only to 0.042%.

Tests were carried out on additional flint aggregates from four sources, two in the Thames Valley, one in Essex and another in Cambridgeshire. None of the flint gravels exceeded 0.02% in these tests, even after 7 years, but the single flint-bearing sand (composed of quartz and flint) exhibited rather more but still comparatively low expansivity, reaching 0.05% after 4 years and 0.06% after 7 years. Once again the expansion bar specimens made with flint aggregates exhibited some evidence of ASR, in the form of gel deposits visible to low-power microscopical examination after testing.

The inherent reactivity of flint, notwithstanding the lack of expansion in these tests, was demonstrated by some chemical tests reported in RP 20 (pat test and rapid chemical test described in RP 25, see below). A series of 'dilutions' using Clee Hill basalt (considered inert) with Thames Valley flint aggregate did not produce any significant expansions in the tests at 20°C, although earlier work in the USA had suggested that at least some British flint could be induced to cause expansion, up to 0.14% at 12 months, when present as around 20% of the total aggregate, at least at the elevated temperature of 38°C[17].

A range of acid to intermediate volcanic rock materials, including rhyolites and andesites, was tested using the same regime, variously as 'whole aggregate' or as 50 : 50 mixtures with Clee Hill basalt (considered inert). None of the aggregates or combinations tested produced any expansions greater than 0.03%, even after 3 years. Chemical tests also showed a 'complete lack of reactivity' for these volcanic rocks.

Malmstone was subjected to extensive testing as it had been found to contain opaline silica, although it was not in practice a source of aggregate for normal concrete. As expected, the chemical tests indicated considerable reactivity of Malmstone, but the expansion bar tests produced no excessive movements, even in various 'dilutions' with flint sand and Clee Hill basalt. This apparent anomaly was explained in RP 20 as being caused by the highly porous nature of Malmstone, rather than by any inherent inadequacy in the expansion bar test procedure.

Jones and Tarleton[7] considered that the findings reported in RP 20 confirmed those given 6 years earlier in RP 17 'on a firmer and wider basis', so that it could be concluded that 'the normal British aggregates so far tested, when used as whole aggregates, are not expansively reactive with high-alkali cements at normal temperatures'. Crucially, however, they added some words of caution in respect of flint aggregates, perhaps partly because of the observation of evidence of reaction in the absence of harmful expansion in the tests:

There remains the possibility therefore that some flints...may be encountered which, under adverse conditions of 'dilution', alkali content, water content and temperature, may cause trouble. So far as present evidence goes, however, this is considered to be rather unlikely.

The final NBS Research Paper, No. 25[8], comprised two parts, both concerned primarily with test procedures and, as such, accessory to the substantive results and conclusions of RP 20, but nevertheless influential over future British testing practice. In the first part (actually part V of the NBS series on alkali–aggregate reaction), the apparent deterioration in bending strength of mortar bars was found to be 'useful supplementary evidence' but was not considered to be a preferable alternative to expansion testing. Also, the observation of gel deposits either within or on the outer surfaces of mortar bars was thought to provide 'at least valuable evidence of potential expansive reactivity', although it was conceded that such observations had not always been accompanied by expansion in the BRS test programme.

The second part of RP 25 (actually part VI of the NBS series) discussed the test methods used at the BRS and gave recommended test procedures. The pat test (or gel pat test), developed from a method earlier described by Stanton et al.[18], was recommended as 'a useful rapid test for reactive material' and was perhaps used subsequently more in the UK than elsewhere. A rapid chemical test was recommended as 'a useful acceptance test', albeit a modified version of the ASTM C289 method[19], and a maximum dissolved silica-alkalinity reduction ratio of 1.5 was suggested, in contrast to the maximum ratio of 1 given in ASTM C33[20]. Finally, the expansion bar test was recommended as 'the standard reliable test for determining the degree of expansive reactivity of an aggregate in various combinations with cements'. A BRS method for the expansion bar test was described which differed in several important details from the ASTM C227 method[21]; the cement alkali content was standardised at 1.2% Na_2O eq., the aggregate–cement ratio was 2 rather than 1.25 and the preferred storage temperature was 20°C rather than 38°C. Using this BRS test, the aggregate was to be classed as 'expansively reactive' if measured expansion exceeded 0.05% at 6 months or 0.1% at 12 months, in contrast to the revised ASTM C33 similar limits at 3 and 6 months respectively. Some mention was made of the possibility of applying the expansion test to concrete specimens, but it would be more than 20 years before such a procedure was developed at the BRS.

5.2.3 Significance and effect

At the time that Dr Jones was finalising his BRS research programme, there were no known examples of ASR affecting structures in the UK, and this will inevitably have been regarded as an important factor bearing upon the emphasis of the conclusions to be drawn from the laboratory results. The laboratory findings were seemingly consistent with the experience in practice, that is British natural aggregates were 'not expansively reactive'. To their considerable credit, Jones and Tarleton recognised that flint aggregates, in particular, had at least a hypothetical potential for expansive reaction in certain circumstances, notwithstanding the lack of expansion produced for flint aggregates in the BRS tests. However, the prophetic cautions contained within the conclusions to the BRS work in respect of British flint aggregates were generally superseded in the readers' appreciation by the overall reassurance which appeared to emerge from an apparently thorough survey.

Petrographically, flint (a distinctive chert of Upper Cretaceous age) is a potentially alkali-reactive form of silica, even without the presence of any opaline material. As early as 1947, flint from Dover cliffs had been shown to be expansively reactive in American tests when forming around 20% of an otherwise non-reactive total aggregate[17]. In the BRS test programme, flint aggregates had shown clear evidence of ASR within expansion bar specimens, as well as in separate chemical tests, although no significant expansions had been recorded. For some reason, the expansion testing approach, modelling normal concrete with small mortar specimens of high cement content, appears to have been regarded as beyond reproach, despite its failure to cause expansion even with opal-bearing 'Malmstone' material. Subsequent events have shown justification for Jones and Tarleton's cautious appraisal of British flint aggregates, but it was to be nearly 20 years before the UK concrete industry would accept that ASR was anything other than a 'foreign problem'.

5.3 False security: complacency in the 1960s

5.3.1 *British Standards*

The British Standards Institution (BSI) is an independent chartered body devoted to the assurance of uniformity and quality of materials and products in the UK. Published British Standards are not legally binding, except when they are called up by a contract specification or similar document, but they are accepted throughout the construction industry as being dependable and authoritative guidance. The constitution of the BSI ensures that all published standards are compiled by representative committees and that they are regularly reviewed and amended or revised when necessitated by new knowledge, recent developments or changed circumstances. British Standards, including Codes of Practice, are therefore important and fairly reliable indicators of the contemporary views within the construction industry at the times of their publication, although they do represent a consensus and may not adequately cover particularly contentious matters.

The materials used in the manufacture of concrete were covered in the 1960s, as now, by BS Specifications, notably BS 12 for Portland cement[22] (ordinary and rapid-hardening), BS 882 for aggregates[23] (natural) and BS 3148 for the mix water[24]. Test methods were given for cement in BS 12 and BS 4550[22,25] and for aggregates in BS 812[26]. Guidance for the making and use of concrete for structures was given in CP 114, CP 115 and CP 116[27-29], later superseded by CP 110[30] (now BS 8110[31]), whilst test methods for concrete were given in BS 1881[32]. With one small exception, none of these British Standard publications made any direct or allusive reference to ASR until the 1980s, and even then only BS 8110 contained any substantive guidance. The single exception was BS 3148[24], in which it was suggested (in an appendix) that the alkali content of some mix waters might be of significance in the presence of alkali-reactive combinations of aggregates and cement, although it was further stated that, 'So far, no naturally occurring expansively reactive aggregates have been found in the United Kingdom.'

The 1954 edition of BS 882, the specification for natural aggregates for concrete, was extensively revised and a new edition published in 1965. A comparison of the 1954 and 1965 editions shows that the standard was rearranged and, in places, completely rewritten. The final results of the BRS research programme had been published in 1958 (see 5.2.2) and the findings must have been familiar to all members of the revising committee; indeed, although the full membership of that committee is no longer available, it is known that the BRS was represented by Teychenné[33]. Therefore, it seems certain that a conscious decision was taken by the BSI committee to omit any reference to ASR in the concrete aggregates specification. One surviving member of that committee has indicated that the matter was discussed and that, because BRS research had only recently concluded that ASR was unlikely to be a problem with British aggregates and no actual examples had been reported, it was considered inappropriate or even unhelpful to raise concern over ASR in the national specification[33]. Even in the 1983 revision of BS 882, mention of ASR is restricted to a non-mandatory note within an appendix, and that is only a cross-reference to the Code of Practice, BS 8110.

From the end of the Second World War to the discovery after 1971 that ASR had damaged the Val de la Mare Dam in Jersey, and by 1977 that concrete on the UK mainland was also affected, there was virtually no reference to the possibility of ASR in British Standards relating to concrete and concrete materials. Therefore, any supplier or user dependent upon the provisions of the relevant British Standards for assuring the quality of concrete would not have been caused to make any allowance whatever for the possibility, however remote, of ASR occurring at some time after construction. It has been argued, albeit retrospectively, that the 'broad-brush' quality requirements of BS 882[23], including the provision that 'aggregates... shall not contain deleterious materials in such a form or in sufficient quantity to affect adversely the strength at any age or the durability of the concrete', could be considered to have embraced ASR, even though ASR was not one of the specific examples of 'such deleterious materials' listed in the standard to clarify the above-quoted requirement. The validity of such an argument depends upon the extent to which alternative authoritative guidance, other than BSI documents, was available in the UK to advise suppliers and users of the possibility of ASR.

5.3.2 Other authoritative guidance and advice

In the UK, apart from BSI, authoritative national guidance on construction materials and practice is available from the Building Research Station (or Building Research Establishment) and, where appropriate, from the Cement and Concrete Association (C&CA, now renamed the British Cement Association, BCA) and the Transport and Road Research Laboratory (TRRL). Occasionally similarly authoritative and useful guidance may be produced on specific topics by the relevant professional institutions, learned societies or even trade associations. Academic publications and certain textbooks can also provide useful guidance, but, by their nature, are not necessarily representative of the consensus viewpoint.

After the publication of Jones and Tarleton's final NBS reports in 1958 (see 5.2.2), the BRS (BRE) did not produce any further official guidance until 1982, when Digest 258 was issued[34], some 6 years after the identification of ASR on the UK mainland. However, BRE researchers had been involved in the investigation into the Val de la Mare Dam in Jersey (see 5.5) and Midgley described a technique arising from this investigation at the 1976 ASR conference in London[35]. The next publication on ASR by BRE staff did not appear until 1979[36]. It is thus apparent that for the whole of the 1960s, and also for a good part of the 1970s, the seeming reassurance provided by the earlier 1946–1958 BRS research programme remained unchallenged. Indeed, a BRE Digest published in 1971[37], on the subject of concrete appearance, contained the following statement:

Fortunately, there are no known deposits of aggregate in the United Kingdom that have any significant degree of alkali–aggregate reactivity, but a designer of concrete buildings *for other parts of the world* might run into serious difficulties if he does not get advice on the suitability of the available aggregates. (present author's emphasis)

Clearly, ASR was still to be regarded as 'a foreign problem'.

Although the C&CA declared an intention to carry out research into ASR in 1948[3], there is no available evidence that any such work was carried out, and the earliest C&CA publication on the subject appeared as late as 1977[38], with their first research work appearing a year later[58]. A C&CA advisory note on the subject of aggregate 'impurities' was published in 1970[40] and made some reference to 'alkali-reactive minerals', but concluded confidently, 'In this country, these minerals are fortunately not present in the aggregates used for making concrete but, in any event, commercially available cements seldom contain enough alkalis to be troublesome. In practice then, the question of aggregate–alkali (*sic*) reaction *concerns only cements and aggregates used abroad*' (present author's emphasis). Roeder carried out some limited research at C&CA into flint reactivity as part of a wider study in 1974–75[41], but the results were not formally published and he was careful to state that, 'More recently interest has centred on the possibility of reaction between the aggregate and the alkalis in the cement but no examples of such reaction are known in this country'. Again, therefore, until the late 1970s, the clear impression given was that ASR, when mentioned at all, was only 'a foreign problem'.

The Sand and Gravel Association of Great Britain (SAGA) organised an important symposium in 1968 on the subject of 'sea-dredged aggregates for concrete', which were enjoying a rapidly increasing rate of use. The various authors, for example Shacklock[39], were primarily concerned with the possible influences of sea salt and shell material, and ASR was only mentioned briefly by two of the contributors. Haigh[42] (from a firm of consulting engineers) stated that, 'Alkali-reactive aggregates, apparently not a problem with aggregates in this country, are to be *found frequently overseas*.' (present author's emphasis). Van de Fliert[43], referring to some Dutch concerns over the use of flint aggregates, said that 'About 10 years ago, simultaneous research in Great Britain (presumably that at BRS) and in the Netherlands showed that this expansion reaction with crypto-crystalline quartz stone (flint) in concrete need not be feared' (present author's parentheses).

Most subjects become covered by numerous textbooks over a period of time, but inevitably some take on a particular authority, either in deference to the eminence of the author and/or his affiliations, or because of popular demand and the consequent reprinting and periodic revising of the work. For illustrative purposes, just three such books will be examined here. Neville's book, *Properties of Concrete*, was first published in 1963; a second edition was produced in 1973, and the third edition is dated 1981[44]. In the 1963 and 1973 editions, a thorough account of ASR is concluded by the statement, 'The alkali–aggregate reaction of the type described has fortunately not been encountered in Great Britain, but is *widespread in many other countries*' (present author's emphasis). The 1981 edition is reworded slightly and makes brief reference to examples identified in Great Britain in 1978, but, for nearly 20 years, this popular and influential book had appeared to support the notion that ASR was 'a foreign problem'.

Lea's important specialist textbook, *The Chemistry of Cement and Concrete*, was reissued as the third edition in 1970[45], being a revision of the second edition by Lea and Desch published in 1956. Since Lea had been Director of Building Research throughout the period of Jones's work at BRS in the 1950s, his rather cautious summary of ASR in respect of the UK is now seen to be significant. Lea states:

A survey of the common British aggregates has failed to reveal any containing alkali-reactive constituents, but care is needed when using aggregates of geological types which might contain such constituents and of which no previous service experience is available. It might have been expected that flint, which is such a large constituent of the Thames Valley gravels, would be reactive. Long experience with this aggregate has shown no such troubles, but it appears that it consists of a microporous mass of silica and does not contain opal.

Although Lea does not imply that ASR is exclusively an overseas problem, and advises caution with some new and untried sources of aggregate, he is nevertheless fairly unequivocal in his assessment of UK flint gravel as a non-reactive aggregate.

Finally, the Road Research Laboratory (later TRRL) published the well-known *Concrete Roads, Design and Construction* in 1955[46], and it remained available for purchase until 1975. The advice given therein in respect of ASR is consistent with the findings emerging from the BRS, and the interpretation embodied by the BSI and other authorities:

Some aggregates, notably those containing opaline silica or chert, may combine with the alkalis in the cement to produce disruption.... This trouble has not been observed in Great Britain, although it is met with in the USA and elsewhere.

5.3.3 *Signs of scepticism*

It is apparent from the preceding sections that the overwhelming impression gained by the UK construction industry during the 1960s and the early 1970s was that ASR was an interesting but, fortunately, a foreign problem. Geologically, this perception was always tenuous, since the inherent reactivity of certain mineral and rock varieties does not change at political boundaries, and in fact proper residual concern was expressed to some extent both by Jones and Tarleton[7] and by Lea[45] in respect of those aggregate materials which were not included in the BRS research programme. By the late 1960s and early 1970s, there are some signs that not everyone in the industry was completely reassured, especially those who had experienced alkali reactivity in other parts of the world.

Idorn was invited to present a lecture to the Concrete Society in London in May 1968, on the subject of 'the durability of concrete'[47]. Idorn had recently completed a substantial doctoral study of ASR and other threats to durability of concrete structures in Denmark[48], and much of his lecture to the Concrete Society concerned ASR. Whilst it is undoubtedly significant that Idorn did not presume to suggest in his lecture that ASR was likely to become recognised as a problem in the UK, it is equally apparent that the materials he had studied were generally similar to many of those found and used in the UK. It was this general similarity with Danish and German flint aggregates, and the uncertain nature of the distribution and provenance of marine aggregates in the North Sea, that caused Fookes to recommend the regular examination and ASR testing of the sea-dredged aggregates to be stockpiled for use in concrete for the prestigious Thames Barrier project in London, preparations for which commenced in the early 1970s[49].

Applied geological research into ASR and other aspects of concrete commenced at Queen Mary College (QMC), London University, at the end of the 1960s, inspired by Fookes (a visiting lecturer and later visiting professor) and also promoted by the growing awareness of special concrete problems being encountered in the unprecedented Middle East construction boom[50]. Doctoral research commenced at QMC in 1972 to include some study of concrete from the Val de la Mare Dam in Jersey and also the alkali–silica reactivity of British flint aggregates[51]. Rapid experience was gained by the investigation of reactive materials from overseas, including Cyprus[52], Bahrain[53] and elsewhere[54]. The Geomaterials Group at Queen Mary College is now rightly regarded as one of the foremost authorities on ASR in the UK and they started out by being unwilling to accept that it was safe to dismiss ASR as being exclusively 'a foreign problem'.

5.4 First warning: the Val de la Mare Dam in Jersey

5.4.1 Background

The Val de la Mare (Figure 5.2) mass concrete gravity dam at St Ouen's in Jersey was built between 1957 and 1961[9,51,55,56]. The structure, which is 29 m maximum height above foundation level and extends for 168 m straight across the valley, retains 950 million litres out of the island's total water storage capacity of 1500 million litres. Construction was in vertical bays or 'blocks', each 6.7 m wide and made up of lift sections each 1.2 m high. The concrete was made using UK Portland cement, together with local aggregates and water from the stream that was to be dammed, and the majority was placed in 1959–60.

Figure 5.2 Val de la Mare Dam, Jersey, in April 1989 (cf. Figure 5.7).

Figure 5.3 Part of the downstream face of the Val de la Mare Dam, Jersey, viewed from the parapet walkway, showing map cracking and exudations. From Ref. 52.

Figure 5.4 Displacement by up to 13 mm of the concrete parapet of the Val de la Mare Dam, Jersey. From Ref. 52.

The Jersey New Waterworks Company first reported in 1971 that four of the concrete sections had become discoloured, dampened and cracked, and the concrete crest handrail was in some places displaced by up to 13 mm (Figures 5.3 and 5.4). This led to investigations, including sonic speed measurements[57], petrographical examinations and aggregate tests, and remedial action including the installation of anchor bars, which was completed by 1975. Monitoring over the following 13 years has demonstrated a slowly continuing expansion at a rate of 50–75 microstrains/year and the failure of an anchor bar, caused by seismic shock rather than ASR, but confidence has grown that 'the dam can now enjoy a working life comparable with other concrete gravity dams'[9].

5.4.2 *Reactive constituents*

The coarse aggregate comprised mainly crushed diorite or granodiorite from a large quarry at Ronez on the north coast of Jersey, mixed with a smaller proportion of beach gravel containing rounded pebbles of 'Jersey shale', chert and various igneous rocks including dacite. The fine aggregate was a fossil beach sand from a pit at St Ouen's Bay, comprising quartz, quartzite and feldspar, with small amounts of chert, shell and various igneous materials. Although the beach aggregates were found to contain some potentially reactive constituents, principally chert and dacite (acid volcanic rock), there is little evidence that these materials reacted within the dam concrete. The reactive silica was present as secondary hydrothermal veining and vugh-fillings within the crushed dioritic rock aggregate source. The veining material, which is associated with shear zones and lamprophyre dyke zones within the quarry, comprised variable mixtures of chalcedony and opal. These vein and chalcedonised lamprophyre materials show reaction within a few days at 20°C in the gel pat test (Figures 5.5 and 5.6)[51], whereas no such rapid reaction could be obtained with the chert and dacite particles from the beach deposits.

Although the chalcedony/opal veining affects much of the Ronez rock mass, it clearly forms much less than 1% of the rock, and contemporary work by BRS considered it to be less than 0.1%[9]. The presence of some beach aggregate in the coarse fraction will have diluted this content of chalcedony/opal still further in the concrete. The 'pessimum' for opal is known to be very low[58], say 1–5%, but expansive reaction seems unlikely with opal contents as low as those computed for the Ronez veining material. It must therefore be presumed that variations during quarrying will occasionally lead to higher concentrations of reactive silica in certain batches of concrete.

The BRS carried out tests on aggregate samples using the mortar bar method, which had been the basis of their NBS research programme in the 1950s. The test details are not available, but it is reported that none of their results exceeded the ASTM C33 guidance criteria[20] at 6 months[9] and that

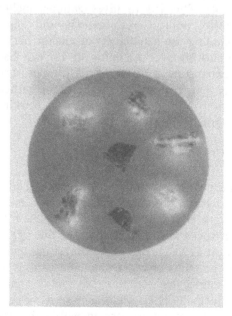

Figure 5.5 Gel pat test of chalcedony/opal vein material, similar to the reactive constituent present in concrete from the Val de la Mare Dam, Jersey, after 14 days of the test at room temperature. From Ref. 52.

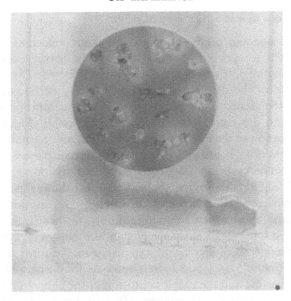

Figure 5.6 Gel pat test of chalcedonised lamprophyre dyke rock, similar to the reactive constituent present in concrete from the Val de la Mare Dam, Jersey, after 14 days of the test at room temperature. From Ref. 52.

results for the Ronez and beach aggregates were similar. In view of the reliance placed upon mortar bar test results in providing reassurance about UK aggregates, this must have been a worrying outcome. It seems likely, from more recent experience, that expansions would only have been produced in the laboratory by the controlled addition of the opal veining material to the aggregate, and perhaps only then using comparatively high alkali contents and concrete rather than mortar specimens.

The source (or sources) of the cement in the UK is not known for certain, and it is possible that cements from various works were supplied over the period of construction of the dam. Approximate values for two concrete samples have been quoted as 0.74% and 0.96% Na_2O eq. by weight of cement[9], and Coombes found that cement supplies during placement of the most severely damaged concrete (June–August 1960) most probably averaged 0.95% Na_2O eq.[55]. Thus, the most affected portions of concrete might have contained higher than typical amounts of opaline vein material in the aggregate and also relatively high concentrations of alkalis from the cement.

Additional reactive alkalis could have been derived from the rock aggregates themselves, or from seawater salt contamination of the beach aggregates, and in any water-retaining structure there is always the possibility of alkali migrations to form localised concentrations.

5.4.3 *Features and damage attributed to ASR*

The visible damage and correspondingly low sonic velocities were almost entirely restricted to concrete lifts cast during the critical period June to August 1960 (Figure 5.7) and block 6 was worst affected. The downstream concrete surface of the dam was stained with heavy carbonate exudation from between the sections, but the most serious discoloration was mainly concentrated into those areas cast in the critical period (Figure 5.3). The overall discoloration in the affected areas was from an original grey–buff to a relatively bright orange–brown. Affected areas also displayed random map cracking, these cracks being frequently bordered by orange–brown discoloration of slightly smoother surface texture than the surrounding concrete. There were white, sometimes porcellaneous, exudations from these cracks, especially those orientated horizontally or nearly so, and these deposits included some alkali–silica gel. No surface 'pop-outs' were observed, but there was evidence of significant expansion in the form of large-scale misalignment (Figure 5.4). Good correlation was found between sonic velocities, date of casting and the visible condition of the concrete. Nearly all of the sonic velocity results for 47 concrete lifts cast during the critical period were less than the overall average of 4.56 km/s for the 77 unaffected lifts measured[57] and, between 1972 and 1983, the maximum apparent 'reduction' in sonic velocity for the affected concrete was 14%.

Apart from measurement of the obvious handrail displacement, the early reports[55,56] do not indicate any attempt to quantify the amounts of any expansive movement which had occurred up to 1971. The BRS is reported

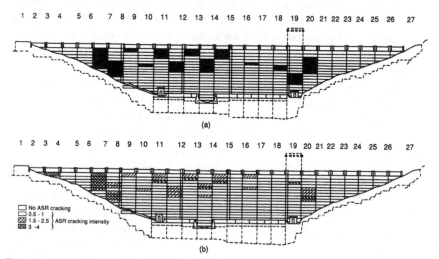

Figure 5.7 Downstream elevation of the Val de la Mare Dam, Jersey, showing (a) the lifts cast June–August 1960 and (b) their cracking intensity. From Ref. 9.

to have carried out some tests for latent expansion using core samples, obtaining results of only about 0.01% after 1 year at 20°C and 100% relative humidity (RH), and similarly after 3 months at 38°C and 100% RH. Although these results are low[14,15] and might indicate that most of the potential for reaction and expansion had been exhausted before drilling and testing of the cores, there is no certainty that the samples were kept moist prior to testing in accordance with modern practice, and also it is not clear whether any of the tests were pursued for storage periods longer than those reported.

The petrographical examinations carried out by the BRS do not appear to have been published, although it is generally believed that microscopic evidence of ASR was identified. Sims[51] obtained four small samples of the concrete and carried out petrographical examinations; he was able to identify some gel deposits within the concrete, variously infilling voids or lining cracks and fracture surfaces, and sporadic potentially reactive aggregate particles, including chalcedony, chert and rhyolite, but he could not find any definite reaction sites in the four samples available. Sims also reported less microcracking in the samples than might have been expected for expansive ASR and suggested that perhaps the main damaging reactions had occurred deep within the innermost parts of the mass concrete, perhaps there promoted by conditions of higher temperatures and increased pore water pressures. This suggestion appeared to be supported by the composition of the gel deposits, in which relatively reduced alkali contents and the presence of calcite and sulphates indicated considerable migration away from the reacting centres. It was further suggested that the conditions prevalent at the site of reaction might have favoured the production of a comparatively fluid gel, able to migrate, rather than a viscous swelling gel. If so, the observed cracking and expansion in the structure may have resulted primarily from the initial reaction, with continued gel-producing reaction not promoting any significant further expansion.

5.4.4 Remedial measures

Although it was assumed that reaction would continue, the investigations in 1971–74 indicated that the concrete would not become incapable of taking the required compressive loads, but that expansive cracking could lead to instability because of increasing internal uplift pressures[9]. Remedial action was therefore restricted to measures to maintain stability of the dam against increased internal uplift, and also it was decided that only the worst-affected parts would be dealt with initially, other areas being subjected to a similar form of repair as and when further deterioration occurred.

Three 40-mm Macalloy high tensile steel anchor bars were installed through the upstream side of the worst-affected block 6 and into the underlying rock; these were post-tensioned to produce a minimum factor of safety against overturning of 1.7 for hydrostatic internal uplift pressure of 100% in the upstream face and zero on the downstream face (Figure 5.8). Since drilling equipment was available on site, it was decided also to drill pressure-relief drainage holes vertically at 3.4-m centres through the central blocks 10–19 to intercept any seepage flow and provide some control over the development of uplift pressures.

A grouting trial was undertaken on block 6 to try to seal any ASR cracking on the upstream side to inhibit water seepage. Although a low-viscosity oil-based chemical grout was used (Polythixon 60/40 DR), very little penetration was achieved; it is not clear whether this was because of some inadequacy in the grouting system or instead because the concrete interior contained fewer cracks and microcracks than had been expected.

As well as the actual remedial measures, facilities were established, especially on block 6, for the future monitoring of movements (Table 5.2). This involved the instrumental monitoring of block 6, including vibrating-wire anchor bar load cells, electrically operated piezometers and 24 mechanical strain gauges across lift and vertical joints. However, as with most condition monitoring, much reliance is still placed upon regular visual inspections, supported by crack surveys and *in-situ* sonic velocity surveys. A further 44 electric piezometers were installed on selected lift joints in 1981, and later removed and reinstalled in 1983 after recalibration and treatment to reduce creep effects, but a new design has proved less effective than the earlier instruments in these conditions, and by 1988 only a third were giving reliable readings[9].

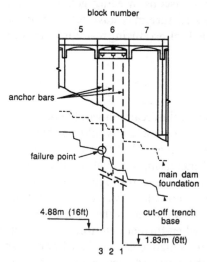

Figure 5.8 Detail of the downstream elevation of the Val de la Mare Dam, Jersey, showing the anchor bar installation to worst-affected block 6. The 'failure point' refers to the 1977 anchor bar fracture due to seismic shock. From Ref. 9.

Table 5.2 Val de la Mare Dam, Jersey: main monitoring activities and frequencies. After Ref. 9.

Monitoring activity	Frequency of records 1975–1988
Piezometer readings	2 weeks
Relief drain flows	Monthly
Anchor bar load cells	Monthly
Crest movement gauges	3 months
Visual inspection and survey or face cracking	3 months (later 6 months)
Other block 6 instrumentation	3 months (later stopped)
Sonic velocity survey	3 years (later 4 years)
All readings plotted on graphs and reviewed	Monthly, unless significant changes occur which are notified immediately
Review reports with recommendations for immediate future work	3 years
Inspection by independent panel alkali reactivity engineer	As engineer considers necessary

5.4.5 Experience to date and the future

It is now 14 years since monitoring of the Val de la Mare Dam commenced, and the findings generally indicate that 'there can be "life after ASR" for mass concrete gravity dams'[9]. Concern had arisen in 1977, when it was discovered that one of the three anchor bars had fractured at a depth of 18.6 m below the top of the dam. However, an intensive investigation revealed that the failure could only have been caused by a high-velocity tensile dynamic load, probably from a seismic shock, and the anchor was not replaced. The two remaining anchors have continued to perform satisfactorily, and in fact act as large extensometers together with the vibrating-wire load cells. Over 7 years (1975–82) the load on anchor 1 increased by 2.5% and, even after controlled destressing in 1982, the load has continued slowly to increase at a rate of around 2.5 kN per annum.

The flows from the pressure-relief drains drilled in 1974 have, with one exception, remained small, and encrustation with calcium carbonate has been the only problem, necessitating periodic cleaning. The exception was a drain in block 12, which did develop unexpectedly high flows in 1980, as a result of leakage through a particular lift joint. This was corrected by the installation of an epoxy-sealed neoprene rubber gasket into the suspect joints on the upstream face of the dam.

There now seems no reason to question the integrity of the Val de la Mare Dam for the foreseeable future, but inevitably it remains of special interest as the first example of ASR recognised in the British Isles. It seems logical that some other concrete structures on Jersey must also be affected by ASR, and a limited survey has been carried out[9,59]. Concretes made using the granodiorite aggregate source (with sporadic chalcedony/opal veins) exhi-

bited worse cracking at all ages than concrete made with only beach aggregates (Figure 5.9); indeed only one example with beach aggregates was more
than a case of 'suspected' ASR. In the case of the granodiorite concretes, the
water-retaining structures, including Val de la Mare Dam, were more seriously affected than the 'dry' concretes. Cole & Horswill[9] claim that, for Jersey
concretes, ASR damage does not worsen significantly after an age of 15–20
years (the dam was 9 years old when ASR was discovered and is now
27 years old), although further deterioration caused by other destructive
processes, including corrosion of embedded steel in reinforced structures,
remains a possibility.

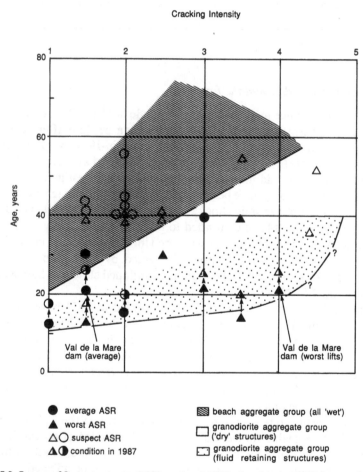

Figure 5.9 Survey of Jersey concretes (1982, updated 1987), showing that ASR damage is largely
restricted to concretes made using granodiorite coarse aggregate and especially water-retaining
structures. The 'cracking intensity' scale is defined by Ref. 59. From Ref. 9 after Ref. 59.

5.5 Fear of epidemic: the growth in mainland examples

5.5.1 *Plymouth—the first examples*

For a while it was possible to regard the dam in Jersey as something of 'a special case': caused by hitherto unsuspected opal-bearing secondary veins in the otherwise well-proven aggregate source and also possibly relating to the quarrying of an unusually intensely veined body of rock and/or the use of a very high-alkali cement. Between 1976 and 1978, any such residual complacency was overcome by the discovery of ASR on the UK mainland, first in the south-west of England and then in the English Midlands and South Wales[60].

The first UK mainland example was found at an electricity substation at Milehouse in Plymouth, Devon, affecting unreinforced concrete bases set in the ground with their upper surfaces exposed to the weather (Figure 5.10). A number of organisations were associated with the investigation into severe cracking of these bases and the identification of ASR, including Queen Mary College, the C&CA, Messrs Sandberg and Plymouth Polytechnic[11,60]. Before long, the Central Electricity Generating Board had identified a number of similar installations in the south-west and South Wales exhibiting broadly similar random cracking patterns, concrete compositions and exposure conditions. At Milehouse, the concrete comprised crushed limestone coarse aggregate, a sea-dredged fine aggregate containing a subordinate proportion of chert and a locally produced high-alkali cement (probably around 1.0–1.2% Na_2O eq.). Reaction sites were observed in the concrete, involving brown

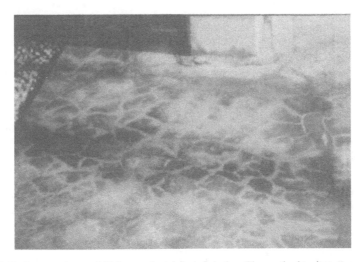

Figure 5.10 Concrete base at Milehouse electricity substation, Plymouth, showing map cracking. From Ref. 11.

and white chert particles, with the formation of reaction rims and associated gel deposits (Figure 5.11). Although some other causes of cracking were apparent[11], it has become generally accepted that ASR was the principal mechanism responsible for the damage. It is now considered that the particular conditions to which these concrete bases were exposed, being buried in the ground except for the upper horizontal surface, which was subject to wetting and drying, was an important factor in the promotion of ASR, but some earlier investigations considered that stray electrical currents might have been involved[61].

In Palmer's 1978 review[60], concrete associated with two reservoirs in the Midlands and another in South Wales were mentioned as examples of ASR, as well as two rather atypical cases respectively involving east coast wartime antitank defences and possible reactivity with greywacke aggregates in Scotland (see 5.6.4). However, it was probably the identification of ASR in the Charles Cross Multistorey Car Park in Plymouth which was mainly responsible for causing practising structural engineers in the UK to become acutely aware of the potential threat of ASR. The car park was built in 1970, using a helical ramp design, and was exhibiting slight to serious cracking of the concrete at an age of just 7 years. Some beams exhibited severe surficial cracking, the patterns of which were influenced by the underlying reinforcement and stress fields, and some column heads exhibited substantial cracking and fragmentation (Figure 5.12). The ground-level retaining walls, some parapet walls and pavements in the top level exhibited classic map-cracking patterns. Laboratory examination of the concrete again revealed the presence of reactive chert particles in a sea-dredged fine aggregate, together with a non-

Figure 5.11 Detail of the concrete from a cracked base at Milehouse electricity substation, Plymouth, showing reactive chert particles. From Ref. 11.

Figure 5.12 Charles Cross Car Park, Plymouth, showing a severely cracked column head.

reactive coarse aggregate (variously crushed limestone or crushed granite) and a high-alkali Portland cement. The car park has been used for a thorough study of concrete condition classification and crack mapping[62]. Load tests were carried out on various concrete units at the car park in 1981, 1982, 1985 and 1986[63]. The results confirmed that the beams 'were still able to carry their design load, and their actual service load plus a significant margin safely'. Consequently, the main *in-situ* structural frame was strengthened in 1983 and the car park has remained in service, being monitored and the structural condition being periodically reviewed.

The reactive combination of a chert (or flint)-bearing fine aggregate, a non-reactive (often crushed limestone) coarse aggregate and a high alkali Portland cement was becoming established as the cause of damaging ASR in the UK and, with only a few exceptions, this has continued to be the case as more and more examples have been identified. Chert or flint had always been recognised as a potentially reactive material (see 5.3.1) and it was the comparatively small proportion of chert in the total aggregate combination which appears to have approached the 'pessimum' for chert, especially in the presence of high alkali concentrations. The combination for concrete of local crushed limestones and crushed granites with sea-dredged sands, imported from further east in the English Channel, was a new and unfortunate development in the south-west of England in the late 1960s or early 1970s, to meet increased construction demand in an area traditionally short of good-quality land-based sand. It was the dilution of the overall chert or flint content of the aggregate by the presence of non-flint coarse aggregate, as opposed to the flint gravel which had more usually been combined with such flint sands in south and south-eastern England, which coincidentally created the reactive mixture. There is no reason to suppose that the use of sea-dredged aggregates, instead of land-based aggregates of similar mineralogical composition, had any adverse influence on the concretes, except of course for any cases where inadequate washing might have left a residual content of sea salt.

5.5.2 *Geographical controls*

Apart from the Jersey ASR caused by opaline veination of an aggregate source and some other possible cases involving greywacke aggregate (see 5.6.4), nearly all UK examples of ASR have involved chert or flint as the reactive constituent, and then only when present in subordinate 'pessimum' proportions and also in the presence of high concrete alkali contents (Table 5.3). Since most concretes are made using locally derived aggregates and cements, it is therefore not surprising that certain regions of the country have tended to yield most of the definite or relatively serious cases of ASR. The mixture of imported flint-bearing marine sand with local limestones and granites in south-west England has already been mentioned. In South Wales, somewhat similarly, sand dredged from the Bristol Channel contains small proportions of chert (probably Carboniferous) and is commonly blended with local crushed limestone aggregate. In the cases of both south-west England and South Wales, at least some of the locally made Portland cements were high-alkali in character. In the English Midlands, especially in the East Midlands, concretes are frequently made using the abundant fluvioglacial sands and gravels commonly termed 'Trent Valley' aggregates. These materials are dominated by quartzites, but both chert and flint are typically present, in varying ratios and proportions, in both the coarse and fine fractions. Again, at least one of the Portland cements made in the region was a high-alkali type, and ASR has sometimes occurred when the local aggregates and cement have been used together. These sands and gravels are somewhat polymictic, and other types of potentially alkali-reactive constituents have also been recognised, including some metamorphic quartzites and glassy rhyolite, although it is uncertain to what extent, if any, such materials have acually caused or contributed to concrete damage.

In other parts of the UK, to date, local combinations of aggregate and cement have not proved to be susceptible to ASR, and only sporadic cases of structures possibly affected by ASR have been reported. However, it would

Table 5.3 Some examples of reactive aggregate combinations identified in UK concretes affected by ASR. Adapted from Ref. 15.

Coarse aggregate	Fine aggregate	Structure locations	Reactive minerals
Granodiorite	Quartz beach sand	Jersey	Chalcedony/opal in granodiorite
Limestone or granite	Sea-dredged quartz/flint sand	SW England	Flint/chert in fines
Limestone	Sea-dredged quartz/chert sand	S Wales	Chert in fines
Trent Valley gravel	Trent Valley sand	Midlands	Flint, chert and possibly quartzite and rhyolite
Granite	Essex quartz sand with flint	Midlands	Flint
Limestone, glass and chert	Limestone, glass and chert	NW England	Glass
Lightweight aggregate	Quartz sand with chert	SE England	Chert
Flint gravel	Quartz sand and siliceous limestone	SE England	Silica in limestone
Greywacke sandstone and argillites	Natural sand mainly quartz and quartzite	Wales, SW England and N England	Cryptocrystalline quartz in greywacke/argillite*

*Additionally, alkali–silicate reaction might be involved.

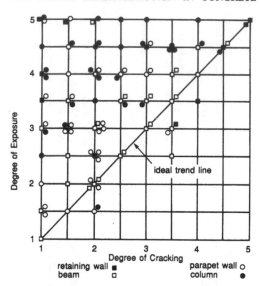

Figure 5.13 Diagram relating the degree of exposure to the degree of cracking at Charles Cross Car Park, Plymouth, showing that exposure helps deterioration to occur. The scales for degrees of exposure and cracking are defined in Ref. 62. From Ref. 62.

not be sensible to regard any geographical region as being completely free from the possibility of ASR and, in particular, the nationwide distribution and use of precast concrete units must always be borne in mind.

As well as the presence of a reactive aggregate combination and of a sufficient amount of alkalis, a supply of moisture is necessary for damaging ASR to occur in concrete. The degree and frequency of wetting has an important environmental control over ASR, and it is believed that cyclic wetting and drying may be the most conducive condition[14]. Consequently the incidence and severity of ASR may sometimes depend upon geographical aspect; for example, Fookes et al.[62] found a sympathetic relationship between exposure and deterioration at the Charles Cross Car Park in Plymouth (Figure 5.13). In some circumstances, regular moisture migration might cause localised concentrations of alkalis, so enabling ASR to occur in affected parts of a structure[64].

5.5.3 The spread of discoveries and allegations

It is not possible to state categorically the number of concrete structures that are affected by ASR in the UK, much less to provide any detailed schedule, because of the persistent climate of secrecy that has bedevilled the growth in awareness of ASR in this country. At first it was hoped that the examples identified in the mid-1970s, such as the electricity substation cases, would prove to be 'special cases' and that ASR would continue to be 'an extremely rare occurrence in the British Isles'[38]. Unfortunately, the number of cases of ASR confirmed or claimed has progressively risen; the 1982 BRE Digest 258[34] stated that about 30 cases had been identified in addition to the dam in Jersey and the Plymouth electricity substation bases, but by 1988 the Digest 330[65], which superseded Digest 258, admitted to more than 170 cases. A majority of these affected examples appear to have been civil engineering structures, rather than buildings, but this might be because such construction is usually subject to a more rigorous programme of inspection and condition assessment.

Nevertheless, it does appear that 'the incidence of damage diagnosed as being caused by ASR in the UK is very small when compared with the total amount of concrete construction carried out'[13]. One major study of a sample of the nearly 6000 highway bridges for which the Department of Transport is responsible in England has apparently identified signs of ASR in between a half and a third of the structures examined[66]. However, it is believed that no more than about 10% were damaged to some extent by ASR; moreover highway structures are liable to be particularly vulnerable to ASR because of the conditions to which they are exposed. It is also important to distinguish between those few cases of ASR in which structural damage has resulted and those more numerous cases where any surficial damage is only cosmetic, or which at worst could provide a facility for later deterioration by other agencies, such as frost, if left untreated.

It is also now recognised[14] that microscopic evidence of the presence of ASR in concrete is not necessarily adequate proof that any damage has been caused or even accentuated by ASR, so that many statistics regarding the scale of the effect of ASR in the UK, or indeed elsewhere, must be treated with some caution. In an admirably thorough review of his observations over a decade, for example, French[67] reported some 40% of 300 concrete structures to be exhibiting evidence of ASR, but these reactions ranged from 'just detectable to serious destructive processes' and he provided no clear indication of the proportion considered to exhibit the latter degree of severity. There is no easy and unequivocal means of quantifying such degrees of damage caused by ASR and, in any case, it is likely to be time-dependent; a minor case of ASR identified today could possibly develop into a more serious case over a period of time. In one appraisal of a group of 93 UK structures, using a core expansion test as a measure of severity, only 32% exhibited expansions greater than about 0.10% and just 6% exhibited expansions greater than 0.15% (Figure 5.14)[15].

It has been noticed that the concrete structures found to be affected by ASR in the UK were built during a period from the 1930s to 1970s, but the number of affected structures appears to show a preponderance in the 15 years 1960 to 1975, which interestingly post-dates the BRS research work of the 1950s (Figure 5.15). Some authors have suggested that this apparent

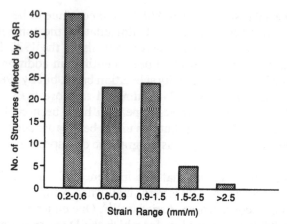

Figure 5.14 Expansion of concrete cores taken from 93 UK structures affected by ASR. From Ref. 15.

upsurge in ASR in the UK might be attributed, in part at least, to changes in the composition and nature of Portland cement being manufactured in the UK[68]. Changes to the cement manufacturing process and greater controls over the emissions into the atmosphere from cement works might have caused the alkali contents to rise, and mineralogical modifications to achieve greater

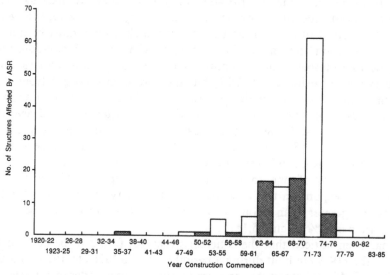

Figure 5.15 Diagram showing the frequency of the identified occurrence of ASR in construction of UK structures between 1920 and 1985. The histogram is *not* weighted to reflect the variations in intensity of construction. From Ref. 15.

rates of strength gain might have led to the formation of a more permeable hydrated product. There seems to be little scientific evidence to support these allegations and, although alkali content records before 1960 are not exhaustive, there do not appear to have been any substantial changes in either the range or the weighted average alkali content of UK cements[69]. A more probable reason for the apparent upsurge of ASR between 1960 and 1975 is that this period also corresponds to a major growth period in UK construction, most notably the rapid expansion of the motorway system. Also, like any other new phenomenon to be discovered, it might be assumed that some earlier examples were misdiagnosed and either successfully repaired or demolished.

Whatever the number of affected UK structures may be, very few case studies have been published, even in part form, and most information remains confidential or on restricted access. Some of the best-publicised examples are structures along the A38 trunk road in Devon, including the Marsh Mills Viaduct on the outskirts of Plymouth, an underbridge at Plympton and the Voss Farm bridge further east. The Marsh Mills Viaduct, which was built in 1969–70, consists of two separate three-lane carriageways, each supported on a series of T-shaped piers. The deck beams were precast in East Anglia and were not affected by ASR[70], but the piers, cross-beams, deck slab and pile caps were found to be damaged to varying extents by ASR[70]. As for other cases in the Plymouth area, the reactive combination consisted of crushed limestone coarse aggregate, a sea-dredged sand containing a subordinate proportion of chert and the local high-alkali Portland cement from Plymstock works. Areas of concrete subjected to severe wetting were most seriously affected, including the cross-beam ends beneath deck joints but most notably the buried pile caps, which were perpetually wet in the marshy ground (Figure 5.16). Loading has been reduced by permanently reducing the traffic flow from three to two lanes on each carriageway, and steel stanchions have been installed beneath the cantilever heads of the piers (Figure 5.17), partly to reduce bending moments but also to provide more direct load transference onto the piles to limit shear stresses in the badly cracked pile caps. It is understood that these measures are intended to prolong the serviceability of the structure for up to 10 years, when replacement is planned.

The Voss Farm overbridge, also built around 1970, was found to exhibit severe cracking attributable to ASR in the columns, the abutments and wing walls and the foundations (Figure 5.18). The concrete mix was similar to that found to be reactive elsewhere in the region, with a minor proportion of chert being the active constituent. The circular columns were heavily reinforced and the crack pattern was wholly orientated according to the stress field, so that only vertical cracks were apparent with no trace of the map cracking often considered typical of ASR (Figure 5.19). Corrective action included the replacement of the columns, and the ground anchoring and refacing of the abutments and wing walls.

Figure 5.16 Marsh Mills Viaduct, near Plymouth, showing an excavation to reveal a severely cracked concrete pile cap, largely below the local water table.

Of the cases of ASR in the English Midlands, probably the Smorrall Lane overbridge on the M6 near to the Corley services has been most widely reported[63]. The two-span bridge, which was built in 1969, exhibited some severe cracking of the central pier-top beam, the deck beams and the soffit of the deck slab; cracking patterns on the beams were reminiscent of flexural

Figure 5.17 A pier of the Marsh Mills Viaduct, near Plymouth, showing the steel stanchions being installed as a measure to support the cantilever heads and modify load transference.

Figure 5.18 Voss Farm Overbridge, Devon, showing a severely cracked wing wall.

Figure 5.19 Voss Farm Overbridge, Devon, showing cracking in a heavily reinforced circular column (steel straps installed as temporary precaution). Photograph taken after removal of the column, showing the development of some horizontal cracking in addition to the dominant vertical cracking. Photograph by courtesy of G.V. Walters of ECC Quarries Limited, Exeter, Devon.

and shear cracking (Figure 5.20). Petrographic examinations established the presence of ASR (Figure 5.21)[71], with the reactive constituent being subordinate chert present in both the coarse and fine aggregates which were otherwise dominated by quartzite. Structural engineers considered that 'the two simply supported spans could become an unpredictable hazard within 5 years',[63] and, consequently, the concrete decks were replaced by new steel constructions; the old concrete beams were removed and retained for continued monitoring and further testing.

Another example of ASR in the Midlands has been described in detail by Nixon and Gillson[72]. In 1982, cracking of concrete bases at an electricity substation were once again reported, this time at Drakelow Power Station, near Burton-on-Trent. The main reactive constituent was thought to be chert, present at around 3% of the otherwise sandstone and quartzite sand and gravel aggregate, although some particles of rhyolite may also have reacted[73]. There were indications that the concrete alkali contents, which were not originally unduly excessive (see 5.6), might have been enhanced by alkalis derived from the aggregates and also be alkali migration into the upper parts of the slabs caused by moisture movements.

This review of known cases of ASR in the UK, as with earlier such reviews[60,74], is incomplete, unavoidably eclectic and destined rapidly to become outdated. Identified cases in south-west England now include foot, road and rail bridges, car parks, office, shop, college and hospital buildings, and concrete blocks at substations, TV transmitters and microwave towers. Jetty, bridge, river wall, reservoir and electricity substation structures have

Figue 5.20 Smorrall Lane Overbridge, Warwickshire, showing regular tensile-style cracking of deck beams across the eastbound carriageway. From Ref. 71.

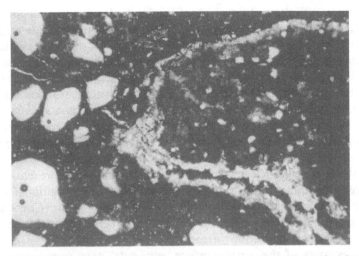

Figure 5.21 Photomicrograph of concrete from Smorrall Lane Overbridge deck slab, showing a reaction centre involving chert. Pln. Pols. (approx. × 15 magnification). From Ref. 71.

all been found to be affected in South Wales. In the Midlands, identified cases have included foot and road bridges, motorway junction complexes, reservoir structures, colliery shafts and electricity substation bases. Sporadic or 'special' cases (see 5.5.4) have been reported from other areas and, regrettably, many other cases probably remain to be discovered. However, it is important to remember that these identified cases still represent a small proportion of the nation's total stock of concrete structures and also that the severity of any damage is frequently not critical.

5.5.4 *Special cases*

Apart from the Val de la Mare Dam in Jersey, UK cases of ASR have almost exclusively involved chert as the reactive constituent, and the physical manifestations of reaction have usually been restricted to surficial cracking, some surface discolorations and, occasionally, small displacements of the concrete. Some 'special' exceptions are described in this section, with the question of artificial aggregates being considered separately in the succeeding section (5.5.5).

As early as 1978, Palmer[60] described a dam in Scotland, built in 1936, which exhibited evidence of an alkali–aggregate reaction involving a greywacke aggregate derived from an excavation on site and apparently not used for any other construction. It appeared that the reaction had not caused any disruption and it was considered that the reaction had proceeded to completion, so that no remedial action was required. More recently, a number of structures in Wales and the north of England, also made using greywacke or similar aggregates, have been identified as exhibiting alkali–aggregate reactivity. The Maentwrog Dam, near Dolgellau in North Wales, which was completed in 1927, exhibits extensive cracking and heavy white exudations on the downstream face and has apparently experienced several phases of repair during its lifetime. A so-called 'alkali–greywacke' reaction has been observed in the concrete under the microscope[14,75], which is at least in part a conventional ASR involving fine-grained quartz in the rock matrix. At present it is uncertain to what extent, if at all, this example might represent a type of 'alkali–silicate' reaction, similar to that originally described in Canada[76]. It is also unclear whether the 'alkali–greywacke' reaction could have been a significant cause of the surface cracking, or the continual leakage through the dam which occurs mainly at construction joints. Construction work has now started to replace the old dam at Maentwrog with a new dam structure a short distance downstream.

In other parts of the world, surface 'pop-outs' have sometimes been a common symptom of ASR, but this has rarely been the case in the UK, where

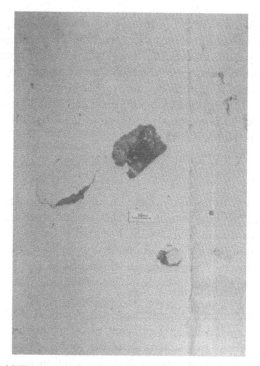

Figure 5.22 Unusual UK example of ASR in which surface gel exudations and pop-outs are the only detrimental features. Chert particles react in a surface zone which is greatly enriched in alkalis. From Ref. 77.

Table 5.4 The concentration of alkalis (Na$_2$O eq., kg/m^3) in the internal surface zone of concrete retaining walls as the result of the continuous migration of moisture; the surface zone is greatly enhanced in alkalis, whilst the centre of the wall is probably depleted in alkalis. After Ref. 77.

Structure location	Core sample	Centre of concrete wall	Internal surface zone of concrete wall*
A	1	1.4	10.1
	2	1.4	12.6
B	1	0.8	15.8
	2	2.0	18.0
	3	0.7	13.1

*Approximately 0–50 mm in case of location A and 0–25 mm in the case of location B.

neither 'pop-outs' nor copious gel exudations are typical features of concrete surfaces on structures diagnosed as being affected by ASR[14]. Sims and Sotiropoulos[77], however, have recently described some 'special' cases in which surface 'pop-outs' and associated gel exudations are the only defects (Figure 5.22), with no cracking, expansion or other damage being apparent. In these cases, of which several different locations were investigated, the reactive particles comprised chert which formed only a very minor percentage of an otherwise wholly oolitic limestone aggregate. Localised ASR was able to occur because of the very substantial concentration of alkalis that was brought about by the continuous migration of moisture though basement retaining walls (Table 5.4), the back sides of which were in direct contact with wet ground and the front sides of which were exposed to an internal drying environment. The alkali content was found to be enhanced by up to five times by this process, and it was thought possible that sealing or painting of the front concrete surface might have exacerbated the effect.

5.5.5 Artificial aggregates

The possibility that some 'artificial' aggregates in use in the UK might prove to be alkali-reactive was first considered by Jones and Tarleton[6,7], who included some air-cooled blast furnace slag in their BRS survey. Jones and Tarleton also utilised Pyrex glass as a known reactive material, and glass has long been recognised as being potentially reactive with Portland cement; glassfibre reinforcement of concrete can only be successful using alkali-resistant glass. Figg[78] has described a case in which decorative glass aggregate in concrete cladding panels for a building in north-east London caused ASR, resulting in cracking, distortion and spalling of the units. Although the non-structural panels had been made using a low-alkali white Portland cement, the glass aggregate itself had contained sufficient reserves of available alkalis to perpetuate the expansive reaction.

Although many of the aggregates that are manufactured or derived from industrial waste products are both siliceous and poorly crystalline or even glassy, there appear to have been no reports of any concrete structures in the UK being damaged as the result of ASR involving artificial aggregates, other than the decorative glass already mentioned. There have also been very few published attempts to assess the ASR potential of such artificial aggregate materials. Sims and Bladon[79] carried out a limited programme to investigate the ASR potential of sintered pulverised fuel ash (PFA), which is widely used as a lightweight aggregate ('Lytag'). No significant expansion of mortar bar and small concrete specimens was recorded, and yet microscopical examination of the concrete at various ages suggested that some gel production might have occurred, impregnating the cement paste matrix but seemingly not giving rise to expansive forces. It is also possible that the voided nature of such lightweight materials might allow a considerable degree of ASR to be accommodated without any disruptive stresses being generated within the mortar or concrete. Further research is needed before artificial aggregates can be safely regarded as being incapable of expansive ASR in concrete over a prolonged period of service in conducive exposure conditions.

5.5.6 *Urgent search for remedies*

As the number of confirmed cases of ASR built up in the UK and it became clear that the problem was no longer restricted to 'foreign' or 'special case' occurrences, it became urgent to identify the main causal factors and so to formulate measures for avoiding the reaction in future construction. At the initiative of the C&CA, an 'Inter-Industry' Meeting was held on 10 June 1981, at which it was agreed to invite Michael Hawkins, as the engineer of the county most afflicted with cases of ASR, to form an independent working party of specialists to co-operate in the production of some guidance notes for the practising engineer. The first report of the 'Hawkins Working Party' was published in 1983[12] (see 5.6.1) and later updated, as a Concrete Society technical report, in 1987[13] (see 5.6.2). BRE Digest 258[34] was published shortly before the first 'Hawkins Report' and heralded a number of the concepts shortly to be endorsed by the Hawkins Working Party, including the naming of aggregate group classifications ('trade groups') considered 'very unlikely to be susceptible to attack by cement alkalis' and the notion of using a threshold limit on concrete alkali content of 3 kg/m^3 as an alternative to the use of 'low-alkali' cement to avoid ASR in new construction.

Cement alkalis could be established by analysis and were, in any case, usually obtainable upon request from the manufacturer. By contrast, BRS Digest 258 maintained that, 'There are as yet no British Standard tests for the susceptibility of aggregates to attack by alkalis and ASTM tests have not been found to predict the reactivity of UK aggregates accurately'. This statement is significant in view of the reliance that had previously been placed on the mortar bar test by the BRE/BRS in the 1950s during their previous investigations into the ASR potential of British aggregates. Appropriately, the baton was taken up by the BSI in 1981, when the aggregates committee, CAB/2, formed a new working group, WG10, to evaluate methods for assessing the ASR potential of aggregates (see 5.6.4).

5.6 Countermeasures: minimising the future risk of ASR

5.6.1 Hawkins Working Party—I

The 'Hawkins Report' was published in 1983[12] and emphasised that the incidence of ASR in the UK was 'small', but the importance of 'minimising the risk of ASR' in future work was embodied by the warning that, 'When ASR has been diagnosed in a structure, there are no methods which can be reliably recommended at present for either preventing further damage or carrying out effective and lasting repairs'. The report was presented as practical guidance, based on contemporary information and intended only to apply to materials, conditions and practice encountered in the UK.

It was considered that ASR causes damage only if the three following factors are all present: sufficient moisture, sufficient alkali and a critical amount of reactive silica in the aggregate. The recommendations given in the Hawkins guidance notes were thus based on ensuring that at least one of these factors was absent. The risk of ASR for a particular concrete in a particular environment could be assessed by reference to these factors (Figure 5.23) and, if a risk was deemed thereby to exist, the same factors provided a range of possible optional precautions to be implemented. Two levels of precaution were envisaged: a level of 'normal concrete construction' which it was supposed would apply in the majority of cases, and a level for 'particularly vulnerable forms of construction', such as concrete 'buried in water-logged ground with the top surface exposed' (e.g. electricity substation bases).

A sufficiency of moisture for damaging ASR to occur was considered to be any exposure to moisture, including relative humidity levels exceeding 75%. In practice this meant any externally exposed concrete was in an environment deemed conducive to ASR.

The amount of 'reactive alkali' available was considered in three main ways according to: (a) the alkali content of the cement, (b) the alkali content of a blend of cement with a certain minimum content of a mineral addition such as ground granulated blast furnace slag (GGBFS) or pulverised fuel ash (PFA), or (c) the alkali content of the concrete. In respect of (a) above, the Hawkins Report stated, 'The use of Portland cement with an alkali content of 0.6% or less is accepted worldwide as the best means of minimising the risk of damage due to ASR'. Only sulphate-resisting Portland cement (SRPC)

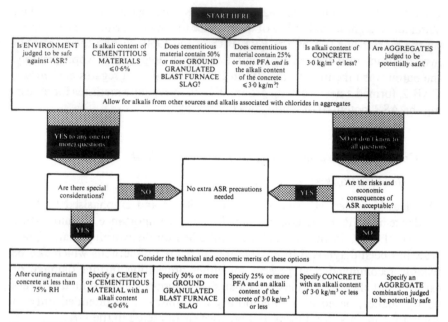

Figure 5.23 Simplified flow chart for the assessment of risks and precautions against ASR, according to the guidance given by the Hawkins Working Party. See Concrete Society Technical Report No. 30, 1987[13], for full details. Redrawn and adapted from Refs. 12 and 13.

was available in the UK with a guaranteed alkali content of not more than 0.6%, but ordinary Portland cement (OPC) could be found to be 'low-alkali' by examination of manufacturer's analytical data making due allowance for variations, or be rendered 'low-alkali' by the controlled replacement of part of the cement by GGBFS or PFA, which were assumed to contribute 'no reactive alkali to the concrete'.

In respect of (b) above, a combination of GGBFS and any UK Portland cement which contained 50% or more GGBFS was considered equivalent to a 'low-alkali' cement. Similarly, a combination of PFA and any UK Portland cement which contained 25% or more PFA was considered equivalent to a 'low-alkali' cement, with the added condition that the alkali content of the concrete provided by the Portland cement in the combination was not more than 3.0 kg/m³.

Finally, in respect of (c) above, it was considered that damage from ASR was unlikely to occur if the concrete had an alkali content of 3.0 kg/m³ or less, which was calculated from the alkali content of the cement and the maximum expected cement content of the concrete, making due allowances for future variations in cement alkali contents and concrete batch weights.

This approach was not considered adequate for 'particularly vulnerable construction'. The value of 3.0 kg/m^3, which was coincidentally equivalent to using a high-alkali cement of say 1.0% Na$_2$O eq. in a typical structural concrete of say 300 kg/m^3 cement content, seems largely to have been based upon research by Hobbs at the C&CA[80] (Figure 5.24), using Beltane opal and mortar bars, although it was considered that the results agreed with earlier findings in Germany and elsewhere[81,82].

The 1983 Hawkins Report was not able to be particularly helpful in the guidance on aggregates. It was stated that, 'There are at present no British Standard tests for the alkali-reactivity of aggregates', but the American (ASTM) chemical and mortar bar methods[19,21] 'commonly used in other countries' were considered not to be 'a practical basis for the specification of British concrete materials'. Modifying the group classification approach put forward in BRE Digest 258[34], the Hawkins Report stated that, 'At present, it seems that aggregates consisting wholly of rock types in the Basalt, Gabbro, Granite, Gritstone, Hornfels, Limestone, Porphyry and Schist trade groups..., would be classified in the UK as "unlikely to be reactive"'. The word 'wholly' was explained as meaning both coarse and fine aggregates, and that the source was not contaminated by reactive silica (such as the secondary opal veining in the granodiorite source in Jersey). Consideration of the com-

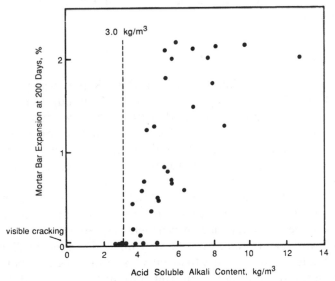

Figure 5.24 Relationship between the expansion of mortar bars at 20°C and the acid-soluble alkali content of the mortar, using opal as the reactive constituent, showing no expansion at alkali levels lower than 3.5 kg/m^3. After Ref. 82.

position of an aggregate could therefore enable an aggregate to be adjudged 'unlikely to be reactive' or 'contains constituents which are sometimes found to be reactive', but, in assessing the amount of reactive silica in the latter case, the Hawkins Report warned that, 'the critical proportion is not capable of easy prediction and in any case will vary with the aggregate combination'.

In addition to the three main controlling factors (moisture, alkali contents and reactive aggregate), the possibilities of alkali migration through moisture movement and the contribution of alkalis from sources other than cement (from other constituents or from external sources) were recognised, although no additional precautions could be recommended, save by tanking the concrete or by the use of definitely non-reactive aggregates. The working party was also anxious to ensure that zealous efforts to achieve the avoidance of ASR did not lead to the neglect, or even act to the detriment, of other considerations:

It is essential that adjustments which may be made to the mix in order to avoid ASR should never lead to the use of materials, cement contents or water/cement ratios which will be inadequate to ensure general durability.

The first Hawkins Report was well received in the industry, but it was only guidance and, in the absence of a BS specification, it was clear that a wide range of approaches to 'minimising the risk of ASR' would arise from different specifiers, including extreme cases where *all* of the 'optional' precautions would be required to apply instead of just *one* of the options chosen according to circumstances. The Hawkins Working Party therefore remained in being, now affiliated to the Concrete Society, and commenced work on 'model specification clauses'.

5.6.2 *Hawkins Working Party—II*

The second substantive edition of the Hawkins Report was published as Concrete Society Technical Report No. 30 in 1987, although an earlier consultation document had been published in 1985. The new edition was styled *Guidance Notes and Model Specification Clauses* and the opportunity had been taken to revise and update the guidance notes, which were otherwise based upon the same principles and followed the same general approach as the original Hawkins Report.

In his foreword to the new edition, Michael Hawkins reaffirmed that damage by ASR affects only a small proportion of concrete construction undertaken in the UK, but added, 'Nevertheless, when it does occur the cost of remedial work can in some cases be very high'. The report still accepted that ASR tended to occur most frequently in certain regions, but warned that, 'The increasing transportation over long distances of both materials and concrete products means that the potential for damage from ASR will not be restricted to particular regions'. For this reason, and others, the simple

UK map depicting the locations of reported cases of ASR which had been included in the first edition was omitted from the revision. The model specification clauses were based upon the revised guidance notes and will not be separately described in this summary, except where convenient to clarify the advice.

The main revisions in the guidance notes concerned the contributions of reactive alkalis from GGBFS and PFA, the derivation of additional alkalis from sodium chloride and other sources, and some changes and improvements to the guidance about aggregates. In the first Hawkins Report, the GGBFS and PFA components were, for the purposes of that report, considered to contribute no reactive alkalis to the concrete system. In the new report, the reactive alkali content of GGBFS and PFA is defined as being the water-soluble alkali content, when determined using the BS 812[26] extraction method for aggregates and the BS 4550[25] analysis method for cements. In consequence, when the concrete alkali content is being calculated, the GGBFS or PFA reactive alkali contents must be taken into account (Figure 5.25). This approach has been disputed, and some authorities maintain that the reactive alkali content of these additions should be based upon a notional proportion of the acid (rather than water) soluble alkali determination: 50% in the case of GGBFS and 17% (or one-sixth) in the case of PFA[65]. The Hawkins Working Party has established a technical subcommittee to carry out further research into this controversy and, until the findings become known and are accepted, opinions will remain divided as to the best approach.

In the first report, it had been recognised that sodium chloride in sea-

---

Site combinations of Portland cement with either ground granulated blast furnace slag (GGBFS) or PFA

When cement to BS 12 is combined on site with either GGBFS or PFA the reactive alkali content of the concrete is calculated from:

$$A = \frac{(C \times a) + (E \times d)}{100}$$

where A = reactive alkali content of concrete (kg/m^3)
       C = target mean Portland cement content of concrete (kg/m^3)
       a = reactive alkali content (%) of cement
       E = target mean content of either GGBFS or PFA in the concrete (kg/m^3)
       d = average reactive alkali content (%) of either the GGBFS or the PFA as provided by the manufacturers.

Figure 5.25 Calculation of the 'reactive alkali content' of concrete in accordance with the Concrete Society TR30 guidance, one of several equation options given in the report. After Ref. 13.

dredged and some other aggregates might be a source of additional alkalis, but it was only suggested that the CP 110[30] limits on chloride for reinforced concrete should be applied. By the time of the revised edition, it had become clear from experiences in Denmark and from work at the BRE[83] that the addition of sodium chloride to the concrete increased the hydroxyl ion concentration and could promote greater ASR expansion. Consequently, provision was made in the calculation of concrete reactive alkali content for the additional sodium associated with any chloride present in the coarse and fine aggregates. No recommendations were made in respect of any reactive alkalis which could possibly be released from some rock types found in aggregates, although the opinion was expressed that, 'No significant amounts of reactive alkali will be derived... (chlorides excepted) from natural aggregates in the UK'.

The guidance on aggregates was amended in two important ways: the classification of aggregates 'unlikely to be reactive' was improved and extended, and a new concept was introduced regarding aggregates comprising largely flint or chert. The list of rocks and minerals considered 'unlikely to be reactive' was revised (Table 5.5) according to the updated BS 812: Part 102 (1984), in which the 'trade groups' had been replaced by a more straightforward list of common rock types found in aggregates, and an appendix was included to provide some guidance on those rock types which had not been included on that tabulated list. Feldspar and (unstrained) quartz were included on the list of constituents 'unlikely to be reactive', enabling many fine aggregates to be assessed more readily.

A dilemma was presented by chert and flint, because although these constituents have been involved with most of the UK examples of ASR, they are extremely common in UK aggregates, and in the highly built-up south-east

Table 5.5 Rocks and minerals 'unlikely to be reactive', according to the guidance notes given in Concrete Society Technical Report No 30, 1987[13]. The list of rock names is based upon that given in BS 812:Part 102 (1984)[26].

Andesite	Feldspar*	Microgranite
Basalt	Gabbro	Quartz*†
Chalk‡	Gneiss	Schist
Diorite	Granite	Slate
Dolerite	Limestone	Syenite
Dolomite	Marble	Trachyte
		Tuff

*Feldspar and quartz are not rock types but are discrete mineral grains occurring principally in fine aggregates.
†Not highly strained quartz and not quartzite.
‡Chalk is included in the list since it may occasionally be a minor constituent of concrete aggregates.

of England they form the major part of most aggregate combinations. The revised Hawkins Report advised that, 'Experience to date from cases of ASR involving sands and gravels indicates that up to a maximum of 5% by mass of flint, chert or chalcedony taken together in the combined aggregate can be tolerated'. However, at the other end of the range, 'It is currently considered that a combination of fine and coarse aggregate which contains more than 60% by mass of flint or chert is unlikely to cause damage due to ASR'. In other words, in the absence of any other potentially reactive constituent, aggregates either containing no more than 5% or alternatively containing more than 60% chert and flint are to be regarded as 'unlikely to be reactive'. The value of 60% was based partly upon observation, whereby concretes made with predominantly flint aggregates were not found to be affected by damaging ASR, and partly upon limited laboratory experiments at the BRE, the results of which have since been confirmed by further work at the BRE and elsewhere (Figure 5.26).

No 'detectable' opal is permitted for aggregates 'unlikely to be reactive' and it is important to realise that opal is considered by the report to be unacceptable under any circumstances:

Sources (of aggregate) known to contain opaline silica should not be used even when the alkali content of either the cementitious material or the concrete is being controlled.

This is because there is some evidence that small amounts of opal in concrete can cause ASR damage even when the cement alkali content is less than 0.6% or the concrete alkali content is less than 3.0 kg/m^3.

No specific guidance is given in respect of alkalis derived from any sources other than the cementitious or aggregate materials, but the report requires that, 'If alkalis in excess of 0.2 kg/m^3 of concrete come from other sources they must be taken into account'. In effect this means that many engineers will require exhaustive alkali audits to be carried out for all the concrete mixes proposed for a given project.

5.6.3 *The Department of Transport and other specifiers*

BS 8110[31] superseded CP 110[30] in 1985 as the Code of Practice for 'Structural Use of Concrete'. Whereas CP 110 had not mentioned ASR, BS 8110 (in clause 6.2.5.4) included recommendations that followed the first Hawkins Report[12] and BRE Digest 258[34]. Other national authorities maintain concrete specifications which are widely recognised and respected, including *inter alia* the Property Services Agency (PSA), British Rail, the Central Electricity Generating Board (CEGB) and the British Airports Authority (BAA), but perhaps the most important in this context is the Department of Transport (DTp). Concrete and concrete materials are included in the DTp *Specification for Highway Works*[84], which was published in 1986 to supersede the earlier *Specification for Road and Bridge Works*[85].

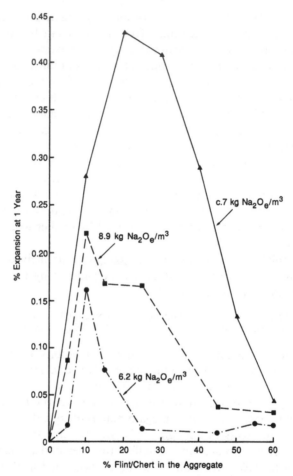

Figure 5.26 Some relationships between flint/chert content of aggregate and concrete prism expansion, indicating a flint/chert 'pessimum' of 10–30% for flint–limestone combinations and no significant expansions for mixtures containing more than 60% flint. Redrawn and adapted from Refs. 103 and 104.

The published *Specification for Road and Bridge Works* did not contain any reference to ASR. However, in July 1982 a new Clause 1618 was issued by the DTp to its regional offices to amend the specification to include 'Measures to control alkali–aggregate reaction in respect of all structural concrete'[86]. The DTp amendment was generally similar to the guidance which appeared in the BRE Digest 268 and the first Hawkins Report. The total alkali content deriving from cement was not permitted to exceed 3 kg/m³ of concrete unless both the coarse and fine aggregates were 'not susceptible to alkali–aggregate reaction' according to criteria stipulated. Low–alkali SRPC or blends of cement and GGBFS (in which GGBFS was at least 50%) were permitted as means of reducing alkali contents that would otherwise exceed 3 kg/m^3. The concrete alkali content was to be calculated from cement manufacturer's data, adding 0.15% to allow for possible future variations, or alternatively adding twice the standard deviation where that would amount to less than 0.15%.

The criteria governing the susceptibility of aggregates were very restrictive, being based upon the classification given in BRE Digest 258 rather than the slightly modified version given in the first Hawkins Report. Only aggregates in which *all* of the 'significant' rock components (i.e. > 5% by mass of the coarse or fine aggregate) were in the trade groups: granite or gabbro or basalt (except andesites) or limestones or porphyry (except dacites, rhyolites and felsites) or hornfels, were to be considered 'not susceptible' to AAR. Of the commonest UK coarse aggregates, only the limestones could satisfy these requirements, and almost none of the natural fine aggregates could be judged 'not susceptible' according to such criteria. Again the presence of opal rendered any aggregate 'susceptible' to AAR whatever the remaining composition of that aggregate.

An extensively revised DTp specification was published in 1986[84], and this included clauses for the 'control of ASR', which were broadly similar to those contained within the 1985 draft revision of the Hawkins Report. The requirements are essentially similar to those provided in the model specification clauses of the revised Hawkins Report later issued in 1987[13] (see 5.6.2), although the DTp is perhaps slightly more restrictive. Although the DTp specification allows that, 'When the proportion of chert or flint is greater than 60% (by weight) of the total aggregate it shall be considered to be non-reactive providing it contains no opal, tridymite or cristobalite', the lower limit for chert, flint or chalcedony taken together is only 2% by weight instead of the 5% implied by the Hawkins Report in the guidance notes. Also, whereas the revised Hawkins Report considers that the use of the 'certified average alkali content' of cements eliminates the need for any deliberate variation factors in the calculation of the reactive alkali content of concrete, the DTp specification continues to require that twice the standard deviation is added to the average of 25 daily determinations of cement alkali content by the manufacturer and also that 10 kg is added to the intended cement content of the concrete to allow for possible batching errors. In practice the difference could sometimes be important, with the Hawkins Report recommendations allowing concrete alkali contents to range up to 3.75 kg/m^3, whilst the DTp requirements seek to make 3.0 kg/m^3 an absolute maximum limit.

The 1986 DTp specification agreed with the draft revision of the Hawkins Report in requiring only the water-soluble alkali contents of GGBFS and PFA to be taken into account. However, it is understood that the DTp specification is shortly to be amended in line with BRE Digest 330, in which

50% and 17% proportions of the acid-soluble alkali contents are preferred[65] (see 5.6.2).

5.6.4 *Quest for a test*

The test methods adopted for the BRS research in the 1950s (see 5.2) had all been derived from American experience, and in the early 1950s several of these had evolved into ASTM standard methods[19,21]. In the UK, presumably because of the apparent reassurance arising from the BRS research programme, there were no similar moves to create any British Standard methods for testing concrete constituents for ASR. Once examples of ASR had been identified on the UK mainland in the mid-1970s, a new urgency was created for finding a predictive test method which would enable the problem to be avoided in new construction. At first it was assumed that the mortar bar expansion test (ASTM C227[21]) was reliable but that the practical problem concerned the long duration of the test: Palmer (C&CA) in 1977[38] said that, 'there is no satisfactory short-term test. . . . Only long-term tests on the aggregate can show with any certainty whether there is any danger in its use', and Gutt and Nixon (BRE) in 1979[36] said 'the mortar bar method is the most generally accepted method of assessment available and if the limitations of time and sampling are borne in mind it has been shown to give a good guide to the reactivity of the aggregate'.

In a thorough review in 1981, Sims emphasised the indicative, rather than definitive, nature of the various tests for ASR and stressed the importance of petrographical examination[87]:

A knowledge of the nature of the materials being considered is central to the assessment programme and some form of petrographic examination should never be omitted. In some cases the petrographic appraisal may obviate the need for any further testing, and in most cases the most pertinent indicative testing sequence will be identified.

The inadequacy of the mortar bar test for UK reactive aggregate combinations had not yet been realised, but Sims did warn that, 'In a few cases, mortar bars produce anomalous results that are difficult to explain' and noted that 'the flint-bearing aggregates of south-east England invariably produce mortar bar results that give no cause for alarm, and yet the deleterious alkali-reactivity in the south-west of England has mostly concerned similar materials which are blended with other rock types and are thereby present in very much smaller proportions'. It was then considered that testing actual cement–coarse aggregate–fine aggregate combinations, rather than the individual aggregate constituents separately, might avoid such a misleading outcome. However, by 1986, Sims[88] could report that:

In the UK the mortar-bar test has not predicted any aggregate combination to be potentially alkali-reactive, including those involved in known cases of ASR. It is possible that the mortar-bar test ... is insensitive to those circumstances encountered in the UK which depend upon the occurrence of a critical combination of constituents and conditions.

A concrete prism test was considered to be more reliable (Figure 5.27).

A BSI working group (WG10) was established in 1981, at the direction of the aggregates committee (CAB/2), and began its work by comparing a range of existing procedures[89,90], using some of the most widely used UK aggregates. There was 'an almost complete lack of correlation between the results of the different methods'[89]. The working group considered it necessary to develop a standardised method for petrographical examination because, 'In the absence of other agreed test methods, the specifications against ASR in the UK are still having to rely on lists of "safe" rock types and identification of suspect minerals'[89] (e.g. BRE Digest 258, the 1983 Hawkins Report and the 1984 DTp amendment to the *Specification for Road and Bridge Works*). Initial work by the group also indicated that 'a test method using concrete stored at 38°C and 100 per cent RH will identify pessimum combinations of aggregates'.

A detailed method for the petrographical examination of aggregates has

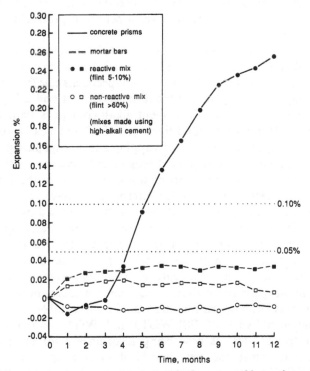

Figure 5.27 Mortar bar and concrete prism test results for comparable reactive and non-reactive UK aggregate combinations. After Ref. 88.

been developed by the BSI working group (latterly by WG3), although this is not specific to ASR and has only recently been issued as a draft for public comment. It is intended that the method will eventually be published as BS 812: Part 104. The all-important task of interpreting the findings of a petrographical examination to evaluate the ASR potential of an aggregate will be the responsibility of the BSI subcommittee charged with reviewing BS 882[23] and, initially at least, it seems that only non-mandatory guidance will emerge. The relatively low proportion of potentially reactive constituents that can sometimes be most critical for ASR in an aggregate has produced some profound problems in formulating the standard petrographical examination method; the requirements that can necessitate the analysis of large samples and also impose complicated statistical controls are likely to cause difficulties in commercial practice. The 1986 DTp specification[84] introduced an additional requirement for aggregate sources to be examined as well as processed aggregate samples, so that the prospects of future variations could be geologically assessed. Although the BSI working group (WG3) has now undertaken an appraisal of such source examination procedures, it seems likely that in practice most considerations of ASR potential will continue to be based upon the examination of representative samples of processed aggregate. Some recent work has suggested that the variations in petrographic composition of processed aggregates over a period of production might be less than expected[91].

A concrete prism test method has now been developed by the BSI working group (WG10) and was recently issued as a draft for public comment[92]. The method employs four $75 \times 75 \times 200$ mm prism specimens cast from a concrete mix made using the aggregate combination under test and a high cement content (700 kg/m^3), in which the cement alkali content is standardised at 1.0% Na$_2$O eq. Normal test conditions are 38°C and nearly 100% RH, utilising a closely specified regime of wrapping and storage in separate containers. A precision trial is under way at present and, given acceptable results, it seems likely that the concrete prism test could become the first British Standard test for ASR. No criteria for interpretation have yet been agreed, however, and again this problem will need to be addressed by the BS subcommittee reviewing BS 882. At present, despite some early optimism, it seems unlikely that any simple limits could be agreed for overall application of the concrete prism test, although it might prove possible to establish a threshold value beneath which an aggregate combination would be regarded as definitely non-expansive for those special circumstances in which no risk of damage from ASR, however remote, would be acceptable.

The development work of BSI/WG10 and WG3 continues and, for example, the 'gel pat test' is now being reconsidered for use as an opal detection test to assist the petrographical examination of aggregates. Recently, a RILEM international committee has been established, under British chairmanship, to investigate the wide range of existing and new tests for ASR. Over the nearly 50 years since ASR was discovered, many have sought to devise a universally reliable test to predict ASR potential, but this 'quest for a test' looks increasingly like the earlier search for El Dorado and we must almost certainly learn how to interpret with confidence the often equivocal but always indicative results of the standardised methods already available.

5.7 Taking stock: diagnosis, prognosis and management

5.7.1 The diagnosis of ASR

It was suggested earlier (see 5.5.3) that some cases of ASR in the UK might have been misdiagnosed prior to the mid-1970s, with the cracking and any other damage being wrongly ascribed to a variety of different mechanisms, perhaps including frost damage, sulphate attack and excessive drying shrinkage. Since the 1970s it has probably also been the case that some examples of concrete cracking have been just as wrongly ascribed to ASR, when they were in fact caused by more familiar processes. What is it that makes such diagnoses so difficult? It is principally because the characteristic map pattern of macrocracking caused by ASR is also a common symptom of several other mechanisms, including drying shrinkage, and some of these mechanisms are statistically more likely to occur than ASR; also, in stressed locations, macrocracking caused by ASR can completely lack the random map pattern so often regarded as diagnostic and instead exhibit a strongly systematic pattern aligned to the stress field. The accurate diagnosis of ASR actually involves the careful and coordinated assessment of both site observations and laboratory findings and, even then, it can be difficult to establish with certainty a causal link between any evidence of ASR and any damage which might have occurred to the concrete structure.

In 1984, the C&CA (now BCA) organised a working party of specialists, from a range of firms and organisations actively involved in the diagnosis of ASR, in order to formulate a standardised approach to diagnosis, and the resultant BCA report was published in 1988[14]. It had been recognised previously that, 'The only certain evidence of alkali–silica reaction is provided by microscopic examination of the interior of concrete to identify positively the presence of gel and of aggregate particles which have reacted'[34,36]. Sims termed these 'reliably diagnostic features'[88], and the recognition of such evidence is an essential part of the BCA diagnosis guidance.

Most considerations of ASR as a possible cause of distress in concrete structures arise because of the presence of surface cracking and the consequent desire or need to establish the cause of cracking prior to any remedial action being implemented. Occasionally concrete is examined for ASR as part of a routine inspection or as part of a reappraisal for change of use or sale purposes, but this is generally to be discouraged in the UK. As with any investigation to establish the cause of cracking in concrete, the age at which cracking first appeared is an important piece of evidence, especially to distinguish between early-age phenomena such as shrinkage and longer term damage caused by ASR, but in practice this information is rarely either available or reliable. The BCA report lists the features possibly indicative of ASR to be observed on site, including environmental conditions, cracking, discoloration, exudations, pop-outs and any displacements or deformations, and requires each feature to be classified according to extent and severity (Table 5.6).

The BCA report devotes a lot of attention to the range of laboratory techniques considered appropriate in the UK for the diagnosis of ASR, usually using core samples drilled from those parts of a structure in which any ASR is considered most likely to have occurred. Direct visual identification of features 'reliably diagnostic' of ASR is regarded as the only positive evidence, and this is generally achieved by the microscopical examination of thin sections and/or sawn sections. Both 'alkali–silica gel deposits' and 'sites of expansive reaction' can usually be recognised if they are present in a concrete sample (Figures 5.28 and 5.29). Other supporting information can be derived from chemical analyses of the concrete (including determination of alkali content), the identification of any exudations, the 'alkali immersion test' and the measurement of compressive and tensile strengths and their ratio. Hobbs has suggested that the microcracking pattern resulting from ASR might be distinctive and has described an assessment procedure, using impregnation of sawn slices of concrete by fluorescent resin and their examination in ultraviolet light[82].

The expansion testing of core samples has been used extensively in the UK

Table 5.6 Classification scheme for assessing the extent and severity of each feature possibly indicative of ASR to be observed during a site inspection. After Ref. 14.

Classification	Index
Extent of feature	
Not significant	1
Slight, up to 5% of area or length as appropriate	2
Moderate, 5–20%	3
Extensive, more than 20%, but not all	4
Total, all areas affected	5
Severity of feature	
Insignificant	1
Minor features of a non-urgent or purely cosmetic nature	2
Unacceptable features requiring attention	3
Severe defects or problems requiring immediate attention	4
Structurally unsafe	5

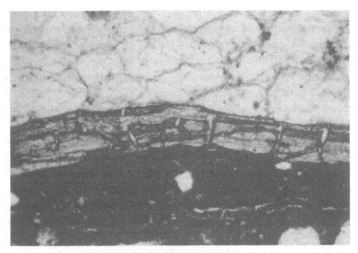

Figure 5.28 Photomicrograph of concrete showing desiccated alkali–silica gel infilling a micro-crack peripheral to a quartzite coarse aggregate particle. Pln. Pols. (approx. × 35 magnification). From Ref. 71.

as a method for diagnosis and also for evaluating any residual capacity for future expansion[34,93]. Some consultants have devised their own versions of this test, but the BCA report includes a standard method of test, using cores of 70–100 mm diameter and with a length/diameter ratio of at least 2. Storage of the core is in a sealed container and the conditions are controlled to ensure

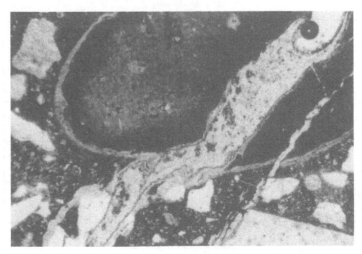

Figure 5.29 Photomicrograph of concrete showing a site of expansive reaction, involving a chert particle. Pln. Pols. (approx. × 45 magnification).

a temperature of $38 \pm 2°C$ and a relative humidity as close as possible to 100%. The measurement of any expansion is achieved by using a mechanical strain gauge periodically to record any length changes between studs affixed systematically to the sides of the core. There is some evidence that certain types of concrete can exhibit ultimately greater expansions at storage temperatures lower than 38°C (say 20°C or even 13°C)[94], but the rate of expansion is significantly reduced in such cases and the prolongation of the test would usually be unacceptable for diagnosis purposes.

Curves of expansion against time of storage in the core test (Figure 5.30) are typically asymptotic, providing moisture loss does not occur, and the maximum expansions recorded at various ages range from about 0.02% (200 microstrains) up to more than 0.25% (2500 microstrains). The comparatively rapid expansion within the first few weeks of the test can help to indicate or confirm the presence within the concrete of existing alkali–silica gel capable of reabsorbing moisture lost and/or absorbing further moisture, causing early expansion. Any expansion at later ages in the test might indicate a potential within the concrete for further, or as yet unrealised, reaction and resultant movement.

Although the core test has been widely used in the UK and has been recommended by the Institution of Structural Engineers in their interim guidance[15] (see 5.7.2), it is as well to note the cautious nature of the BCA working party's advice:

At present, the precision of the test methods and the lack of information on the correlation between the results of core expansion testing and expansion in the structure, do not permit firm guidance to be given on the interpretation of expansion test results.

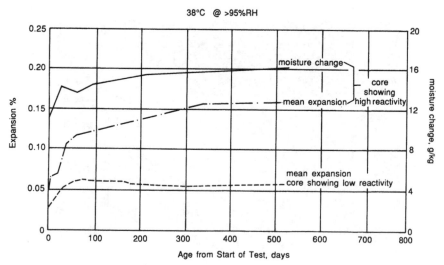

Figure 5.30 Typical levels of expansion recorded at 38°C in tests carried out on cores drilled from structures affected by ASR. Redrawn and adapted from Ref. 14.

The boundary between diagnosis and prognosis is not clear-cut, and the identification of the most likely cause of cracking is of limited usefulness without also assessing the risk of continued expansion and cracking in those areas already affected and the risk of expansion and cracking in areas currently unaffected. The BCA report[14] introduces a novel 'risk classification system', in which the overall risk can be appraised (low, medium or high risk) from the 'materials risk', in turn assessed from the site observations and laboratory findings, and the 'environmental risk' assessed from the exposure conditions involved.

5.7.2 Assessing structures and the future risks

Once it became apparent that ASR was affecting a sizeable number of structures and buildings in the UK, many of them publicly owned, it became necessary to adopt a systematic approach to the appraisal of structural integrity and future performance.

Initially, such engineering appraisals were frustrated by the paucity of research, in the UK or elsewhere, concerning the influence of ASR on engineering properties, although it was starting to become apparent from overseas experience that many structures severely cracked and disfigured by ASR had not actually suffered any engineering failure even over a long period of time[2]. In broad terms, once ASR had been diagnosed as a major cause of damage, the structural engineer had to arrive at two judgements: the existing engineering condition of the structure, and the projected future performance and serviceability. The consequential course of action, whether to demolish, support, repair or monitor further, would depend upon the conclusions reached from structural analyses and a variety of laboratory and site tests.

The issues to be decided may be simply stated[82,93,95]:

(1) Does the ASR damage significantly impair the strength, stiffness, stability and serviceability of the structure; is the structure safe?
(2) Is expansion continuing and will damage continue to develop; will the reaction reach an equilibrium or can the reaction be arrested; what will the ultimate expansion and damage be?
(3) Can the damage lead to secondary deterioration due, for example, to frost or corrosion of reinforcement?
(4) Does the structure need to be repaired or strengthened, need attention from the point of view of its appearance, or need to be monitored and managed?

These considerations were the subject of an *ad hoc* committee set up by the Institution of Structural Engineers in 1986, under the chairmanship of David Doran, and the first full report of the 'Doran Committee' was published in late 1988 as 'Interim technical guidance on the appraisal of existing structures'[15]; a more comprehensive treatment is planned for around 1990–91. The UK approach to issues (1–3) above will now be briefly considered in accordance with the ISE interim guidance; issue (4) is covered in 5.7.3 and 5.7.4.

Research results on the effects of ASR on the mechanical properties of concrete are limited and variable, but useful work has been conducted in the UK by Swamy at Sheffield University. Swamy and Al-Asali[96,97], using opal and fused silica reactive constituents, have found that tensile and flexural strengths were appreciably more affected by ASR than compressive strength (Table 5.7), with the ratio of tensile to compressive strength being reduced by about 45% at an ASR expansion of 0.1%, and by about 60% at an expansion of 0.3%. There was also a substantial reduction in modulus of elasticity as ASR expansion increased, possibly because of microcracking. However, in appraising a concrete structure, the effect of ASR on the strength or load-bearing capacity of a structural member is much more important than the properties of the concrete material alone, and for this much is dependent upon the restraints derived from internal steel reinforcement or from externally applied stresses related to the loading patterns. Some testing of reinforced concrete members in Japan and Denmark has indicated that, in some circumstances, the induced compressive stress arising from the restraint of ASR expansion by reinforcement can lead to unimpaired or even increased shear capacity[82]. According to Hobbs[82], quoting Courtier, 'overall expansion' of structural components is the dominant parameter, with expansions less than about 0.12% or 0.15% having the effect of 'prestressing the member without significantly affecting its load-carrying capacity'. In practice, the structural adequacy of any particular construction at a particular time can probably only be ascertained by full-scale proof-loading, and any deterioration in the reserves of stiffness and strength can be determined by repeated proof-loading wherever this is practicable[63].

The approach to appraisal and the determination of subsequent actions

Table 5.7 Percentage reduction (compared with control at same age) in engineering properties of laboratory concrete specimens after expansion caused by ASR involving either opal or fused silica as the reactive constituents. Compiled and adapted from Ref. 97.

| | Concrete expansion (%) | | | |
| | 0.05 | | 0.10 | |
Concrete property	4.5% opal	15% fused silica	4.5% opal	15% fused silica
Compressive strength	9	12	11	11
Modulus of rupture	—	30	—	48
Indirect tensile strength	—	27	—	29
Dynamic modulus of elasticity	16	11	37	20
Ultrasonic pulse velocity	7	2	11	6

adopted by the ISE 'interim guidance' is based upon a 'structural severity rating' scheme, which in turn utilises assessment of 'expansion index', reinforcement detailing, the site environment, and the consequence of further deterioration (Table 5.8). An arbitrary scale of expansion indices (I–V) is defined, being computed from determinations of 'current expansion' based upon crack width measurements, plus 'potential additional expansion' based upon core expansion testing including an allowance for long-term enhancement. Three classes (1, 2 and 3) of reinforcement detailing are defined, according to whether the cage is one-, two- or three-dimensional and also to the nature of the anchorage; attention must also be paid to particularly sensitive structural details. Three levels of site environment are defined (dry, intermediate, wet) and are broadly equivalent to the three conditions of exposure described in the BCA report on diagnosis[14]. Each 'structural severity rating' is then separately considered according to whether the consequence of further deterioration is deemed 'slight' or 'significant'. The consequential action or 'management procedure' is then defined by the ISE 'interim guidance' according to the determined 'structural severity rating', ranging in five steps from 'very mild' requiring only routine inspections at normal frequency, up to 'very severe' demanding immediate action and detailed investigation. The suggested scheme appears to have some merit in standardising the approach

Table 5.8 Scheme for establishing the 'structural severity rating' given in Ref. 15. The 'expansion index' is computed from crack width measurements and core expansion test results. The site environment and reinforcing detailing classes are defined in the Institution of Structural Engineers interim guidance.

		Expansion index of ASR								
		I		II		III		IV		V	
Site environment	Reinforcement detailing class	Consequence of further deterioration									
		Sli.	Sig.	Sli.	Sig.	Sli.	Sig.	Sli.	Sig.	Sli.	Sig.
Dry	1	VM	VM	VM	M	M	M	S	S	SV	VSV
	2	VM	VM	VM	M	M	M	S	SV	VSV	VSV
	3	VM	VM	VM	M	M	S	S	SV	VSV	VSV
Intermediate	1	VM	VM	VM	M	M	S	S	SV	VSV	VSV
	2	VM	VM	M	M	S	S	SV	VSV	VSV	VSV
	3	VM	VM	M	M	S	SV	SV	VSV	VSV	VSV
Wet	1	VM	VM	M	S	S	SV	SV	VSV	VSV	VSV
	2	VM	VM	S	S	SV	SV	VSV	VSV	VSV	VSV
	3	VM	M	S	S	SV	SV	VSV	VSV	VSV	VSV

Sli, the consequence of structural failure is either not serious or is localised; Sig, there is risk to life and limb or a significant risk of serious damage to property.
Structural severity ratings: VM, very mild; M, mild; S, serious; SV, severe; VSV, very severe.

and decision-making, but it seems to founder on the imprecision of the 'expansion index' which combines two determinative methods (crack width measuring and core expansion testing) which are either suspect or at least not yet capable of reliable interpretation.

Whatever the effect on structural strength, ASR leads to cracking of the concrete surface, and this raises concerns over the possible susceptibility to damage from other causes, including frost attack, carbonation and reinforcement corrosion. Although there seems to be some empirical evidence[95] that corrosion may not occur as readily as might otherwise be expected when the concrete cover is cracked by ASR, the ISE interim guidance rightly advises, 'In the absence of proven experience, ASR cracking must be regarded as introducing a risk of corrosion'.

5.7.3 Monitoring and management

Internationally, it is extremely rare for structures affected by ASR to require replacement or substantial repair, and in most cases 'management' by monitoring and selective remedial actions is an adequate response. According to Wood, for example, 'for many of the structures we investigate, the robustness of the reinforcement and the mildness of the reaction enable them to be managed on the basis of routine inspections and some action to reduce the moisture and salt supplied to the structure'[98]. A good example of such structure management has already been described, the Val de la Mare Dam (see 5.4). In those cases where ASR can be shown definitely to have terminated, or reached an equilibrium, it might be possible to consider a lasting remedial solution that will not require long-term management and monitoring.

The scope of monitoring and management varies according to the 'structural severity rating' as determined according to the ISE interim guidance (see 5.7.2). In one study of some 63 structures[98], only 10% were rated 'very severe', 24% were 'mild' or 'very mild' and just less than half were rated as 'serious', which suggests annual engineering inspection, crack monitoring and occasional core testing. Schemes for the inspection of structures are described in the BCA report[14] on diagnosis and by Fookes et al.[62], but with repetitive inspections it is essential to ensure that reliable comparisons can be made between inspections carried out at different times and perhaps even by different engineers; good-quality notated photography and/or scale drawings can assist. Crack monitoring involves plotting the evolution of crack patterns as well as measuring fluctuations in crack widths or local displacements, and it is important to monitor initially uncracked locations as well as the more obviously affected areas. Various instrumental techniques are available for measuring movements across cracks or elsewhere on a structure, but it is important properly to control the veracity of such readings and to make appropriate corrections for background environmental factors.

5.7.4 *The prospects of repair*

It used to be thought that ASR was the cause of an irreversible deterioration of concrete, ultimately leading to structural failure unless action was taken to replace or to 'manage' the affected units. Some researchers considered that attempted repairs could, in some circumstances, make the damage worse, for example by sealing cracks that would otherwise have provided a non-expansive escape route for newly generated gel. It has been agreed for some time that protection of the concrete from exposure to moisture must have some beneficial influence, at least in reducing the rate of reaction even if the reaction could not be completely arrested, but the performance of coating systems appears to have been extremely variable. It was these uncertainties over the prospect of successful repair that emphasised the importance of implementing the recommendations of the Hawkins Working Party in the UK (see 5.6).

In the UK up to the present, most 'repairs' have consisted of substantial replacements or extensive reconstruction, such as at the Val de la Mare Dam, Charles Cross car park, Voss Farm Bridge and Smorrall Lane Bridge (see 5.4.4 and 5.5.3), or temporary support works prior to planned replacement in the foreseeable future, such as the Marsh Mills Viaduct (see 5.5.3). Measures to deflect or reroute water migration, or physically to protect concrete surfaces with shelters or temporary claddings, have usually been regarded as only emergency measures to retard deterioration whilst investigations and structural appraisals are completed. There is little British experience to date, and only discouraging international experience, to suggest that surface coating systems can be other than negligibly effective at preventing further ASR and cracking from occurring, although some success has been claimed for silanes[99], especially if they are periodically reapplied.

The most hopeful experience in the UK has been the observation, over a period of some years of repetitive re-examination, that a level of stability of ASR damage can be reached at some stage, often at between 10 and 15 years after construction, after which further crack development and movements are negligible and very slow. It is not clear why this should occur, but in most cases it seems more likely to represent a physical equilibrium rather than an exhaustion of one of the chemically reactive constituents. Limited trials have suggested that crack sealing and some coating types might prove to be long-term effective treatments for affected concretes which have reached such a stage of physical equilibrium.

5.8 New reassurance: life after ASR

5.8.1 *Back in proportion*

At the height of the publicity over ASR discoveries in south-west England, one enterprising journalist coined the term 'concrete cancer' to describe ASR to lay readers[100]. Although the term is misleading, because ASR is a scarring rather than a terminal 'disease', and engineers rightly deprecate the use of such emotive terminology, it is interesting, because it reveals the level and nature of the public response to the apparently epidemic spread of ASR damage to concrete. Like all human fears of 'diseases' or other phenomena of unknown cause and uncertain remedy, the public response to ASR exaggerated the actual incidence of occurrence and the severity of damage to affected structures. However, the causation is now much better understood, there is confidence that future construction should be largely free of ASR, and it is now becoming realised in the UK that most structures disabled by ASR can remain in service, with many being able to satisfy the originally intended lifespan. Fear of 'concrete cancer' has therefore been reduced and the responsible engineering concern is now more in proportion with the real scale of the problem.

5.8.2 Emphasis on survival

It has already been explained (see 5.7.3) that most UK structures found to be affected by ASR can be 'managed' by monitoring and selective repair; demolition and reprovision is rarely a necessary option in the absence of other prejudicial factors. In 1985, Somerville suggested that, 'Experience from other countries—which have faced the problem for a much longer period—should be drawn upon; there, demolition is virtually never considered'[95]. More recently, Hobbs has asserted that, 'no concrete structure or part of a structure anywhere in the world has...collapsed due to ASR alone'[82]. Considerations other than ASR which could lead to a decision prematurely to dismantle an ASR-affected structure would include, *inter alia*, the discovery of inadequate structural detailing in critical positions leading to an unacceptable reduction in the safety margin, secondary deterioration such as reinforcement corrosion facilitated by surface cracking caused by ASR, or simply an economic aspect whereby the cost of replacement might be less than the cost of long-term management, especially if replacement also enables desirable improvements in the facility to be achieved. In summary, few structures affected by ASR in the UK will need to be replaced or substantially reconstructed unless other non-ASR factors are involved and, even then, it appears that only a very few cases would necessitate demolition for purely engineering reasons: most will survive the discovery of ASR.

5.8.3 A more confident future

Although it is nearly 50 years since ASR was first discovered, it is less than 15 years since it started to become apparent in the UK that ASR was not wholly a 'foreign problem' and that many British concretes were damaged to some extent by the reaction. In that comparatively short time, the British construction industry has identified the means of prevention, established the means of reliable diagnosis and made considerable progress in the development of a dependable cure. The Hawkins Working Party (see 5.6) has determined the various concrete design characteristics which will minimise the risk of ASR in future construction, and there is ample evidence that the appropriate precautions are being widely implemented in normal concrete production and are also being vigorously specified for major building and civil engineering projects. A BS concrete prism test for assessing the expansivity of aggregate combinations may soon be available, and an important 4-year programme of research being sponsored by the Mineral Industry Research Organisation (MIRO) might enable criteria for interpreting the concrete prism test results to be established. Another part of the MIRO programme concerns research into the reactivity of British flint aggregates.

The guidance provided by the BCA Working Party on Diagnosis should ensure that, over the forthcoming years, the occurrence of ASR in the existing stock of concrete structures is properly identified and correctly assessed; there ought to be fewer misdiagnoses or gross overstatements of severity. Even in the more contentious area of prognosis and structural assessment, there is a growing understanding of the ways to cope with the effects of ASR, and the recent interim guidance from the Institution of Structural Engineers indicates that positive advice on accurate appraisal and restorative measures may soon become available. In the meantime, the BRE through an 'ASR Forum' is coordinating research, at various centres in the UK, into the structural effects of ASR.

Of course, it would be naive to suppose that no previously unforeseen difficulties will be identified as time goes by, and already awareness is developing of some factors which complicate approaches to prevention, diagnosis and cure. One important such factor involves the diverse provenance of the alkalis in concrete, which, it is now realised, can be derived from many common sources other than just the cement. The possible alkali contributions from aggregates or from mineral additives have been discussed earlier (see 5.6.2), but recent work by Swamy[101], using highly reactive types of aggregate, has suggested that externally derived alkalis, especially in the form of de-icing salts, can 'have the quickest and most devastating effects on hardened concrete undergoing ASR expansion': the resultant damage was reported to be significantly more serious than for concretes containing comparable amounts of reactive alkali incorporated into the original mix. This might imply that UK concrete road surfaces and highway structures in general might be particularly at risk and, whilst it is true that highway bridges would feature strongly in a schedule of UK structures affected by ASR, recent limited work at the Transport and Road Research Laboratory (TRRL) has provided some early reassurance about concrete road pavements[102].

There will be increasing confidence in the ability of materials engineers reliably to distinguish between potentially reactive and non-reactive aggregate combinations and, in due course, there will be standardised British methods for the examination and testing of aggregates (see 5.6.4). However, 'special' or atypical types of occurrence might still be encountered from time to time, and new situations might arise as improved transportation or even the single European market causes materials to be used together in hitherto untried combinations. Particular difficulties might arise with artificial aggregates, because less is known generally about their long-term performances and, in any case, new and unproven varieties are always likely to be offered to the construction industry.

Whatever developments might be ahead, at present the occurrence and results of ASR in the UK are now being understood and dealt with in a manner more commensurate with the true scale of the problem. There is now considerable and justifiable confidence in the UK that vigilance will cause ASR to be a much less frequent occurrence in the future and that most of the existing stock of affected structures will be successfully enabled by management or repair to achieve something approaching their originally intended service lives.

5.9 Concluding remarks

ASR continues to be a matter of intense, and perhaps sometimes disproportionate, concern in the UK. At first it had been regarded as an exotic, rare and, above all, foreign problem; later the exaggerated response to the discovery of ASR in UK structures became reflected in the perception of a 'concrete cancer' epidemic as portrayed to the general public by the media. The occurrence of ASR is now accepted by most engineers as one amongst many durability threats to concrete construction, but the risk of any resultant damage being of serious structural concern is now recognised to be comparatively small. Nevertheless, damage caused by ASR can occasionally be very significant, and reliable means of arrest and repair cannot always be applied with confidence, so that, for the foreseeable future, critical or particularly vulnerable structures will need to be designed to negate the risk of ASR and precautions will continue to be taken with most normal construction at least to minimise the risk of ASR. The search for dependable predictive tests will go on, and hopefully knowledge and understanding will improve so that the likelihood of ASR occurring in a given concrete in a particular location can be more accurately evaluated. At present it would seem that most of the existing stock of concrete structures will be able to survive the experience of ASR, and methods of treatment are being developed to try to ensure that, in the future, ASR can be regarded as a tolerable wound and not as a terminal condition.

Acknowledgements

The author recognises and appreciates the support of the partners of Messrs Sandberg and also the opportunities arising from the commissions received from their various clients, without whom this chapter could not have been written. Thanks are also due to Miss Alison Douglas for typing the manuscript and to several colleagues and other authors for their assistance with the diagrams and photographs. Finally, the author gratefully acknowledges the enduring patience and understanding of his wife and daughter, who were frequently abandoned during the hours of writing and editing.

References

Note: In the case of British and American Standards, the modern edition is cited and reference to any earlier versions is made in parentheses.

1a. Stanton, T.E. (1940) Influence of cement and aggregate on concrete expansion. *Engng. News Rec.* **124**(5), 59–61.
1b. Stanton, T.E. (1940) Expansion of concrete through reaction between cement and aggregate. *Am. Soc. Civ. Engrs. Pap.* December 1781–1811.
2. Tuthill, L.H. (1982) Alkali–silica reaction—40 years later. *Concrete International* 4(4), 32–36 [and Discussion (1983) **5**(2), 65–67].
3. Cement and Concrete Association (1948) *Thirteenth Annual Report.* 30th June.
4. Jones, F.E. (1952) *Reactions Between Aggregates and Cement. Part I, Alkali–Aggregate Interaction: General.* National Building Studies Research Paper No. 14, DSIR/BRS. HMSO, London.
5. Jones, F.E. (1952) *Reactions Between Aggregates and Cement. Part II, Alkali–Aggregate Interaction: British Portland Cement and British Aggregates.* National Building Studies Research Paper No. 15, DSIR/BRS. HMSO, London.
6. Jones, F.E. and Tarleton, R. D. (1952) *Reactions Between Aggregates and Cement. Part III, Alkali–Aggregate Interaction: The Expansion Bar Test and Its Application to the Examination of Some British Aggregates for Possible Expansive Reaction with Portland Cements of Medium Alkali Content.* National Building Studies Research Paper No. 17, DSIR/BRS. HMSO, London.
7. Jones, F.E. and Tarleton, R.D. (1958) *Reactions Between Aggregates and Cement. Part IV, Alkali–Aggregate Interaction: The Expansion Bar Test and its Application to the Examination of Some British Aggregates for Possible Expansive Reaction with Portland Cements of High Alkali Content.* National Building Studies Research Paper No. 20, DSIR/BRS. HMSO, London.
8. Jones, F.E. and Tarleton, R.D. (1958) *Reactions Between Aggregates and Cement. Part V, Alkali–Aggregate Interaction: Effect on Bending Strength of Mortars and the Development of Gelatinous Reaction Products and Cracking. Part VI, Alkali–Aggregate Interaction: Experience with Some Forms of Rapid and Accelerated Tests for Alkali–Aggregate Reactivity: Recommended Test Procedures.* National Building Studies Research Paper No. 25, DSIR/BRS. HMSO, London.
9. Cole, R.G. and Horswill, P. (1988) Alkali–silica reaction: Val de la Mare Dam, Jersey, case history. *Proc. Instn Civ. Engrs., Part 1* 84 (December), 1237–1259 (*Paper 9372*).
10. Poole, A.B. (Ed.) (1976) *Proceedings of a Symposium on the Effect of Alkalies on the Properties of Concrete, London.* Cement and Concrete Association, Wexham Springs, Slough, UK.
11. Sandberg, Messrs. (1977) *132/33 kV substation, Milehouse, Plymouth. Investigation of Cracking in Concrete Bases.* Confidential report to South Western Electricity Board. Messrs Sandberg Ref. MWO/60.
12. Hawkins, M.R. (Ed.) (1983) *Alkali–Aggregate Reaction: Minimising the Risk of Alkali–Silica Reaction, Guidance Notes.* Report of a Working Party. Cement and Concrete Association, Ref. 97–304, Wexham Springs, Slough, UK.
13. Concrete Society (1987) *Alkali–Silica Reaction: Minimising the Risk of Damage to Concrete.*

Guidance Notes and Model Specification Clauses. Concrete Society Technical Report No. 30, London.

14. British Cement Association (1988) *The Diagnosis of Alkali–Silica Reaction.* Report of a Working Party. British Cement Association, Ref. 45.042, Wexham Springs, Slough, UK.

15. Institution of Structural Engineers (1988) *Structural Effects of Alkali–Silica Reaction—Interim Technical Guidance on Appraisal of Existing Structures.* Institution of Structural Engineers, London.

16. Midgley, H.G. (1951) Chalcedony and flint. *Geol. Mag.* **88** (May–June), 179–184.

17. McConnell, D., Mielenz, R.C., Holland, W.Y. and Greene, K.T. (1947) Cement aggregate reaction in concrete. *J. Am. Concr. Inst.* **19**(2), 93–128.

18. Stanton, T.E., Porter, O.J., Meder, L. C. and Nicol, A. (1942) California experience with expansion of concrete through reaction between cement and aggregate. *J. Am. Concr. Inst.* **13**(3), 209–236.

19. American Society for Testing and Materials (1987) *Standard Test Method for Potential Reactivity of Aggregates (Chemical Method).* ASTM C289–87 (ASTM C289-52T, ASTM Standards, 1952, Part 3, 943).

20. American Society for Testing and Materials (1986) *Standard Specification for Concrete Aggregates.* ASTM C33-86 (ASTM C33-55T, ASTM Standards, 1955, Part 3, 1145–1150 and Revised in ASTM Reprint 12, 1956).

21. American Society for Testing and Materials (1987) *Standard Test Method for Potential Alkali-Reactivity of Cement–Aggregate Combinations (Mortar-Bar Method).* ASTM C227-87 (ASTM C227-52T, ASTM Standards, 1952, Part 3, 44–51).

22. British Standards Institution (1978) *Specification for Ordinary and Rapid-Hardening Portland Cement.* BS 12:1978 (BS 12:1958, BS 12:Part 2:1971).

23. British Standards Institution (1983) *Specification for Aggregates from Natural Sources for Concrete.* BS 882:1983 (BS 882:1954, BS 882 and 1201:1965, BS 882,1201:Part 2:1973).

24. British Standards Institution (1980) *Methods of Test for Water for Making Concrete (Including Notes on the Suitability of the Water).* BS 3148:1980 (BS 3148:1959).

25. British Standards Institution (1970 and 1978) *Methods of Testing Cement.* BS 4550:Part 2:1970 and BS 4550:Parts 0 and 1 and 3 to 6:1978.

26. British Standards Institution (1984 and 1985) *Testing Aggregates.* BS 812:Parts 101 and 102:1984, BS 812:Parts 103, 105, 106 and 119:1985 (BS 812:1960, BS 812:1967, BS 812:Parts 1 to 3:1975, BS 812:Part 4:1976).

27. British Standards Institution (1969) *The Structural Use of Reinforced Concrete in Buildings.* CP 114:Part 2:1969 (CP 114:1957).

28. British Standards Institution (1969) *The Structural Use of Prestressed Concrete in Buildings.* CP 115:part 2:1969 (CP 115:1959).

29. British Standards Institution (1969) *The Structural Use of Precast Concrete.* CP 116:Part 2:1969 and Addendum No. 1, 1970 (CP 116:1965).

30. British Standards Institution (1972) *The Structural Use of Concrete.* CP 110:Parts 1 to 3:1972.

31. British Standards Institution (1985) *Structural Use of Concrete.* BS 8110:Parts 1 to 3:1985.

32. British Standards Institution (1983 and 1986) *Testing Concrete.* BS 1881:Parts 101 to 122:1983, BS 1881:Part 125 and 201 to 203 and 205 to 206:1986 (BS 1881:1952, BS 1881:Parts 1 to 5:1970, BS 1881:Part 6:1971).

33. Teychenné, D.C. (1987) Personal communication.

34. Building Research Establishment (1982) *Alkali–Aggregate Reactions in Concrete.* BRE Digest 258. HMSO, London.

35. Midgley, H.G. (1976) The identification of opal and chalcedony in rocks and methods of estimating the quantities present. *Proceedings of a Symposium on The Effect of Alkalies on the Properties of Concrete, London,* pp. 193–201.

36. Gutt, W. and Nixon, P.J. (1979) Alkali aggregate reactions in concrete in the UK. *Concrete (J. Concr. Soc.)* **13**(5), 19–21.

37. Building Research Station (1971) *Changes in the Appearance of Concrete on Exposure.* BRS Digest 126. HMSO, London.

38. Palmer, D. (1977) *Alkali–Aggregate (Silica) Reaction in Concrete.* Cement and Concrete Association, Advisory Note, Ref. 45-033, Wexham Springs, Slough, UK.

39. Shacklock, B.W. (1969) Durability of concrete made with sea-dredged aggregate. *Pro-*

ceedings of a Symposium on Sea Dredged Aggregates for Concrete, 9th December 1968. Sand and Gravel Association of Great Britain, Ref. 97.101, London, pp. 31–33.

40. Keen, R.A. (1970) *Impurities in Aggregates for Concrete*. Cement and Concrete Association, Advisory Note No. 18, Wexham Springs, Slough, U.K.

41. Roeder, A.R. (1975) *Some Properties of Flint Particles and Their Behaviour in Concrete*. Unpublished individual project, Advanced Concrete Technology Course, 1974–1975, Cement and Concrete Association, Wexham Springs, Slough, UK.

42. Haigh, I.P. (1969) General principles of control of aggregates. *Proceedings of a Symposium on Sea Dredged Aggregates for Concrete*, 9th December 1968. Sand and Gravel Association of Great Britain, Ref. 97.101, London, pp. 47–50.

43. Van de Fliert, C. (1969) Experience with sea-dredged aggregates in the Netherlands. *Proceedings of a Symposium on Sea Dredged Aggregates for Concrete*. 9th December 1968. Sand and Gravel Association of Great Britain, Ref. 97.101, London, pp. 57–60.

44. Neville, A.M. (1963, 1973 and 1981) *Properties of Concrete*, 1st, 2nd and 3rd Edns. Pitman, London.

45. Lea, F.M. (1956 and 1970) *The Chemistry of Cement and Concrete*, 2nd Edn (with Desch, C.H.) and 3rd Edn, Edward Arnold, London.

46. Road Research Laboratory (1955) *Concrete Roads, Design and Construction*. DSIR. HMSO, London.

47. Idorn, G.M. (1969) *The Durability of Concrete*. Concrete Society Technical Paper PCS 46, London.

48. Idorn, G.M. (1967) *Durability of Concrete Structures in Denmark, a Study of Field Behaviour and Microscopic Features*. *DSc Thesis*, Technical University of Denmark, Copenhagen.

49. Fookes, P.G. (1984) Discussion on the design and construction of the Thames Barrier coffer dam. *Proc. Instn. Civ. Engrs, Part 1* 76 (May), 567–568.

50. Fookes, P.G., Collis, L., French, W.J. and Poole, A.B. (1977) *Concrete in the Middle East* (reprints of five articles from *Concrete*). Cement and Concrete Association, Viewpoint Publication, Ref. 12.077, Wexham Springs, Slough, U.K.

51. Sims, I. (1977) *Investigations into Some Chemical Instabilities of Concrete*. PhD Thesis, University of London (Queen Mary College).

52. Poole, A.B. (1975) Alkali–silica reactivity in concrete from Dhekelia, Cyprus. *Proceedings of a Symposium on Alkali–Aggregate Reaction Preventive Measures, Rejkyavik, Iceland*.

53. French, W.J. and Poole, A. B. (1974) Deleterious reactions between dolomites from Bahrain and cement paste. *Cem. Concr. Res.* **4**, 925–937.

54. Sims, I. and Poole, A.B. (1980) Potentially alkali-reactive aggregates from the Middle East. *Concrete (J. Concr. Soc.)* **14**(5), 27–30.

55. Coombes, L.H., Cole, R.G. and Clarke, R.M. (1975) Remedial measures to Val de la Mare Dam, Jersey, Channel Islands, following alkali–aggregate reactivity. Symposium, *British National Committee on Large Dams* (BNCOLD), University of Newcastle upon Tyne, Proceedings Paper 3.3, pp. 1–10.

56. Coombes, L.H. (1976) Val de la Mare Dam, Jersey, Channel Islands. *Proceedings of a Symposium on the Effect of Alkalies on the Properties of Concrete, London*. Cement and Concrete Association, Wexham Springs, Slough. UK, pp. 357–378.

57. Sherwood, D.E., Marriott, M. and Smith, J. (1975) Non-destructive testing of concrete dams by sonic speed measurement. Symposium, *British National Committee on Large Dams* (BNCOLD), University of Newcastle upon Tyne, Proceedings Paper 3.2, pp. 1–7.

58. Hobbs, D.W. (1978) Expansion of concrete due to alkali–silica reaction: an explanation. *Mag. Concr. Res.* **30**(105), 215–220.

59. Fookes, P.G., Cann, J. and Comberbach, C.D. (1984) Field investigation of concrete structures in South-West England, Part 3. *Concrete (J. Concr. Soc.)* **18**(11), 12–16.

60. Palmer, D. (1978) Alkali–aggregate reaction. Recent occurrences in the British Isles. *Proceedings of the Fourth International Conference on the Effects of Alkalies in Cement and Concrete*, pp. 285–298.

61. Moore, A.E. (1978) Effect of electric current on alkali–silica reaction. *Proceedings of the Fourth International Conference on the Effects of Alkalies in Cement and Concrete*, pp. 69–72.

62. Fookes, P.G., Comberbach, C.D. and Cann, J. (1983) Field investigation of concrete

structures in South-West England, Parts 1 and 2. *Concrete (J. Concr. Soc.)* **17** (3 and 4), 54–56 and 60–65.

63. Wood, J.G.M., Johnson, R. A. and Abbott, R.J. (1987) Monitoring and proof loading to determine the rate of deterioration and the stiffness and strength of structures with AAR. Institution of Structural Engineers/Building Research Establishment, Seminar, *Structural Assessment—Based on Full and Large-Scale Testing*, 6–8 April.

64. Nixon, P.J., Collins, R.J. and Rayment, P.L. (1979) The concentration of alkalies by moisture migration in concrete—a factor influencing alkali–aggregate reaction. *Cem. Concr. Res.* **9**, 417–423.

65. Building Research Establishment (1988) *Alkali Aggregate Reactions in Concrete.* BRE Digest 330. Department of the Environment, London.

66. Montague, S. (1989) Most DTp bridges in danger from chloride. *New Civil Engineer*, 2nd March, 8 (referring to a report *The Performance of Concrete in Bridges: a survey of 200 highway bridges*. Published by DTp, HMSO, London 1989).

67. French, W.J. (1987) A review of some reactive aggregates from the United Kingdom with reference to the mechanism of reaction and deterioration. *Proceedings of the Seventh International Conference on Concrete Alkali–Aggregate Reactions*, pp. 226–230.

68. Buttler, F.G., Newman, J.B. and Owens, P. (1980) Pfa and the alkali–silica reaction. *Consulting Engineer*, November, 57–62.

69. Corish, A.T. and Jackson, P.J. (1982) Portland cement properties—past and present. *Concrete (J. Concr. Soc.)* **16**(7), 16–18.

70. Mott Hay and Anderson (1986) *Marsh Mills Viaduct Appraisal and Recommendations. Report on Full Appraisal and Recommendations for Future Management of Marsh Mills Viaduct.* Vol. 1 Report, Vol. 1 Appendices. Department of Transport, London.

71. Sandberg, Messrs (1984) *Examination of Concrete for Alkali–Reactivity and Chloride Content, Smorrall Lane Overbridge, M6 Motorway.* Confidential report to Warwickshire County Council. Messrs Sandberg Ref. L/4045/G/5.

72. Nixon, P.J. and Gillson, I.P. (1987) An investigation into alkali–silica reaction in concrete bases at an electricity substation at Drakelow Power Station, England. *Proceedings of the Seventh International Conference on Concrete Alkali–Aggregate Reactions*, pp. 173–177.

73. Sandberg, Messrs (1986) *Examination of Aggregate and Concrete Samples. Drakelow Substation, Alkali–Silica Reactivity Study.* Confidential Report to the Central Electricity Generating Board. Messrs Sandberg Ref. L/7067/G.

74. Allen, R.T.L. (1981) Alkali–silica reaction in Great Britain—a review. *Proceedings of the Fifth International Conference on Alkali–Aggregate Reaction in Concrete*, S252/18.

75. Anon (1986) Full inspection follows Welsh dam AAR alert. *New Civil Engineer*, 13th November, 8.

76. Gillott, J.E., Duncan, M.A.G. and Swenson, E.G. (1973) Alkali–aggregate reaction in Nova Scotia. IV. Character of the reaction. *Cem. Concr. Res.* **3**, 521–535 (also National Research Council of Canada, RP 567).

77. Sims, I. and Sotiropoulos, P. (1989) A pop-out forming type of ASR in the United Kingdom. *Eighth International Conference on Alkali–Aggregate Reaction*, Kyoto, Japan: poster-session paper (submitted for publication to *Cem. Concr. Res.*).

78. Figg, J.W. (1981) Reaction between cement and artificial glass in concrete. *Proceedings of the Fifth International Conference on Alkali–Aggregate Reaction in Concrete.* S252/7.

79. Sims, I. and Bladon, S. (1984) An exploratory assessment of the alkali-reactivity potential of sintered pfa in concrete. *Proceedings of the Second International Conference on Ash Technology and Marketing.*

80. Hobbs, D. W. (1980) *Influence of Mix Proportions and Cement Alkali Content upon Expansion due to the Alkali–Silica Reaction.* Cement and Concrete Association, Technical Report 534, C&CA Ref. 42.534, Wexham Springs, Slough, UK.

81. Palmer, D. (1981) Alkali–aggregate reaction in Great Britain—the present position. *Concrete (J. Concr. Soc.)* **15**(3), 24–27.

82. Hobbs, D.W. (1988) *Alkali–Silica Reaction in Concrete.* Thomas Telford, London.

83. Nixon, P.J., Page, C.L., Canham, I. and Bollinghaus, R. (1988) Influence of sodium chloride on alkali–silica reaction. *Advances in Cement Research* **1**(2), 99–106.

84. Department of Transport (1986) *Specification for Highway Works*, Parts 1–7 and *Notes for guidance*, Parts 1–6, HMSO, London.

85. Department of Transport (1976) *Specification for Road and Bridge Works*, 5th Edn (and Supplement No. 1, 1978). HMSO, London.

86. Department of Transport (1982) Control of alkali–silica reaction. Interim amendments to specifications for all the Department's contracts that include structural concrete usage. Clause 1618 addition to *Specification for Road and Bridge Works* (see Ref. 85), July.

87. Sims, I. (1981) Application of standard testing procedures for alkali-reactivity, Parts 1 and 2. *Concrete (J. Concr. Soc.)* **15**(10 and 11), 27–29 and 29–32.

88. Sims, I. (1987) The importance of petrography in the ASR assessment of aggregates and existing concretes. *Proceedings of the Seventh International Conference on Concrete Alkali–Aggregate Reactions*, pp. 358–367.

89. Nixon, P.J. (1986) Testing the alkali silica reactivity of UK aggregates. *Chemistry and Industry* 21st July, 488–489.

90. Nixon, P.J. and Bollinghaus, R. (1983) Testing for alkali reactive aggregates in the UK. *Proceedings of the Sixth International Conference on Alkalis in Concrete, Research and Practice*, pp. 329–336.

91. Sims, I. and Miglio, B.F. (1989) Compositional uniformity of some UK sand and gravel aggregates for concrete. The Institution of Geologists/The Geological Society/The Institution of Mining and Metallurgy/Minerals Industry Research Organisation, Joint Conference, *Extractive Industry Geology*, Birmingham, April (submitted for publication in *Quart. J. Engng. Geol.*).

92. British Standards Institution (1988) Draft for public comment. *Testing Aggregates. Alkali–Silica Reactivity: Concrete Prism Method*. BS 812: Part 123. BSI Document Ref. 88/11922 DC, June.

93. Wood, J.G.M. (1985) Engineering assessment of structures with alkali–silica reaction. One-day conference, *Alkali–Silica Reaction, New Structures—Specifying the Answer, Existing Structures—Diagnosis and Assessment*, London, November. The Concrete Society, London.

94. Wood, J.G.M., Young, J. S. and Ward, D.E. (1987) The structural effects of alkali–aggregate reaction on reinforced concrete. *Proceedings of the Seventh International Conference on Concrete Alkali–Aggregate Reactions*, pp. 157–162.

95. Somerville, G. (1985) *Engineering Aspects of Alkali–Silica Reaction*. Cement and Concrete Association, Interim Technical Note 8, Wexham Springs, Slough, UK.

96. Swamy, R.N. and Al-Asali, M.M. (1986) Influence of alkali–silica reaction on the engineering properties of concrete. *Alkalies in Concrete*, ASTM STP 930, 69–86. American Society for Testing and Materials, Philadelphia, PA.

97. Swamy, R.N. and Al-Asali, M.M. (1988) Engineering properties of concrete affected by alkali–silica reaction. *American Concrete Institute (ACI) Materials Journal* **85**(5), 367–374.

98. Wood, J.G.M., Johnson, R.A. and Norris, P. (1987) Management strategies for buildings and bridges subject to degradation from alkali–aggregate reaction. *Proceedings of the Seventh International Conference on Concrete Alkali–Aggregate Reactions*, pp. 178–182.

99. Idorn, G.M. (1988) Concrete durability in Iceland. *Concrete International—Design and Construction* **10**(11), 41–43.

100. Coates, J. (1983) 'Concrete cancer' plague spreading. *The Sunday Times*, 6th February.

101. Swamy, R.N. (1988) Alkali–silica reaction—sources of damage. *Highways and Transportation (The Journal of the Institution of Highways and Transportation)* **35**(12), 24–29.

102. West, G. and Sibbick, R. (1988) Alkali–silica reaction in roads. *Highways* **56** (1936) 19–24.

103. Walters G.V. (1989) ECC Quarries Limited, Exeter, Devon. Personal communication.

104. Nixon, P.J., Page, C.L:, Hardcastle, J., Canham, I. and Pettifer, K. (1989) Chemical studies of alkali–silica reaction in concrete with different chert contents. *Proceedings of the Eighth International Conference on Alkali–Aggregate Reaction*, Kyoto, Japan, 129–134.

6 Alkali–silica reaction— Danish experience

S. CHATTERJEE, Z. FÖRDÖS and N. THAULOW

Abstract

In Denmark research on alkali–silica reactivity and preventive measures has been carried out in two distinct phases. In the first phase, flint and opaline limestone were identified as the main reactive components in Danish aggregates. Suggested preventive measures were the use of low-alkali cement and/or the use of aggregates containing less than 2% reactive components. At this stage the basic assumptions were that the alkali content of concrete is determined by that of its cement and the alkalis are evenly distributed.

The second phase of research, on the other hand, assumed that a real-life structure may acquire extra alkalis from its environment and there is always a concentration distribution of alkalis in it. An accelerated test method of storing mortar prisms in a saturated NaCl solution at 50°C was devised. New and detailed mechanisms have been proposed for alkali–silica reaction and the associated expansion.

As regards the preventive measures, the following innovations have been made:

(1) The assumptions of a constant and uniform distribution of alkali in a concrete structure have been dropped.
(2) The environment to which a structure may be exposed has been classified as regards its severity.
(3) The acceptance criteria of aggregates, both fine and coarse, have been made very strict, and have been collated with the environmental classes.
(4) The permissible total alkali content of a concrete mix has been collated with the environment to which the concrete would be exposed.
(5) A fly ash Portland cement has been introduced to the Danish market.

6.1 Historical introduction

As far as alkali–silica reaction (ASR) is concerned, Denmark is in an unfortunate situation; most of the aggregate sources contain reactive components in one or another size fraction, i.e. there is always a risk of ASR. It is hardly surprising that the first description of the effects which could be attributed to ASR was reported by Poulsen in 1914[1], long before Stanton's work in the 1940s. However, systematic research on ASR did not start until Nerenst's report in 1951[2]. The research on ASR in Denmark can be broadly divided into two phases—that carried out before the 1970s and that afterwards—with fairly clear differences in their research philosophies.

6.1.1 *First phase*

The earlier research was carried out under the auspices of the Danish Committee on Alkali Reaction in Concrete. The results of this extensive programme were published in 23 reports[3-25]. In this programme basically the American methodologies were used for the detection of alkali–silica reactivity of aggregates[26,27]. Subsequently, petrographic and X-ray diffraction techniques were employed both to characterise the reactive aggregate types and to detect ASR in concrete structures. Flint and opaline limestone were identified as the most prevalent reactive aggregate types in Denmark. The reactive components were found in both sand and gravel fractions of many aggregate sources. One of the important results of this programme was verification of the concept of 'pessimum proportioning'. This concept was based on the assumption of a fixed alkali content of a mix composition and the observation that in mortar bar tests expansion maxima occur within a rather narrow range of alkali to reactive silica ratio. It was considered desirable to avoid the condition of maximum expansion. Another contribution was the classification of flint types, relating their porosity and reactivity. It was also inferred that, if the total flint content of a sand type is less than 2%, it is less liable to cause alkali–silica expansion. The practical implication of this investigation was that the risk of alkali–silica expansion could be avoided by using either a flint-free aggregate or a low-alkali Portland cement or a pozzolan. The research was summarised by Bredsdorff *et al.*[28]. The culmination of the research was Idorn's dissertation in 1967[29] in which a systematic approach to the assessment of field concrete deterioration was presented.

Implicit assumptions of this phase of research were that the alkali content of a concrete structure is determined by the alkali content of the cement and remains constant and uniformly distributed throughout its life. Moreover, Powers and Steinour's hypothesis of ASR, which is consistent with the above assumptions, formed the theoretical basis of this phase[30].

6.1.2 *Second phase*

The second phase of alkali–silica research started with an investigation of concrete roads which had deteriorated within 4 years of their inauguration. During this investigation, it was observed that extensive ASR had occurred in the concrete roads. Since non-reactive granite was used as coarse aggregates and the cement contained about 0.6% Na_2O eq. alkali, it was inferred that the extensive reaction was a result of an interaction between the de-icing salt NaCl and reactive silica present in the sand, i.e. NaCl acted as an accelerator. This inference was then verified by laboratory experiments[31]. The above road investigation also revealed that the cement-stabilised sub-base of the roads (containing about 6% cement) though it used the same reactive sand as the roads, suffered no ASR. A subsidiary investigation indicated that this absence of ASR in the sub-base may be because of absence of free $Ca(OH)_2$ in the sub-base material[32]. At about the same time, the electron probe microanalytical technique was applied to determine the chemical composition of alkali–silica gel[33]. A high content of CaO in the gels analysed, up to 20%, was found. The gel was analysed *in situ* within the cracks at known and increasing distances from the perimeter of the original reacted particle. The CaO content of the gel increased with increasing distances. This paper, which remained somewhat dormant for a time, played a very important role in the elucidation of the ASR mechanism.

The explicit assumptions of this second phase of research were that the real-life structures often receive alkalis from outside sources, e.g. from the groundwater, from de-icing salt, etc., and that there is nearly always a concentration gradient of alkalis in any structure.

Most of the subsequent research in Denmark could be treated as a follow-up to the above-mentioned three papers. The subsequent research could be divided into the following subgroups:

(1) That directed to the elucidation of mechanisms of ASR and attendant expansion.
(2) That directed to the development of methods for the measurement of the expansion capacity of reactive aggregates.
(3) That directed towards the development of preventive methods.

Beside these systematic studies, there were some isolated studies of the effects of ASR on reinforced concrete beams, life expectancy of a structure suffering from ASR reaction, occurrence of ASR in swimming pools, etc.

6.2 Fundamental studies

6.2.1 *Mechanisms of ASR and attendant expansion*

The first investigation to be carried out was on the role of $Ca(OH)_2$ in the expansive ASR[34]. From Ref. 32, one may reason that expansive ASR may be avoided, even when cement content is high, by using a cement which does not form crystalline $Ca(OH)_2$ during hydration, e.g. high-slag Portland cement. The results of this investigation confirmed the above expectation of no expansion in prisms made with slag Portland cement. The X-ray diffraction diagram of the prism made with slag Portland cement and stored in an NaCl bath showed that prisms contained no $Ca(OH)_2$ but had substantial amounts of Friedel's salt, indicating that NaOH had formed in those prisms by the following reaction:

$$4CaO + Al_2O_3 + 2NaCl + 13H_2O = 3CaO \cdot Al_2O_3 \cdot CaCl_2 \cdot 12H_2O + 2NaOH$$

A point to note is that no expansion occurred in spite of this extra NaOH formation. This confirmed the importance of crystalline $Ca(OH)_2$ in the expansive ASR.

The next investigation addressed the mechanisms by which NaCl and $Ca(OH)_2$ accelerate the alkali–silica expansion[35]. In that investigation, prisms made from reactive sand and Portland cement containing varying amounts of diatomites were exposed to saturated NaCl solution at 50°C. The expansion of the prisms was measured up until about 1 year, and at the end of that period the prisms were further examined by petrographic, X-ray diffraction and electron probe microanalytical techniques. Figure 6.1 shows that the alkali–silica expansion decreases with the increase in the diatomite content in the cement. The X-ray diffraction diagrams showed that crystalline $Ca(OH)_2$ content decreased with the increase in diatomite in cement.

The most interesting information came from the electron probe micro-analyses of reactive grains and their surrounding pastes. Normalised major oxide analyses of reacted grains and their surrounding areas are shown in Figures 6.2 to 6.5. The figures show that (i) the entry of CaO and Na_2O into reacted grains decreases with the increase in diatomite content in cement, (ii) more CaO and Na_2O enter the grains when the prisms are stored in NaCl solution than when stored in water, (iii) CaO enters reactive grains

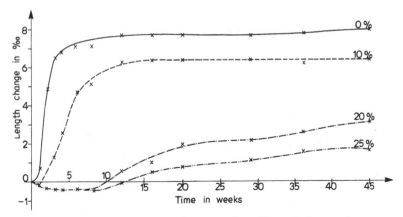

Figure 6.1 Expansion characteristics of mortar prisms. Figures indicate moler contents.

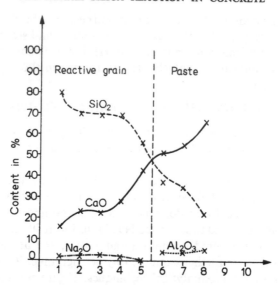

Points of analysis are progressively
away from the core of the grain.

Figure 6.2 The distributions of oxides in and around a reactive grain. Mortar prism stored in water bath. No moler in cement.

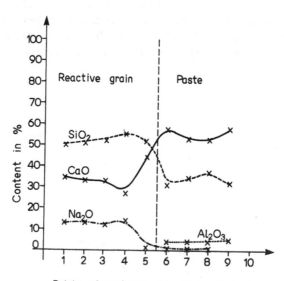

Points of analysis are progressively
away from the core of the grain.

Figure 6.3 The distribution of oxides in and around a reactive grain. Mortar prism stored in an NaCl bath. No moler in cement.

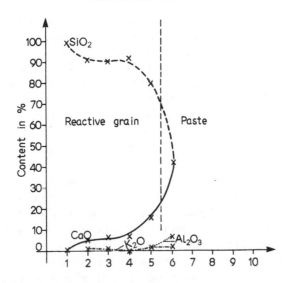

Points of analysis are progressively
away from the core of the grain.

Figure 6.4 The distribution of oxides in and around a reactive grain. Mortar prism stored in water bath. Cement contains 25% moler.

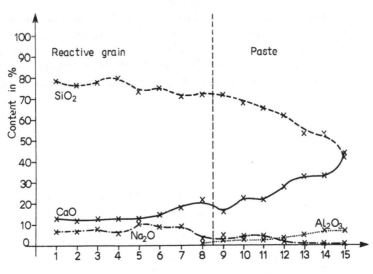

Points of analysis are progressively
away from the core of the grain

Figure 6.5 The distribution of oxides in and around a reactive grain. Mortar prism stored in an NaCl bath. Cement contains 25% moler.

irrespective of the nature of the bath, (iv) in samples free of diatomite, SiO_2 concentration drops sharply at the grain boundary, whereas in samples containing diatomite SiO_2 concentration decreases slowly. A detailed analysis of the microanalytical data indicated that some SiO_2 diffused out of the reactive grains, and the amount of SiO_2 that diffused out increased with decreasing $Ca(OH)_2$ content in the paste. It is obvious that calcium and sodium must have entered the grains as hydrated ions and have been accompanied by OH^- ions so as to maintain electroneutrality.

From the results, it can be seen that expansion of mortar prisms, extent of reaction and penetration of hydrated Na^+, Ca^{2+} and OH^- decreased with the decrease in crystalline $Ca(OH)_2$ in the paste. However, the diffusion of SiO_2 out of a reactive grain increased with the decreasing $Ca(OH)_2$ content.

The observed high Na_2O content of reacted grains of prisms stored in an NaCl bath indicates that a reaction of the type shown in Figure 6.6 occurs during ASR. Availability of both NaCl and $Ca(OH)_2$ near a reactive grain is the controlling factor in the penetration of hydrated Na^+ and OH^-. The above reaction explains how NaCl and $Ca(OH)_2$ accelerate ASR.

In ASR, $Ca(OH)_2$ has at least four roles: it accelerates the penetrations of hydrated Ca^{2+}, Na^+, K^+ and OH^- in a reactive grain; it promotes the reaction shown in Figure 6.6; it hampers the diffusion of SiO_2 out of a reactive grain; and the presence of solid $Ca(OH)_2$ acts as a buffer to maintain a high OH^- concentration.

In a cement paste environment, expansion will occur if the rate of penetration of matter in a reactive grain exceeds that of the SiO_2 diffusing out; the process is shown schematically in Figure 6.7. If the alkali content is low, only a limited amount of hydrated Ca^{2+} and OH^- can penetrate a grain; this is because of the large size of hydrated Ca^{2+}. If the alkali concentration is high, smaller hydrated Na^+, K^+ and OH^- will be able to penetrate a reactive grain unhindered. This penetration of Na^+, K^+ and OH^- will cause a break-

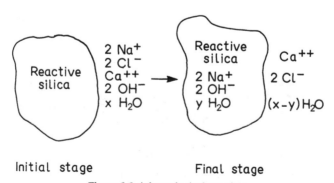

Initial stage   Final stage

Figure 6.6 A hypothetical reaction.

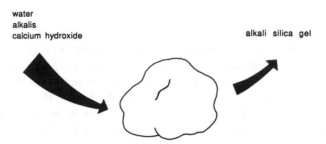

Figure 6.7 A model for the alkali–silica reaction. If the amounts of water, alkali hydroxide and calcium hydroxide entering the reactive particle are larger than the amount of alkali–silica gel seeping out, the particle expands and cracks the surrounding cement paste.

down of Si–O–Si bonds, thereby opening the grain for further penetration of Ca^{2+}, Na^+, K^+ and OH^-. In the presence of excess $Ca(OH)_2$ and high alkali concentration, only a limited amount of SiO_2 can diffuse out, but more materials are pumped in. This generates the pressure necessary for expansion. Note that according to this hypothesis, the chemical reaction and accompanied expansion are not directly related.

The above roles of $Ca(OH)_2$ are quite different from that assigned to it in the hypothesis of Powers and Steinour[30]. According to the Powers and Steinour hypothesis the expanding grain should be low in CaO, whereas according to the newly proposed hypothesis the expanding grain should contain substantial amounts of CaO. The near universally high content of CaO in gels can be seen in Figure 6.8[36].

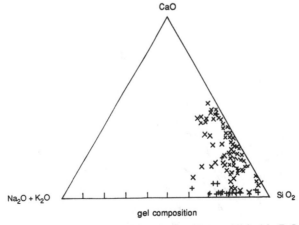

gel composition

Figure 6.8 Composition of alkali–silica gel in concrete. Data published in Ref. 20 and micro-analyses in Ref. 36.

6.2.2 Testing of reaction mechanisms

From the newly proposed hypothesis, it follows that all alkali metal salts are capable of accelerating alkali–silica expansion, and for a given alkali metal salt and reactive aggregate at a given temperature the expansion will decrease with decreasing salt concentration. A recent investigation shows that the above expectations are valid[37]. Figures 6.9 and 6.10 show the results of this investigation. Figure 6.9 shows that all alkali salts accelerate expansion. Alkali hydroxide, which depresses the solubility of $Ca(OH)_2$, causes less expansion than other salts. Figure 6.10 shows that expansion decreases with decreasing KCl concentration. Figure 6.10 shows that a 0.5 N KCl solution, i.e. a *ca* 3% solution, does not cause any expansion; this is consistent with the result of a 10-year submersion of concrete prisms in seawater[38].

The reaction shown in Figure 6.6 constitutes another test for the newly proposed hypothesis. The points to note are that the above reaction should occur in a reactive sand–$Ca(OH)_2$–KCl solution system and during reaction the liquid phase is enriched with $CaCl_2$. This newly formed $CaCl_2$ will in turn depress the solubility of $Ca(OH)_2$ in the liquid phase. High reactivity of a sand creates a high concentration of $CaCl_2$ and a low concentration of $Ca(OH)_2$ in the liquid phase. This inference has been tested recently[39]. Table 6.1 shows the results of this investigation. From Table 6.1, it can be seen that the suspensions containing more reactive sand types have lower OH^- concentrations.

From the results of Figure 6.10, it is possible to get an order of magnitude estimation of the maximum tolerable alkali content. Figure 6.10 shows that in the case of a highly expansive sand a 0.5 N alkali salt solution does not give rise to any expansion, but a 1 N solution does. One may consider that

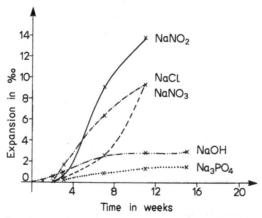

Figure 6.9 Expansion characteristics of mortar bars stored in different sodium salt solutions.

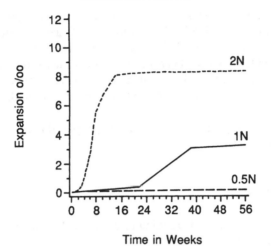

Figure 6.10 Expansion characteristics of mortar bars stored in KCl solutions of varying normality.

the maximum tolerable alkali salt solution will be around 0.75 N. After some time, the concentration in the pore liquid and in the salt bath will be the same. An estimate of the amount of alkali needed to trigger expansive reaction may be estimated from the following considerations. Consider a concrete mix containing 350 kg cement per cubic metre and a water–cement (w/c) ratio of 0.5, i.e. an initial water content of 175 litres. Further assume that, after hardening, the free water content is 85 litres. The maximum tolerable alkali content can be calculated as:

$$\frac{0.75 \times 62 \times 85}{2 \times 1000} = 2 \text{ kg Na}_2\text{O eq. per cubic metre of concrete}$$

For a less expansive sand, more Na_2O could be tolerated. In the above

Table 6.1 The OH^- concentration of the filtrates in milligrams per litre of solution for sand tested by a simple chemical method[39].

Sand no.	Expansivity	Set 1		Set 2		Set 3
		50°C	70°C	50°C	70°C	50°C
1	Nil	590	603	601	580*	593
2	Very low	550*	565	567	561	
3	Low	541	527	549	522	
4	High	499	463	510	457	500
5	High	486	458	488	457	

* Have some difficulties during filtration.

estimation, it has been assumed that no alkali salts may migrate from an outside source and that alkali salt is uniformly distributed throughout the mass of the concrete mix. If alkali salts migrate from an outside source or local evaporation takes place at certain parts, then the local concentration of alkali salt may go above the critical value and give rise to alkali–silica expansion even though the overall alkali content may be lower than the critical value.

Various practical implications of this newly proposed hypothesis have been discussed in a paper[40].

6.3 Prevention of ASR

The development of ASR and cracking of concrete depends primarily on the following parameters:

The alkali content of the concrete
The type and content of the reactive silica in the aggregate
The quality of the concrete (strength, density, air content)
The environmental condition (temperature, humidity, salt).

6.3.1 Alkalis in Danish cement

The alkalis in the concrete normally originate from the cement, but alkalis can also be transported from external sources, for example from de-icing salts. However, the total alkali content of the concrete depends, on most occasions, on the alkali content of the cement and the cement content of the concrete.

The alkali content of the most commonly used Danish types of cement varies between 0.6% and 0.8% by weight of cement, and that is generally expressed as equivalent acid-soluble sodium oxide ($Na_2O + 0.658\ K_2O$) (Table 6.2).

From Table 6.2 it can be seen that the acid-soluble alkali content of the Portland cement is close to 0.6% Na_2O eq. According to the American Standard ASTM C 150-72, a Portland cement with an alkali content of less than 0.6% is defined as a low-alkali cement. In Denmark two special cements with low alkali contents are produced:

Low-alkali sulphate-resisting Portland cement (less than 0.4% alkali)
White Portland cement (less than 0.2% alkali).

6.3.2 Danish Code of Practice for the prevention of ASR

The Danish Code of Practice for the structural use of conctrete[41] requires that the cement should be Portland cement or Portland fly ash cement com-

Table 6.2 Danish types of cement: chemical composition.

		Portland cement 1	Portland cement 2	Portland cement 3	Portland cement 4
SiO_2	(%)	28.52	21.33	24.29	24.31
Al_2O_3	(%)	8.98	4.86	2.38	1.85
Fe_2O_3	(%)	3.47	2.86	2.72	0.34
CaO	(%)	50.03	63.74	65.90	69.32
MgO	(%)	1.07	1.00	0.62	0.55
SO_3	(%)	2.38	2.87	2.31	1.89
L.O.1	(%)	2.10	1.83	0.97	1.14
K_2O	(%)	0.43	0.55	0.19	0.03
Na_2O	(%)	0.23	0.28	0.17	0.12
Na_2O eq.	(%)	0.51	0.64	0.29	0.13
Free CaO	(%)	1.41	1.76	0.96	2.92
Flyash content	(%)	22	2.0	0	0
Water-soluble alkalis					
K_2O	(%)	0.16	0.22	0.05	0
Na_2O	(%)	0.04	0.05	0.02	0
Na_2O eq.	(%)	0.15	0.20	0.06	0

1, Standard cement (Portland fly ash cement); 2, Rapid cement; 3, Low-alkali sulphate-resisting Portland cement; 4, White cement.

plying with the requirement of DS 427, Portland cement and Portland fly ash cement, 2nd Edn, April 1983. In DS 427 there is no regulation of the alkali content of the cement. However, in May 1986, a new Code of Practice was issued. This is the so-called BBB (In Danish: *Basis Betonbeskrivelsen for Bygningskonstruktioner*)[42]. This is a fundamental concrete specification to be used in all public building construction works. A system is set out to prevent harmful alkali–silica reaction.

The parameters used in this system are as follows:

Alkali content of the concrete
Reactivity of the aggregate
Environmental conditions.

6.3.2.1 *Cement and its requirements.* The cement falls into one of four groups of alkali content:

EA extra low-alkali $\leqslant 0.4\%$ eq. Na_2O
LA low-alkali $\leqslant 0.6\%$ eq. Na_2O
MA medium-alkali $\leqslant 0.8\%$ eq. Na_2O
HA high-alkali $> 0.8\%$ eq. Na_2O.

The alkali contributions of the cement and all other alkali sources in the mix are summed up as eq. Na_2O kg/m^3 of concrete.

6.3.2.2 *Aggregate classification*. The aggregates are classified into three groups:

Class P   for use in passive environments
Class M   for use in moderate environments
Class A   for use in aggressive environments.

The classification of the sand with regard to alkali–silica reactive component (porous flint) is shown in Table 6.3. The methods used are the petrographic thin-section point-count method TI-B 52[43] and the mortar bar expansion test in saturated sodium chloride solution TI-B 51[44]. The TI-B 51 method has recently been accepted as a common Nordic test method: Nordtest Build NT 295[45].

For the coarse aggregate (Table 6.4), amount of reactive aggregate is limited by the allowable amount of particles with a density below 2400 kg/m^3. Furthermore, there is a limit on the absorption of the flint with density larger than 2400 kg/m^3. This value is determined on 10% of the flint with porous crust.

6.3.2.3 *Environment classification*. The environmental classes are set out in DS 411[41]. The requirements refer to three environmental classes, characterised by different degrees of aggressiveness commonly found in Denmark.

(1) *Aggressive environmental class*. Comprises environment containing salt or flue gases, seawater or brackish water.
(2) *Moderate environmental class*. Comprises moist, unaggressive outdoor and indoor environment, and flowing or standing fresh water.
(3) *Passive environmental class*. Comprises dry, unaggressive environment, i.e. particularly an indoor climate.

6.3.2.4 *Requirements for the fresh concrete from BBB*. The combination of requirements for the aggregate, the cement and the concrete as a function of the environmental conditions from BBB[42] are given in Table 6.5.

Table 6.3 Classification of sand.

	Class P	Class M	Class A
Volume of reactive flint (%)	No demand	Max. 2%	Max. 2%
Mortar bar expansion at 8 weeks	No demand	Max. 0.1%	Max. 0.1%

Table 6.4 Classification of coarse aggregate.

	Class P	Class M	Class A
Particles with density less than 2400 kg/m³ (%)	No demand	Max. 5.0%	Max. 1.0%
Absorption (%)	No demand	Max. 2.5%	Max. 1.1%

6.3.2.5 *Requirements for hardened concrete.* The requirements for the hardened concrete after BBB[42] are found in Table 6.6.

6.3.3 *Experimental results with fly ash and fly ash cement*

The effect of fly ash on the expansion due to ASR of mortar or concrete samples can be evaluated either by the use of closed systems in which the total alkali content of the systems does not change with time, which is the case when testing is carried out according to, for example, ASTM C441 or the modified ASTM method or by the use of open systems in which samples are in contact with an alkali salt solution, as is the case with the so-called NaCl bath method. The evaluation can also be carried out by an accelerated test method or by a long-time storage method[46–49]. In Denmark both open and closed systems as well as accelerated and long-time storage methods have been used to evaluate fly ash or Portland fly ash cement systems.

In normal circumstances (closed systems), the use of a Danish Portland cement, which has an alkali content of less than 0.8%, in conjunction with a reactive aggregate does not give rise to any destructive ASR. However, in an open system, in which alkali salts from an outside source can migrate into a structure, destructive ASR can develop very quickly.

It is common practice (in Denmark) to use either a low-alkali Portland cement or a pozzolanic cement (Portland fly ash cement) as a safety factor against ASR.

In recent years papers[50,51] have been published by the Nordic Concrete Research, from which it appears that:

(1) Long-time storage tests show that addition of fly ash to concrete mixes has no adverse effect even if its alkali content is as high as 2.34% Na₂O eq.
(2) Accelerated tests according to ASTM show that the addition of fly ash, even of high alkali content, to Portland cements reduces expansions due to ASR.
(3) Accelerated tests with unlimited supplies of alkali salt show that the addition of fly ash to Portland cement reduces, at least, the rate of

Table 6.5 BBB requirements for fresh concrete.

	Environment		
	Passive	Moderate	Aggressive
Water/cement max.*	No demand	0.55	0.45
Cement types	EA LA MA	EA† LA MA ‡	EA LA MA ‡
Filler content minimum (kg/m³ of mortar)	No demand	No demand	650
Max. content of PFA and MS (weight % of C+PFA+MS)	PFA+MS 35% MS 10%	PFA+MS 35% MS 10%§	PFA+MS 35% MS 10%§
Max. content of eq. Na₂O kg/m³ (exclusive content in PFA and MS)	No demand	3.0† ‡	3.0 ‡
Sand class	P	M†	A
Coarse aggregate class	P	M**	A
Max. content of acid-soluble chloride (weight % of C+PFA+MS)††	1.0	0.2	0.2
Min. air content (% of *kitmassevolumen*)‡‡	No demand	15 §§	15
Plasticiser	Required if MS is used	Required if MS is used	Required

*Silica fume is included in calculation of the water–cement ratio by an activity factor of 2.0, fly ash by 0.5.

† Class P sand is allowed in moderate environments if type EA or LA cement is used and the total equivalent Na_2O is less than 1.8 kg/m³ concrete.

‡ In moderate and aggressive environments HA cement (high-alkali) may be used and the requirement to a maximum Na_2O content omitted provided the sand is class A with the further requirements:

Max. 1.0% reactive aggregate
Max. 0.1% expansion of mortar bars after 20 weeks.

and the coarse aggregate is type A with the further requirement:

Max. 1.0% aggregate under 2500 kg/m³ density.

§ MS should not be used together with Portland fly ash cement in moderate and aggressive environments.

** The project may specify less than 1.0% aggregate under 2200 kg/m³ in order to reduce the number of pop-outs.

†† It is not allowed to add chloride to reinforced concrete. For unreinforced concrete the limit is 1.5%.

‡‡ The *kitmassevolumen* is calculated as the total concrete volume minus the volume of aggregates. The requirement may be omitted for structural elements not exposed to frost.

§§ For exposed aggregate finish, the requirement of air content may be omitted.
C, cement; PFA, fly ash; MS, silica fume, microsilica.

Table 6.6 Requirements for hardened concrete.

	Environment		
	Passive	Moderate	Aggressive
Characteristic compressive strength minimum (MPa)	15	20	35

expansion and perhaps also the ultimate expansion. Therefore, the use of the Portland fly ash cement reduces the risk of expansion due to ASR even in an open system in which alkali salts from an outside source may penetrate into a concrete structure. For the case of a moderately reactive sand, this reduction in expansion by using Portland fly ash cement may make the sand acceptable for concrete-making.

(4) In an open system the use of a low-alkali sulphate-resistant Portland cement has very little advantage over an ordinary Portland cement. The most promising results are obtained with the use of Portland fly ash cement (Figures 6.11 to 6.13).

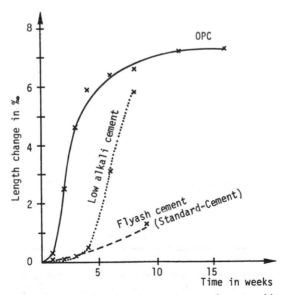

Figure 6.11 Expansion characteristics of sand–cement mortar bars stored in a saturated NaCl bath at 50°C (ordinary Portland cement compared with low-alkali sulphate-resistant cement and with Portland fly ash cement (Danish standard cement).

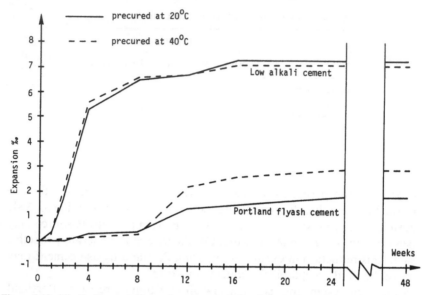

Figure 6.12 Alkali–silica test. Expansion characteristics of mortar bars of Nymølle sand together with low-alkali cement and Portland fly ash cement.

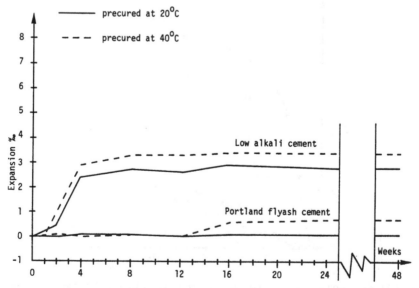

Figure 6.13 Alkali–silica test. Expansion characteristics of mortar bars of Kallerup sand together with low-alkali cement and Portland fly ash cement.

6.4. Conclusions

Denmark is a small country with many of its aggregate sources containing alkali-reactive components. This has led to extensive research programmes on ASR, on both its fundamental aspect and its avoidance in concrete structures. As regards the preventive measures, the following innovations have been made:

(1) The assumptions of a constant and uniform distribution of alkali in a concrete structure have been dropped.
(2) The environment to which a structure may be exposed has been classified as regards its severity.
(3) The acceptance criteria of aggregates, both fine and coarse, have been made very strict, and have been correlated with the environmental classes.
(4) The permissible total alkali content of a concrete mix has been correlated with the environment to which the concrete would be exposed.
(5) A fly ash Portland cement has been introduced to the Danish market.

It is hoped that as a result of the above innovations the incidence of ASR will be very much restricted in the future.

References

1. Poulsen, A. (1914) *Betons Holdbarhed* (lecture in the Institution of Danish Civil Engineers, 21 January 1914). *Ingeniøren*, **31**, 293 300 (discussion pp. 300–304, in Danish).
2. Nerenst, P. (1952) *Betonteknologiske Studier i USA—Rapport over ECA Studierejse*, 14 November 1950 to 15 February 1951 (in Danish). The Danish National Institute of Building Research, Copenhagen, Series SBI Studie No. 7.
3. Nerenst, P. (1957) *Alment om Alkali Reaktioner i Beton* (with an English summary). The Danish National Instutute of Building Research and the Academy of Technical Sciences, Committee on Alkali Reactions in Concrete, Progress Report A1, Copenhagen.
4. Idorn, G.M. (1958) *Concrete on the West Coast of Jutland, Part I*. The Danish National Institute of Building Research and the Academy of Technical Sciences, Committee on Alkali Reactions in Concrete, Progress Report B1, Copenhagen.
5. Idorn, G.M. (1958) *Concrete on the West Coast of Jutland, Part II*. The Danish National Institute of Building Research and the Academy of Technical Sciences, Committee on Alkali Reactions in Concrete, Progress Report B2, Copenhagen.
6. Jeppesen, A. (1958) *Durability and Maintenance of Concrete Structures on Danish Railways*. The Danish National Institute of Building Research and the Academy of Technical Sciences, Committee on Alkali Reactions in Concrete, Progress Report B3, Copenhagen.
7. Tovborg Jensen, A., Wøhlk, C. J., Drenck, K. and Krogh Andersen, E. (1957) *A Classification of Danish Flints etc. Based on X-Ray Diffractometry*. The Danish National Institute of Building Research and the Academy of Technical Sciences, Committee on Alkali Reactions in Concrete, Progress Report D1, Copenhagen.
8. Grey, H. and Søndergaard, B. (1958) *Flintforekomster i Danmark* (The Occurring of Flint in Denmark. With an English Summary). The Danish National Institute of Building Research and the Academy of Technical Sciences, Committee on Alkali Reactions in Concrete, Progress Report D2, Copenhagen.
9. Søndergaard, B. (1959) *Petrografisk Undersøgelse af Daske Kvartære Grusaflejringer*. (Petrographic Investigation of Quaternary Danish Gravel Deposits. With an English Summary).

The Danish National Institute of Building Research and the Academy of Technical Sciences, Committee on Alkali Reactions in Concrete. Progress Report E1, Copenhagen.

10. Haulund Christensen, K.E. (1958) *Evaluation of Alkali Reactions in Concrete by the Chemical Test*. The Danish National Institute of Building Research and the Academy of Technical Sciences, Committee on Alkali Reactions in Concrete. Progress Report H1, Copenhagen.

11. Jones, F.E. (1958) *Investigations of Danish Aggregates at Building Research Station*. The Danish National Institute of Building Research and the Academy of Technical Sciences, Committee on Alkali Reactions in Concrete, Progress Report I1, Copenhagen.

12. Bredsdorff, P., Poulsen, E. and Spøhr, H. (1966) *Experiments on Mortar Bars Prepared with Selected Danish Aggregates*. The Danish National Institute of Building Research and the Academy of Technical Sciences, Committee on Alkali Reactions in Concrete, Progress Report I2, Copenhagen.

13. Bredsdorff, P., Poulsen, E. and Spøhr, H. (1967) *Experiments on Mortar Bars Prepared with a Representative Sample of Danish Aggregates*. The Danish National Institute of Building Research and the Academy of Technical Sciences, Committee on Alkali Reactions in Concrete, Progress Report I3, Copenhagen.

14. Efsen, A. and Glarbo, O. (1960) *Experiments on Concrete Bars. Expansions During Storage in Climate Room*. The Danish National Institute of Building Research and the Academy of Technical Sciences, Committee on Alkali Reactions in Concrete, Progress Report K1, Copenhagen.

15. Trudsø, E. (1958) *Experiments on Concrete Bars. Freezing and Thawing Tests*. The Danish National Institute of Building Research and the Academy of Technical Sciences, Committee on Alkali Reactions in Concrete, Progress Report K2, Copenhagen.

16. Andreasen, A.H.M. and Haulund Christensen, K.E. (1957) *Investigation of the Effect of Some Pozzolans on Alkali Reactions in Concrete*. The Danish National Institute of Building Research and the Academy of Technical Sciences, Committee on Alkali Reactions in Concrete, Progress Report L1, Copenhagen.

17. Poulsen, E. (1958) *Preparation of Samples for Microscopic Investigation*. The Danish National Institute of Building Research and the Academy of Technical Sciences, Committee on Alkali Reactions in Concrete, Progress Report M1, Copenhagen.

18. Idorn, G.M. (1956) *Disintegration of Field Concrete*. The Danish National Institute of Building Research and the Academy of Technical Sciences, Committee on Alkali Reactions in Concrete, Progress Report N1, Copenhagen.

19. Meyer, E.V. (1958) *The Alkali Content of Danish Cements*. Meyer, E.V. (1958) *A New Danish Alkali Resistant Cement*. Andersen, J. and Ditlevsen, L. (1958) *Methods for the Determination of Alkalis in Aggregate and Concrete*. The Danish National Institute of Building Research and the Academy of Technical Sciences, Committee on Alkali Reactions in Concrete, Progress Reports F1, 2 and 3, Copenhagen.

20. Idorn, G.M. (1961) *Studies of Disintegrated Concrete, Part I*. The Danish National Institute of Building Research and the Academy of Technical Sciences, Committee on Alkali Reactions in Concrete, Progress Report N2, Copenhagen.

21. Idorn, G.M. (1961) *Studies of Disintegrated Concrete, Part II*. The Danish National Institute of Building Research and the Academy of Technical Sciences, Committee on Alkali Reactions in Concrete, Progress Report N3, Copenhagen.

22. Idorn, G.M. (1964) *Studies of Disintegrated Concrete, Part III*. The Danish National Institute of Building Research and the Academy of Technical Sciences, Committee on Alkali Reactions in Concrete, Progress Report N4, Copenhagen.

23. Idorn, G.M. (1964) *Studies of Disintegrated Concrete, Part IV*. The Danish National Institute of Building Research and the Academy of Technical Sciences, Committee on Alkali Reactions in Concrete, Progress Report N5, Copenhagen.

24. Idorn, G.M. (1964) *Studies of Disintegrated Concrete, Part V*. The Danish National Institute of Building Research and the Academy of Technical Sciences, Committee on Alkali Reactions in Concrete, Progress Report N6, Copenhagen.

25. Plum, N.M. (1961) *Alkaliudvalgets Vejledning. Foreløbig Vejledning i Forbyggelse af Skadelige Alkali-Kiselreaktioner i Beton* (in Danish). The Danish National Institute of Building Research, Copenhagen, Series SBI.

26. ASTM C-227: *Test Method for Potential Alkali Reactivity of Cement–Aggregate Combinations (Mortar Bar Method)*.

27. ASTM C-289 *Test Method for Potential Reactivity of Aggregates* (*Chemical Method*).
28. Bredsdorff, P., Idorn, G.M., Kjær, A., Plum, N.M. and Poulsen, E. (1962) *Chemical Reactions Involving Aggregate. Chemistry of Cement. Proceedings of the Fourth International Symposium, Washington 1960,* Vol. 2. National Bureau of Standards, Washington, pp. 749–806 (with bibl.). Series: National Bureau of Standards Monograph 43, Vol. 2.
29. Idorn, G.M. (1967) *Durability of Concrete Structures in Denmark.* DABI, Holte, Denmark.
30. Powers, T.C. and Steinour, H.H. (1955) An interpretation of some published researches on the alkali–aggregate reaction". *ACI Journal Proc. V* **51,** 497–516.
31. Chatterji, S. (1978) An accelerated method for the detection of alkali silica reactivities of aggregates. *Cement Concr. Res.* **8,** 647–650.
32. Chatterji, S. (1979) The role of $Ca(OH)_2$ in the breakdown of Portland cement concrete due to alkali silica reaction. *Cement Concr. Res.* **9,** 185–188.
33. Knudsen, T. and Thaulow, N. (1975) Quantitative microanalyses of alkali–silica gel in concrete. *Cement Conc. Res.* **5,** 443–454.
34. Chatterji, S. and Clauson-Kaas, N. (1984) Prevention of alkali silica expansion by using slag Portland cement. *Cement Concr. Res.,* **14,** 816–818.
35. Chatterji, S., Jensen, A.D., Thaulow, N. and Christensen, P. (1986) Studies of alkali silica reaction. Mechanisms by which NaCl and $Ca(OH)_2$ affect the reaction. *Cement Concr. Res.* **16,** 246–254.
36. Thaulow, N. and Thordal Andersen, K. (1988) Ny viden om alkalireaktioner (in Danish). *Dansk Beton* **1,** 14–19.
37. Chatterji, S., Thaulow, N. and Jensen, A.D. (1981) Studies of alkali silica reaction. 4. Effects of different alkali salt solutions on expansion. *Cement Concr. Res.* **17,** 777–783.
38. Bonzel, J., Krell, J. and Siebel, E. (1986) Alkalireaktion im Beton. *Beton* 345–348, 385–389.
39. Chatterji, S. and Jensen, A.D. (1978) A simple chemical method for the detection of alkali silica reaction. *Cement Concr. Res.* **18,** 654–656.
40. Chatterji, S., Thaulow, N. and Jensen, A.D. (1988) Studies of alkali silica reaction. 6. Practical implications of a proposed reaction mechanism. *Cement Concr. Res.* **18,** 363–366.
41. Danish Code of Practice for the Structural Use of Concrete, 3rd Edn, March 1984. Danish Standard DS 411.
42. *Basisbetonbeskrivelsen for Bygningskonstruktioner* (in Danish). Byggestyrelsen, Copenhagen, 1987.
43. Anon (1985) *Test Method TI-B52 Petrographical Investigation of Sand.* Technological Institute, Copenhagen.
44. Anon (1985) *Test Method TI-B51 Alkali Silica Reactivity of Sand* (in Danish). Technological Instutute, Copenhagen.
45. NORDTEST NT Build 295 (1985) *Sand–Alkali–Silica Reactivity Accelerated Test.* Nordtest, Finland.
46. Brotschi, J. and Mehta, P.K. (1978) Test methods for determining potential alkali–silica reactivity in cements. *Cement Concr. Res.* **8,** 191–200.
47. Fördös, Z. (1981) *Undersøgelse af Alkalikiselreaktioner i Beton Tilsat flyveaske* (in Danish). Aalborg Portland, CBL Internal Report No. 29.
48. Chatterji, S. (1978) An accelerated method for the detection of alkali–aggregate reactivities of aggregates. *Cement Concr. Res.* **8,** 647–650.
49. Justesen, C.F. and Nepper-Christensen, P. (1985) *Portlandflyveaske cement* (in Danish). Aalborg Portland, CBL Internal Report No. 36.
50. Chatterji, S. and Fördös, Z. (1985) *Effect of Flyash Addition on Alkali Silica Expansion.* Nordic Concrete Research Publication No. 4.
51. Chatterji, S. and Nepper-Christensen, P. (1987) *Evaluation of Portland Flyash cement as a preventive against alkali–silica reaction.* Nordic Concrete Research Publication No. 6.

7 Alkali–silica reaction— Icelandic experience

H. ÓLAFSSON

Abstract

Since the early 1960s there has been awareness of potential alkali–silica reaction (ASR) problems in Iceland, some aggregates being reactive and Icelandic cement containing a high amount of alkalis. Preventive measures were taken during construction of large concrete structures but not of housing. Research on ASR was carried out, so that when extensive ASR damage to houses in Reykjavík and the vicinity was established in 1979 rational preventive actions, such as using silica fume in Icelandic cement, could be taken. Since 1979 no ASR damage has been found in concrete. Much effort has been devoted to research on remedial measures on damaged houses, the last 10 years with noticeable success.

7.1 Introduction

Cement- and concrete-making in Iceland is unique in many ways. Geologically, the country is so young that calcareous raw materials are only to be found as offshore seashell deposits; and for argillaceous material rhyolitic rock must be used[1]. Because of this, Icelandic cement has a low silica content and is high in alkalis with an Na_2O equivalent of around 1.5%.

Icelandic concrete aggregates often contain much glass in their crystalline structure as a result of rapid cooling and quenching of the volcanic rocks under formation. Hence, these are frequently reactive, even if they are basaltic in composition: basalt is normally considered an inactive rock form. It must also be borne in mind that driving rains are exceedingly harsh and frequent in some parts of the country. These conditions need to be counteracted[2,3].

The three parameters governing alkali–aggregate reactions, namely high alkali content, reactive aggregates and moisture, are therefore to be found in Icelandic concretes, and alkali–silica expansion is also a well-known cause of concrete deterioration in the country. Commonly such expansions result in a network of surface cracks, which expose the surfaces to increased soaking and consequent frost actions. Annual freeze–thaw cycles in the Reykjavík area are around 80, and shifts from driving rain to frost often occur in a few hours as the low-pressure areas pass the island. Therefore the possibility of deleterious ASR in all exposed concrete must be heeded and counteracted. The principles for counteraction in Iceland are the same as elsewhere; but the harsh climatic conditions have produced solutions not as yet applied elsewhere.

The cases of deleterious reactions which have been experienced have almost entirely been in individual, privately owned houses and multistorey apartment buildings and other buildings in the Reykjavík region. This city now houses approximately 110 000 inhabitants, of the total of about 250 000 in the country.

There are a great many sources of natural pozzolana to be found in the country, like tuffs, scorias and pumices, but the most practical source for cement production is a metamorphic rhyolitic rock found in the vicinity of the cement plant[4]. Material from this source was used prior to the start of silica fume replacement, both for production of a pozzolanic cement, used in massive constructions, and as a means for suppressing alkali–aggregate expansion in ordinary Portland cement. The glassy rhyolitic rock is actually the same argillaceous material as is used in the clinker production[5].

7.2 Historical background

Since concrete is the only building material produced in Iceland, it is widely used in house-building as well as for other constructions, such as harbours, bridges and dams. In 1958 an Icelandic cement factory started production, its capacity of 130 000 tons a year fulfilling the requirement for cement in the country. The Icelandic cement is high in alkalis, its content being Na_2O eq. ~1.5%, compared with 0.8% in the cement used before that time.

Being aware of the potential danger of alkali–aggregate reaction in structures subjected to a high degree of moisture, such as harbours, dams and bridges, precautions were taken, using innocuous aggregates, imported low-alkali cement or pozzolanic cement. No serious damage due to alkali–aggregate reactions has been found in such constructions.

In house construction no such precautions were taken. This was partly because it was believed that such constructions were not vulnerable to alkali–aggregate reactions because of their low moisture content; also no knowledge existed of damage to such constructions anywhere else, and special precautions would have increased building cost.

The ASR damage in Iceland is to a great extent due to one aggregate source, Hvalfjord sand. To meet the increasing demand for houses and buildings in the Reykjavík area in the 1970s, it became feasible to exploit seabed deposited aggregates. These were used without washing, in other words they were an additional supply of alkalis to the concrete. The use of these aggregates increased steadily from 1962 to 1972 and from 1972 to 1979 they were predominant. When these aggregates were first used, mortar bar tests carried out by the Icelandic Building Research Institute (BRI) showed that they were reactive, and the municipal authorities were warned of the potential danger. However, no direct actions were taken, but a committee was established to study disintegration of concrete. This committee, which is still active, has sponsored research at the BRI concerning ASR, and this made it possible to take decisive action when in 1979 it first was clear that ASR damage to housing concrete was a serious problem.

In 1975 the Third International Conference on ASR was held in Iceland, and the focus was mainly on preventive measures. At that time no cases of serious ASR damage were known in Iceland. The next year, the first evidence was found, leading to an extensive field survey[1]. In the report published at the beginning of 1979 it was concluded that ASR damage was a major problem in housing concrete in the Reykjavík area. That same year preventive actions were taken including:

Limiting the use of the gravel source in question
Blending silica fume into the Icelandic Portland cement
Changing the criteria for reactive materials
Requesting washing of sea-dredged gravel materials.

These measures seem to have been effective, since now, 10 years later, no cases of ASR damage have been found in structures built after 1979 in spite of extensive field surveys[7–9].

The principal test method used for evaluating alkali–silica reaction is the well-known ASTM C227 mortar bar method. As an indication of the reactivity of Icelandic sands, 8 out of 14 sand types included in a research project were classified as highly reactive. These sands originated from various parts of the country. The classification criteria used were based on 1 year's expansion; less than 0.1% were classified as innocuous and over 0.2% as highly reactive[10]. The chemical composition of Icelandic ordinary Portland cement and silica fume is shown in Table 7.1.

Cores from the first confirmed case of housing seriously damage by ASR were sent for further examination and structural analyses to the Concrete Research Laboratory in Karlstrup, Denmark. In the report received in September 1976 it was concluded that the gel was high in CaO containing:

SiO_2 70–77%
CaO 19–26%
Na_2O 2–8%
K_2O 0.9–1.6%

The reactive aggregate particles contained a large amount of SiO_2, the chemical composition being:

Table 7.1 Chemical analysis of Icelandic Portland cement without any replacement of pozzolana and silica fume presently in use.

Constituent	OPC (%)	Silica fume (%)
SiO_2	20.0	91.8
Al_2O_3	4.3	0.9
Fe_2O_3	3.5	1.5
CaO	63.2	0.0
MgO	2.3	0.8
SO_3	3.2	0.5
Na_2O	1.24	0.43
K_2O	0.23	1.21
Free lime		
LOI	1.2	2.1
Insoluble residue	0.4	35.8

SiO_2 60–73%
Al_2O_3 14–16%
Fe_2O_3 10–13%

and lesser amounts of Na_2O, K_2O, CaO and TiO.

Figure 7.1 shows typical results from SEMEX microanalyses on the gel.

7.3 Field surveys

Several field surveys have been carried out in order to obtain reliable statistical information on the condition of concrete structures. The first field surveys

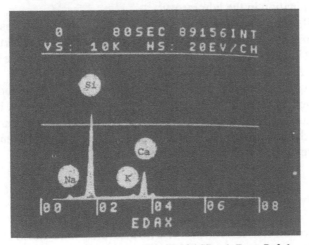

Figure 7.1 SEMEX microanalyses of ASR gel. From Ref. 6.

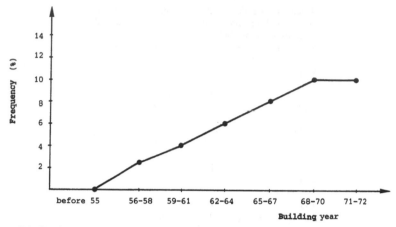

Figure 7.2 Cautious estimate of serious ASR damage in houses in Reykjavík and vicinity from a field survey in 1977–88. From Ref. 1.

concerned all types of concrete damage, but the newer ones have been limited to ASR.

The first field survey concerning houses in the Reykjavík area was carried out in 1977–78 and included 50 houses per year built in the period 1955–72. It was followed by two other field surveys in other parts of the country—one in northern Iceland, Akureyri, where the aggregates are reactive but the climate is drier with less driving rain, and the other close to the international airport, Keflavík, where the aggregates are classified as innocuous but where the climate is wet with heavy driving rain. The initial examination was visual, but cores were taken from houses with map cracking and possible ASR for further examination in the laboratory[1].

The results from these first field surveys showed that ASR was a serious and increasing problem in Reykjavík, while it was very rare in Akureyri and non-existent in Keflavík. In Figure 7.2 a cautious estimation of the frequency of extensive ASR in Reykjvík is shown.

Later surveys carried out in Reykjavík in 1981 and 1984 included newer houses as well as older ones, and in a 1987 survey only houses built after 1979 were studied. These surveys have not shown any ASR damage after 1979, when definite preventive measures were taken, but until then the ASR was widespread, as is seen in Figure 7.3[9].

7.4 Preventive measures

As mentioned in 7.3, several preventive measures were taken in 1979 to prevent ASR in new constructions. These measures will now be discussed.

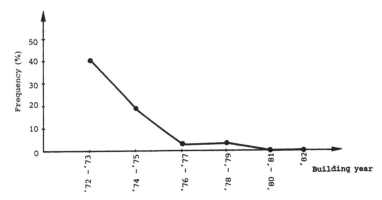

Figure 7.3 Serious ASR damage in houses in Reykjavík according to a survey in 1984. From Ref. 9.

As previously mentioned the principal test method used for evaluating alkali–silica reaction is the well-known ASTM-C227 mortar bar method. The ASTM 6-month criterion of 0.1% expansion was at first used for guidance. However, in 1979, the Icelandic Building Code demanded a stricter criterion of 0.05% in 6 months and 0.1% in 12 months. It still remains to be determined which magnitude of expansion is acceptable in this country and elsewhere.

Interest in silica fume in Iceland dates back to 1972, when the first samples of the material were tested as soon as a ferrosilicon plant was conceived in the country. At first silica fume was considered a horrifying pollutant, but it has proven in all the tests to be a very effective suppressor of alkali–silica expansion[7,11].

Figure 7.4 shows what may be termed the primary results of this endeavour. Based on these results the Icelandic building authorities enforced the strict limit on allowable mortar bar expansion mentioned above. In line with this the State Cement Works has since 1979 only produced cement intermixed with silica fume, replacement at first being 5%, but since 1983 7.5%. Research at BRI has shown that even such a small amount is highly effective in preventing deleterious expansion. Such a level of silica fume has none of the serious negative side effects, such as higher water requirement or increased drying shrinkage, associated with higher levels.

Prior to silica fume, finely ground rhyolite pozzolana was intermixed with OPC, starting in 1973 with 3% and increasing to 9% in 1975–79.

Another building code restriction introduced since 1979 is that unwashed sea-dredged aggregates are now banned. The reason for this ban is easily seen from Figure 7.5, and the explanation for this effect of seawater is that NaCl from the sea may exchange ions with $Ca(OH)_2$ liberated during the cement hydration and form NaOH, which increases the alkalinity of the concrete and its alkali reactivity[7,10–12].

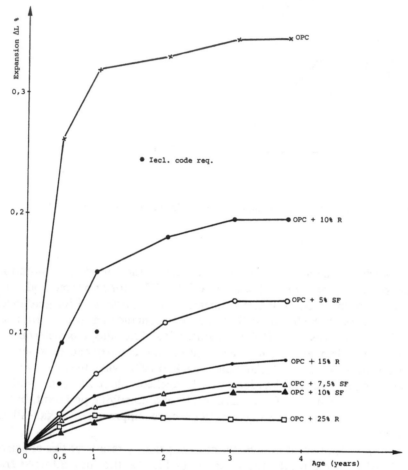

Figure 7.4 Rhyolite (R) and silica fume (SF) *vs* mortar bar expansion (Hvalfjörd sand). From Ref. 7.

7.5 Protection of ASR-damaged concrete

To repair damage caused by alkali–aggregate reaction in concrete it is essential to be able to stop the deleterious reactions. Of the influencing factors that have to be present simultaneously for the reactions to take place, the only factor that can be removed is moisture. By analysing the origin of moisture in exterior walls it has been shown that the cause of the high moisture content is driving rain combined with capillary forces of the concrete. Thus by preventing external water penetrating the concrete while keeping the surface open enough for moisture to escape it should be possible to lower the moisture

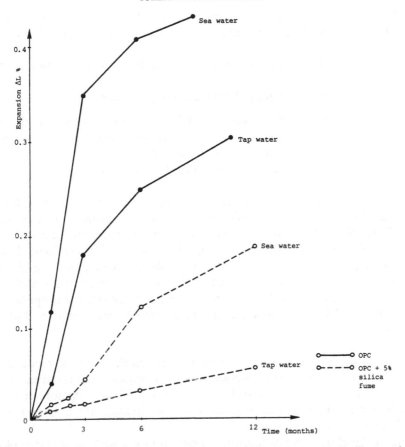

Figure 7.5 The influence of sea water on mortar bar expansion due to alkali–silica reactions (Hvalfjörd sand). From Ref. 7.

content of the concrete and thereby reduce the rate of reactions or stop them altogether.

A number of full-scale repair projects were carried out on exterior walls of ASR-damaged houses in the period 1979–82, aimed at lowering the moisture content in the concrete and stopping the deleterious reactions. Among the methods used were the following[13–15]:

Ventilated panels or claddings with and without insulation.
Rendering applied on insulation.
Impregnation of silicones and silanes.

The moisture content of the concrete was measured by three different methods:

(a) By use of an electrical moisture meter (Protimeter Concrete Master) which registers moisture content close to the concrete surface (Figure 7.6).
(b) By drilling cores without cooling water and drying and weighing in the laboratory.
(c) By use of relative humidity (RH) meters (e.g. Vaisala).

The RH measurements were unstable and not considered reliable and were therefore not used. The protimeter shows relative changes in moisture but not necessarily always exact absolute moisture content.

In the following the main results obtained will be presented:

Ventilated claddings. Twelve houses with degree of damage ranging from moderate to extensive were included in the project. The principal results were as follows: the concrete moisture decreases considerably, regardless of the use of insulation, if care is taken to remove impermeable layers of paint from the concrete surface. Typical test results are shown in Figures 7.7 and 7.8. Figure 7.7 shows moisture variation in the concrete close to the surface, measured by the Concrete Master. The readings are adjusted for temperature effect. Figure 7.8 shows moisture content in drilled cores before and after saturation.

Renderings on insulation. Two basic types were used, hydraulic renderings with steel reinforcement and so-called thin renderings with reinforcement of fine-mesh fibreglass fabric. Two houses were treated with the former type and three with the latter type. A gradual decrease in moisture in the concrete has been registered in all cases.

Figure 7.9 shows typical moisture variation in the surface (0–5 cm) of a concrete wall, and Figure 7.10 shows moisture in the cross-section of the wall measured in drilled cores. No accumulation of moisture in the insulation was registered.

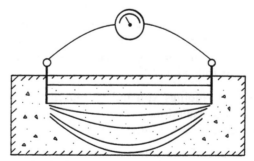

Figure 7.6 Moisture meter (Protimeter Concrete Master).

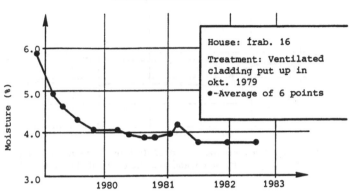

Figure 7.7 Moisture development in concrete surface (0–5 cm) after ventilated cladding was erected. From Ref. 13.

Impregnation with silicones and silanes. Making the surface of the concrete hydrophic by applying silicones or silanes to the walls has given the most promising results. In all experiments, using different types of silicones/silanes, a marked decrease in moisture has been registered, even on surfaces with fine map cracking. In cores drilled with a dry method a dry alkali gel has been observed, indicating that the moisture content has decreased below the critical point. Figure 7.11 shows typical results.

In the last 5 years all described methods have been used for renovation and protection of ASR-damaged houses with seemingly good results. The

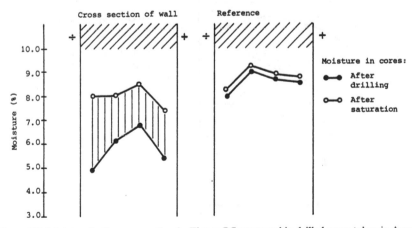

Figure 7.8 Moisture in the same wall as in Figure 7.7 measured in drilled cores taken in August 1982. From Ref. 13.

Figure 7.9 Moisture development in concrete surface (0–5 cm) after thin rendering was applied on $1\frac{1}{2}$-inch polystyrene insulation and in a reference wall. From Ref. 14.

use of silane impregnation is most interesting since, if it is applied before extensive damage is visible, it appears to be possible to prevent or at least delay further damage. An important factor in Iceland is the fact that when the moisture saturation is reduced below 90% freeze–thaw damage will be reduced considerably. Laboratory tests and practical experience have also proved that monosilane impregnation of concrete improves the cohesion of the acrylic paints which are commonly used in Iceland.

When considering the great effect of silanes on the moisture content of exterior walls one should bear in mind the following[15]:

(1) The concrete in question commonly has a water–concrete (w/c) ratio of

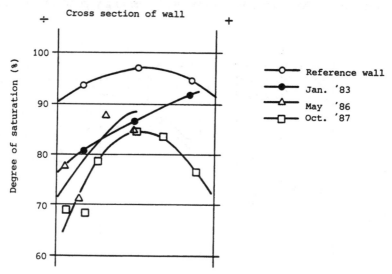

Figure 7.10 Moisture in the same wall as in Figure 7.9 measured in drilled cores taken at different times. From Ref. 14.

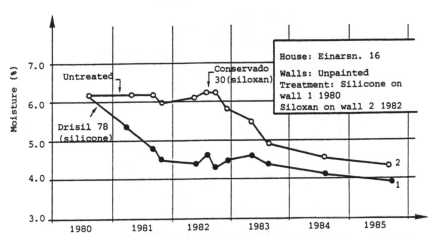

Figure 7.11 Moisture development in concrete after impregnation with silicone (Drisil 78) and siloxane (Conservado 30). From Ref. 15.

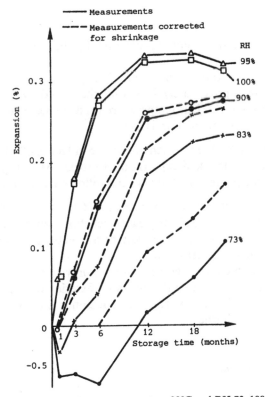

Figure 7.12 Expansion of mortar bars at temperature 38°C and RH 73–100%. From Ref. 16.

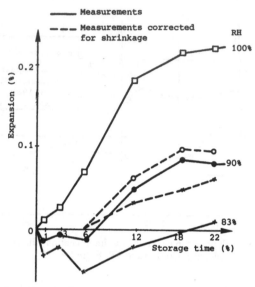

Figure 7.13 Expansion of mortar bars at temperature 23°C and RH 83–100%. From Ref. 16.

0.55–0.7 and air entrainment of 2–3% and thus has a relatively high water absorption.

(2) There is a temperature gradient out through the wall, the inside temperature being approximately 23°C and the outside temperature in the range −10 to +15°C. This temperature gradient seems to be of importance when silane impregnation is used.

A research project has been carried out on the influence of relative humidity and temperature on the alkali expansion of mortar bars. Figures 7.12 and 7.13 show the principal results. It draws the following conclusions:

(1) By reducing the relative humidity from 100% to below 90% the total expansion caused by alkali–aggregate reaction can be reduced[16]. The lower the RH, the greater is the reduction in expansion. Decreased rate of expansion delays deterioration.

(2) Expansion is a function of temperature. By preventing high temperatures in concrete, e.g. by using light colours on exterior walls instead of dark ones or by shielding them from the sunshine, expansion and rate of expansion are diminished.

7.6 Current development

At present, the main activities concerning ASR are as follows:

(1) Field surveys. Every third year a field survey is undertaken on houses built after 1979. Cores are taken from all houses with map cracking for further examination in the laboratory. The cores are tested not only for ASR but also for air entrainment and freeze–thaw resistance.

(2) Research. One research project is under way dealing with the interaction between ASR and freeze–thaw resistance. This project started in 1986 and will be completed in 1990. Another research project deals with petrographic classification of gravels concerning ASR.

(3) Remedial measures on ASR damaged concrete. This includes following up on the projects described in 7.5 as well as testing new methods and materials.

7.7 Future prospects

It is a general opinion, based on laboratory tests and practical experience, that ASR will not cause extensive damage in concrete produced after 1979. Isolated cases of ASR damage cannot be totally excluded since the amount of silica fume in the cement is theoretically not enough to bind all the OH^- in the concrete, and traces of alkali–silica gel have been observed in a few cases.

In the near future the main emphasis will be on the following:

Quality control of concrete production and concrete-making materials.
Increased use of structural analyses of concrete; until now such analyses have been carried out abroad.
Remedial measures on ASR-damaged concrete.

References

1. Kristjánsson, Ríkharður (1979) *Steypuskemmdir-Ástandskönnun* (an outline in Icelandic). The Icelandic Building Research Institute.
2. Gudmundsson, G. and Ásgeirsson, H. (1975) *Some Investigation on Alkali Aggregate Reaction, Cement and Concrete Research*, Vol. 5. New York, pp. 211–220.
3. Gudmundsson, G. (1975) *Investigation on Icelandic Pozzolans*, Symposium on AAR-Preventive Measures. The Icelandic Building Research Institute.
4. Sæmundsson, K. (1975) *Geological Prospecting for Pozzolanic Materials in Iceland, Symposium on AAR-Preventive Measures.* The Icelandic Building Research Institute.
5. Gudmundsson, G. (1971) *Alkali Efnabreytingar í Steinsteypu.* The Icelandic Building Research Institute.
6. Thaulow, N. (1976) *Undersögelse af Beton Borekærne fra Reykjavík.* Aalborg Portland.
7. Ásgeirsson, H. (1986) Silica fume in cement and silane for counteracting of alkali–silica reactions in Iceland. *Cement Concr. Res.* **16**, 423–428.
8. Ólafsson, H. and Helgason, Th. (1983) *Alkalívirkni Steypuefna á Íslandi og Áhrif Salts og Possolana á Alkalívirkni í Steinsteypu.* The Icelandic Building Research Institute.
9. Kristjánsson, R., Ólafsson, H., Þórðarson, B., Sveinbjörnsson, S. and Gestsson, J. (1979–1987). *Field Surveys of Houses.* The Icelandic Building Research Institute.

9a. Ólafsson, H. (1983) Repair of vulnerable concrete. *Proceedings of the Sixth International Conference on Alkalis in Concrete.*

10. Ólafsson, H. and Thaulow, N. (1983) Alkali–silica reactivity of sands, comparison of various test methods. *Proceedings of the Sixth International Conference on Alkalis in Concrete.*

11. Gudmundsson, G. and Ásgeirsson, H. (1979) *Pozzolanic Activity of Silica Dust, Cement and Concrete Research*, Vol. 9. Pergamon Press, New York, pp. 249–252.

12. Ólafsson, H. (1982) *Effect of Silica Fume on Alkali–Silica Reactivity of Cement, Condensed Silica Fume in Concrete.* Division of Building Materials, The Institute of Technology, Trondheim, Norway.

13. Ólafsson, H. and Iversen, K. (1981) *Viðgerðir á Alkalískemmdum í Steinsteypu.* The Icelandic Building Research Institute.

14. Ólafsson, H., Iversen, K and Ankerfeldt, P. (1988) *Udvendig Facadeisolering—Demonstration í Island.* The Icelandic Building Research Institute.

15. Ólafsson, H. and Gestsson, J. (1988) *Áhrif Vatnsfæla á Steinsteypu.* The Icelandic Building Research Institute.

16. Ólafsson, H. (1986) The effect of relative humidity and temperature on alkali expansion of mortar bars. In: *Concrete Alkali–Aggregate Reactions.* Noyes Publications, Far Ridge, NJ, pp. 461–465.

17. Idorn, G.M. (1987) *Concrete Durability in Iceland*, notes from a visit, G.M. Idorn Consult A/S.

8 Alkali–silica reaction— Canadian experience

P.E. GRATTAN-BELLEW

Abstract

The first case of alkali–silica reaction (ASR) in Canada was discovered in the mid 1950s, by which time most of the investigative techniques currently used had already been developed. Three types of reactive aggregates are recognised in the current Canadian Standards:

(1) Alkali–silica reaction, occurring with opal, chert, chalcedony, volcanic glass, cristobalite and tridymite.
(2) Slow/late-expanding alkali silicate/silica reaction, occurring with grey-wacke, argillite, quartz arenite, phyllite, some granitic rocks and gneisses.
(3) Alkali–carbonate reaction, occurring with some dolomitic limestones.

Only types (2) and (3) will be discussed in this chapter.

New test methods for determining the potential reactivity of aggregates are described; the most promising is the accelerated mortar bar test in which the length change of mortar bars stored in 1 M NaOH at 80°C is monitored for 2 weeks. Current investigations into the development of methods for evaluating the effectiveness of supplementary cementing materials (pozzolans) with different types of aggregates are also discussed.

8.1 Historical introduction

Alkali–silica reactivity was first described in the USA by Stanton in 1940[1], but a gap of over a decade was to elapse before, in 1953, an investigation was commenced into what proved to be the first documented case of ASR in Canada in a bridge in Montreal, Quebec, by Swenson[2]. The description of the investigation by Swenson is of considerable interest because it shows that by the late 1950s the techniques for investigating the cause of deterioration of concrete which are used today had already been established; these include drilling concrete cores from damaged structures, measurement of pulse velocity, making polished and thin sections from the cores, and measurement of residual strength of the cores. The observation of gel filling cracks and voids in the concrete, the presence of cracks in aggregate particles and in the paste and the development of reaction rims around reacted particles had already been established as evidence of ASR in the concrete.

Samples of aggregates thought to have been used in the construction of the bridge deck were evaluated for alkali expansivity in concrete prisms which were stored at 55°C and 100% relative humidity (RH). The concrete from the north deck of the bridge in Montreal expanded by 0.082% in 2 years. If the concrete prism test results were evaluated by the criteria given in the current Canadian Standard[3], the aggregate used in the north deck would be classed as reactive as the expansion is over the 0.04% limit specified in Table 8.1. It should, however, be noted that the test was run at 55°C and not at 38°C as specified in the current Canadian Standard; the increased temperature may have enhanced the expansion. It is interesting to note that expansion and deterioration of the bridge in Montreal occurred despite the fact that the alkali content of the cements used at the time of its construction was about 0.61% Na_2O eq., a level which would normally be considered low enough to prevent expansion due to ASR.

Following the initial discovery of alkali reactivity in Montreal, an investigation was carried out into the potential reactivity of cherty limestones of the Trenton Formation in the Montreal area. Chauret[4] concluded that, based on the ASTM Specifications C33-54[5], the limestone was non-reactive; however when his results are evaluated according to the specifications in the current Canadian Standard[3], the aggregate is designated as reactive because, although expansion was less than 0.1% at 6 months, the trend of the expan-

Table 8.1 Suggested maximum expansion values for various test methods for alkali–aggregate reactivity. From CAN3-A23.1 Appendix B.

Prism exposure class test (A23.1-M77, Table 7)	Concrete prism expansion test (A23.2-14A)	Mortar bar expansion test (ASTM C227)	Accelerated concrete expansion (A23.2-14A, Clause 2.5)
A, B, C	0.01% at 3 months* 0.025% at 1 year	0.05% at 3 months† 0.10% at 6 months‡	0.04% at 1 year
D	0.04% at any age (1 year)	0.05% at 3 months 0.10% at 6 months	0.075% at 1 year

* In critical structures such as those used for nuclear containment, a maximum expansion limit of 0.02% at 1 year is recommended.

† As in the above note except that the limits are 0.05% at 6 months and 0.10% at 1 year.

‡ It is recommended that, when used to evaluate carbonate rocks containing chert, the alkali content of the cement used to make the mortar bars be increased to 1.25% by the addition of alkali to the mix water. This is necessary because some limestones containing finely divided chert and assessed as acceptable on the basis of the mortar bar test rsults, ASTM C227, caused deterioration in field concretes.

sion was still upward at the end of the experiment (Figure 8.1). The results of recent mortar bar experiments by Fournier *et al.*[6] for a similar Trenton limestone from Quebec City are also shown in Figure 8.1. Extensive cracking of concrete structures in Quebec City confirms the reactivity of the aggregate. It is unfortunate that the reactivity of the aggregate from Montreal was not identified in the late 1950s because this could possibly have resulted in precautionary measures being taken, which might have prevented much of the concrete deterioration due to ASR which is now common in the Montreal and Quebec City areas.

Apart from one investigation involving ASR[7] most of the research done in Canada from the late 1950s until the mid 1960s involved the alkali–carbonate reaction[8,9]. The alkali–carbonate rock reaction occurs in dolomitic limestones which are found in the Kingston region of south-eastern Ontario. The reaction is thought to involve dedolomitisation and the subsequent expansion of clay minerals[10]; a detailed discussion of it is outwith the scope of this chapter.

In the mid 1960s an extensive investigation was launched into the potential reactivity of aggregates in Nova Scotia and the adjacent parts of New Brunswick (Figure 8.2), where concrete deterioration of highway bridges had been observed[11]. During the course of this investigation, many new rock types were added to the list of potentially reactive aggregates[12–14].

From the mid 1970s onwards there was a large increase in the number of publications on alkali–aggregate reactivity (Figure 8.3) paralleled by an increase in the number of reports of structures affected by the reaction. Examples of concrete deterioration due to alkali–aggregate reaction have

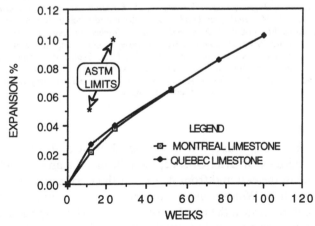

Figure 8.1 Expansion of mortar bars in test ASTM C227 containing silicious limestones from Montreal and Quebec City.

Figure 8.2 Map of Canada showing locations of sources of documented reactive aggregates. 1, *Chilliwack, BC*, acid volcanic rock in gravel deposits; 2, *Near Calgary, Alberta*, metaquartzite rock and chert; 3, *South Saskatchewan River, Saskatchewan*, opaline shale; 4, *Nelson River, Manitoba*, dolomitic limestone (alkali–carbonate reaction); 5, *Sioux Lookout, north-western Ontario*, chert in gravels; 6, *Wawa, Ontario*, silicified volcanic glass and tuff from gravel; 7, *Sudbury area of Ontario*, argillite, greywacke, quartz arenite and quartz sandstone from gravel deposits; 8, *south-western Ontario*, potentially reactive cherts; 9, *Windsor, Ontario*, chert in fine gravel; 10, *Ottawa area, Ontario*, silicious limestone; 11, *Kingston and Cornwall, Ontario*, alkali–carbonate reaction; 12, *Montreal area, Quebec*, sandstone and silicious limestone; 13, *Quebec City, Quebec*, silicious limestone; 14, *Beauceville region, Quebec*, andesitic tuff; 15, *James Bay region of Quebec*, biotite gneiss; 16, *southern New Brunswick*, gravel deposits containing quartz arenites, argillites, greywackes and volcanic rocks; 17, *Nova Scotia*, quartzites, greywackes, phyllites quartz arenites, argillites and rhyolite from several areas; 18, *eastern Newfoundland*, silicified volcanic rock in gravel deposits; 19, *Alert, Ellesmere Island*, greywacke and chert (silicified volcanics).

now been found from Chilliwack in the south-west of British Columbia, through the Prairies[7], Ontario, Quebec, New Brunswick and Nova Scotia to Newfoundland in the east and from Alert[15] at the northern tip of Ellesmere Island in the north to Windsor in south-western Ontario (Figure 8.2).

The apparent increase in the number of structures showing damage due to alkali–aggregate reaction both in Canada and in other countries is probably the result of a number of factors, such as changes in the cement manufacturing

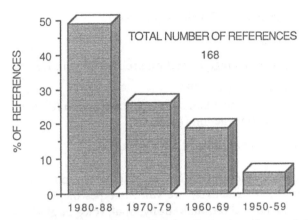

Figure 8.3 Chart showing increase in the number of publications on all types of alkali–aggregate reactivity from 1950 to 1988. Data from Ref. 9.

process, increased alkali content of the cement, increased fineness of some cements, increased cement contents of some concretes, the now widespread use of de-icing salts on highways and, last but not least, an increased awareness of the problem of alkali–aggregate reactivity and a realisation that good concrete should not crack after 5 or 10 years. Many of the structures now showing distress due to ASR were built in the late 1960s to early 1970s, a period when the average alkali content of cements in eastern Canada reached its maximum (Figure 8.4). The use of low-alkali cements in western Canada has resulted in few cases of ASR; even when potentially reactive cherty gravels have been used as aggregate in concrete.

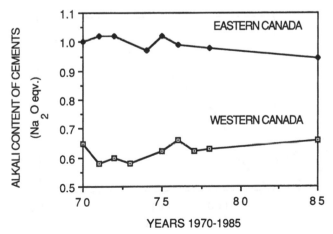

Figure 8.4 Variation in alkali levels in cements in eastern and western Canada from 1970 to 1985. Data from Canadian Portland Cement Association.

8.2 Types and distribution of ASR aggregates

8.2.1 *Types of alkali reactivity*

Three types of alkali–aggregate reaction are identified in Ref. 3:

(1) Alkali–silica reaction (ASR)
(2) Slow/late-expanding alkali silicate/silica reaction (SLASSR)
(3) Alkali–carbonate reaction (ACR).

Only types (1) and (2) are relevant to this chapter.

8.2.1.1 Alkali–silica reaction (ASR). The reactive components of aggregates exhibiting this type of reactivity are various types of poorly crystalline or metastable silica such as chert, chalcedony, opal, volcanic glass, some artificial glasses, tridymite and cristobalite. The term alkali–silica is rather too restrictive as it does not encompass volcanic and synthetic glasses which are silicates, however the term alkali–silica was retained in the standard because it is the one that was used by Stanton in 1940[1] and is widely accepted.

Some alkali–silica reactive materials such as opals and cherts exhibit the pessimum effect[16], which results in maximum expansions in mortar bars or concrete prisms occurring when the reactive component is present in small amounts, typically 2–10% of the aggregate.

Mortar bars made with alkali–silica reactive aggregates and stored under laboratory conditions typically show a rapid onset of expansion followed by a gradual falling off in the rate of expansion after about 1 year (Figure 8.5).

8.2.1.2 Slow/late-expanding alkali–silicate/silica reaction (SLASSR). The term slow/late-expanding alkali–silicate/silica reaction was coined to describe

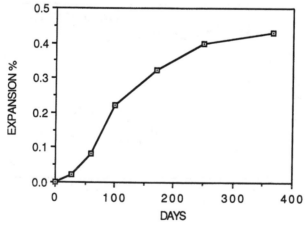

Figure 8.5 Expansion of mortar bars made with silicious limestone reference aggregate from Ottawa. From Ref. 17.

those silicious aggregates which caused delayed onset of expansion in concrete prisms stored at 38°C and 100% RH in the Canadian Standard test, Can3-A23.2-14A. Typical expansion curves for late-expanding silicious aggregates are shown in Figure 8.6.

Not all slow/late-expanding alkali–silicate/silica reactive aggregates cause a 3-month delay in the onset of expansion in concrete test prisms; some begin to expand after 20 or 30 days. Rocks which exhibit delayed expansion in concrete test prisms are, typically, quartz arenites, quartz sandstones, grey-wackes, argillites, phyllites and gneisses in which there is no evidence for the existence of typical alkali–silica reactive materials, cryptocrystalline silica or volcanic glass. The reactive component in these rocks is thought to be microcrystalline quartz[18]. The slow/late-expanding alkali–silicate/silica reactive aggregates commonly show less expansion in laboratory tests than alkali–silica reactive aggregates; typical expansions were tabulated by Grattan-Bellew[19].

The term silicate/silica was used to describe the slow/late-expanding aggregates so as to include aggregates such as some phyllites which contain micro-crystalline quartz and also vermiculite-like minerals, which Gillott et al.[18] showed exfoliate in alkaline solution. It is possible, however, that despite the exfoliation of vermiculite expansion of concrete may be due to reaction of microcrystalline quartz.

The late onset of expansion of concrete made with slow/late-expanding alkali–silicate/silica reactive aggregates occurs only in samples stored under laboratory test conditions; in the field, cracking is observed after 4 or 5 years,

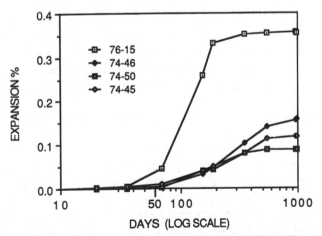

Figure 8.6 Typical expansion graphs of slow/late-expanding alkali–silicate/silica reactive aggregates in the concrete prism test CSA A23.2-14A, run at 38°C and 100% RH, showing late onset of expansion which is typical of these types of aggregates. Samples: 76-15, argillite; 74-46, quartz wacke; 74-45, feldspathic quartzite; 74-50, quartz arenite.

which is about the same time as it takes for cracking to occur in concrete containing alkali–silica reactive aggregates such as silicious limestone.

8.2.2 *Discussion*

Silica is the dominant reactive component in both ASR and SLASSR aggregates, but in some acid volcanic rocks which exhibit ASR the reactive component is volcanic glass, a silicate. The main difference between the two types of reactive aggregates, ASR and SLASSR, is that in the first most of the material, e.g. opal, is soluble in alkali, while in the second only a very small quantity is soluble; this is reflected in the relatively low amounts of dissolved silica observed when SLASSR aggregates are tested in ASTM C289, the chemical method. The low amount of soluble silica in SLASSR aggregates may also account for the late onset of expansion of concrete prisms stored at 38°C and 100% RH.

8.2.3 *Distribution of reactive aggregates in Canada*

Potentially reactive aggregates are of widespread distribution in many parts of Canada (Figure 8.2). The apparent confinement of reactive aggregates to a zone close to the border with the USA is the result of this being the most populated area and hence where most of the aggregate is used.

8.2.3.1 *Volcanic rocks.* Silicified volcanic rocks are found in gravel deposits in southern British Columbia (BC). The only documented case of concrete deterioration due to ASR in BC occurred at Chilliwack[20]. The gravels in the Chilliwack area typically contain between 20 and 40% volcanic rocks[21]. Potentially reactive silicified volcanics are found in gravel deposits near Wawa on the north shore of Lake Huron, but there are no documented cases of deterioration of concrete containing these gravels. The rock is of varied composition, consisting of a silicified tuff composed of feldspar porphyroblasts in a matrix of microcrystalline quartz, silicified, altered andesite in which phenocrysts of altered feldspar are distributed in a fine devitrified matrix and silicified pitchstone. The pitchstone consists of fine-grained microcrystalline to cryptocrystalline quartz (Figure 8.7).

Rhyolitic tuff occurs in a quarry in Beauceville, located some distance south of Quebec City. Use of this aggregate has caused severe deterioration of the nearby Sartigan Dam[22]. The tuff varies from a fine-grained rock consisting of a devitrified glassy matrix in which occur angular to subrounded quartz fragments, to a medium coarse-grained porphyritic material containing angular fragments of quartz, feldspar and sometimes shales and siltstones. The reactive material is probably the fine-grained matrix which consists predominantly of microcrystalline to cryptocrystalline quartz.

The volcanic rocks are found in gravels in Newfoundland; these are of

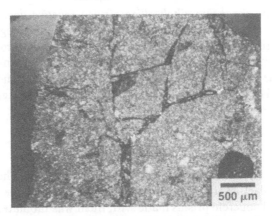

Figure 8.7 Optical micrograph of thin section of devitrified pitchstone showing characteristic curved cracks (C).

variable composition. There is a considerable amount of devitrified glass which appears in thin section as very featureless fine-grained material similar to that shown in Figure 8.7. Andesite also occurs in the gravels. Expansion of a concrete prism made with a gravel containing volcanic rocks is shown in Figure 8.8. Cracking due to ASR is observed in several bridges in eastern Newfoundland. The concrete in the bridges was made with a gravel aggregate containing volcanic rocks.

The silicified tuff and devitrified volcanic glass referred to as cherts[15] occur at Alert, Ellesmere Island. They are broadly similar to those from Newfoundland.

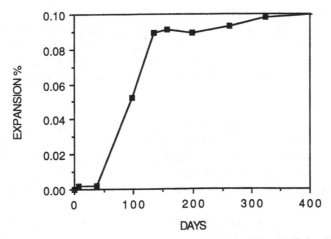

Figure 8.8 Expansion of concrete prism made with a gravel containing silicified volcanics from Come-By-Chance, Newfoundland.

8.2.3.2 *Slow/late-expanding alkali–silicate/silica reactive aggregates.* Rocks falling into the slow/late-expanding alkali–silicate/silica category are widespread in the Precambrian areas of northern Manitoba, Ontario and Quebec, and comprise much of the Palaeozoic metamorphosed sediments which occur in the Maritime provinces of eastern Canada. Extensive studies of the reactivity of slow/late-expanding alkali–silicate/silica reactive aggregates were made in two areas:

(a) Sudbury, Ontario
(b) Nova Scotia.

(a) *Potentially reactive aggregates from Sudbury, Ontario.* The composition of potentially reactive rocks in the Sudbury gravels and the expansion of mortar bars and concrete prisms made with them was tabulated by Grattan-Bellew[19,23]. A stockpile of reactive gravel from Sudbury has been made for use as a reference material. It is available from the Ministry of Transportation of Ontario[17]. The gravels are derived from metasediments of Huronian age which outcrop to the north and southwest of Sudbury[24]. Potentially reactive aggregates in the gravels consist predominantly of argillites, quartz arenites, quartz wackes and feldspathic quartzites. Optical micrographs of four typical potentially reactive aggregates from the Sudbury gravel are shown in Figure 8.9, and graphs of the expansion of concrete prisms made with these aggregates are shown in Figure 8.6; expansion of the four samples is roughly proportional to their content of microcrystalline material (Figure 8.10). The microcrystalline material is composed of crystals of quartz and clay with a grain size of 2–5 μm (Figure 8.9). The clay minerals are predominantly illite and chlorite.

(b) *Potentially reactive aggregates from Nova Scotia.* Greywackes and quartzites are the dominant reactive rocks in the aggregates which were studied by Duncan *et al.*[12]; phyllites and argillites comprise only a small percentage of the gravels. Expansion curves for concrete prisms are similar to those observed with the Sudbury aggregates, with most of the reactive aggregates showing expansions in the range of 0.044–0.186% after 2 years of moist storage at 38°C. The greywackes are similar in appearance to the quartzwacke shown in Figure 8.9.

8.2.3.3 *Granitic rocks.* Some granite aggregates have been found to cause deterioration of concrete. A viaduct in Toronto shows moderately severe cracking which is presumed to be due to alkali–aggregate reaction[17]. The reactivity of some granite is thought to be due to the occurrence of microcrystalline quartz along the boundaries of the large quartz grains[25].

8.2.3.4 *Chert.* The mineralogical composition of cherts varies from very fine-grained opaline silica, through cryptocrystalline silica and chalcedony to microcrystalline quartz. Cherts containing opaline silica are thought to be

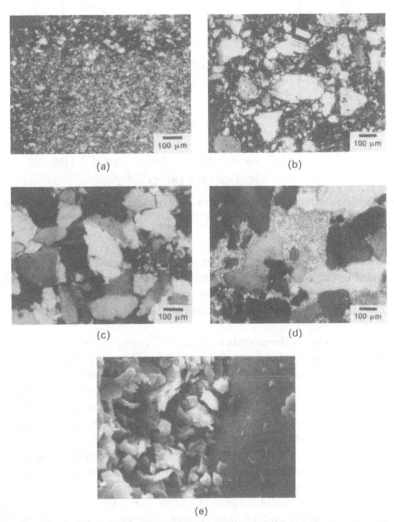

(a)   (b)   (c)   (d)   (e)

Figure 8.9 Micrographs of aggregates from Sudbury, Ontario. (a) Thin section of argillite 76-15 comprising predominantly microcrystalline quartz and clay minerals. (b) Thin section of quartz wacke 74-46, consisting of quartz grains (Q) and feldspar (F) in a matrix of micro-crystalline quartz and clay minerals. (c) Feldspathic quartzite 74-45 consisting predominantly of quartz grains (Q), feldspar (F) and patches of microcrystalline quartz (M). (d) Thin section of quartz arenite 74-50 consisting of large quartz grains (Q) and a small amount of micro-crystalline quartz (M). (e) SEM micrograph of microcrystalline quartz and clay minerals (M) and a large quartz grain (Q).

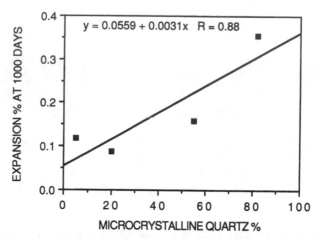

Figure 8.10 Graph showing correlation between microcrystalline quartz content of aggregates shown in Figure 1.6, and expansion of concrete prisms containing these aggregates.

the most reactive[26], however, Kneller *et al.*[27] concluded that the reactivity was related to the pore structure of the chert. There is probably not sufficient information relating expansion of concrete containing chert to its micro-structure to permit accurate prediction of the reactivity of chert from a petrographic evaluation.

Cherts are found in the gravels in western Canada in British Columbia[20] and the adjacent parts of Alberta[28] (Figure 8.2). These cherts are potentially expansive but no case of concrete deterioration involving them has been recorded; this may be because of the use of low-alkali cement in this region.

Palaeozoic cherts occur in south-western and northern Ontario (Figure 8.2). The deterioration of a bridge near Sioux Lookout in northern Ontario due to alkali–silica reactivity has been ascribed to the presence of a small percentage of leached chert in the gravel aggregate[17]. Optical micrographs of some chert from northern Ontario are shown in Figure 8.11. The chert consists predominantly of very fine-grained, featureless silica with bands of chalcedonic silica. Some fragments of the chert are evidently silicified limestones, as bivalves and silicified sponge spicules are visible in the thin sections. Examination of the chert in the scanning electron microscope (SEM) shows that it is composed of irregularly shaped silica particles of 1–4 μm grain size (Figure 8.11). X-ray diffraction analysis showed a strong diffraction maximum at 0.55 nm resulting from amorphous silica and weak reflections due to the presence of a small quantity of quartz.

Concrete deterioration has also been observed in highway structures in Windsor in south-western Ontario, where the fine aggregate contains about 5% leached chert[17].

Chert of Precambrian age is found in northern Ontario, where it is associ-

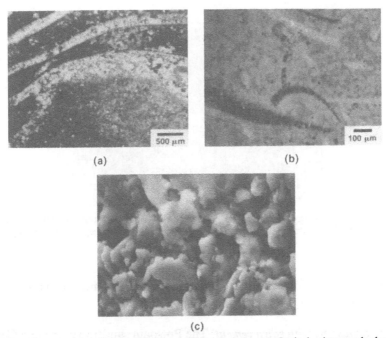

(a)   (b)

(c)

Figure 8.11 Micrographs of chert from northern Ontario. (a) Optical micrograph showing cryptocrystalline silica (C) and amorphous silica (A). (b) Optical micrograph of fine-grained chert derived from limestone, showing silicified bivalves (B) and sponge spicules (S). (c) SEM micrograph of chert illustrated in (b), showing that it is composed of irregularly shaped silica particles in about 5 μm grain size.

ated with iron formations. One case of concrete deterioration due to the use of waste rock containing Precambrian chert from an iron mine was recorded by Rogers[17].

8.2.3.5 *Silicious limestone.* Silicious limestones of Middle Ordovician age are found in eastern Ontario and southern Quebec. A stockpile of reactive cherty limestone from Ottawa has been made for use as a reactive reference aggregate; it is available from the Ministry of Transporation of Ontario[17]. The silicious limestone which belongs to the Trenton group[29] outcrops in a narrow band stretching from Ottawa south-eastwards to Montreal and then east to Quebec City. The expansiveness of the limestone in mortar bars varies with the percentage of insoluble residue in the rock and with the facies to which it belongs[6]. The limestone from Ottawa contains 8% insoluble residue which is comprised of microcrystalline silica and clay minerals[30]. In thin section the reactive limestone is of variable appearance depending upon which horizon it belongs to. The rock varies from a moderately coarse-grained fossiliferous biosparite containing about 5% insoluble residue to a very fine-grained micrite with some fossil fragments; it contains 5–15% insoluble

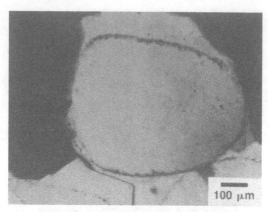

Figure 8.12 Optical micrograph of quartz grain from Potsdam sandstone showing secondary quartz overgrowth (R) surrounding original grain. A band of inclusions separates the overgrowth from the centre of the grain.

matter. The expansion of mortar bars containing this limestone is shown in Figures 8.5 and 8.15.

8.2.3.6 *Sandstone with silica cement.* The Potsdam sandstone of upper Cambrian age outcrops in the vicinity of the St Lawrence Seaway just west of Montreal. Concrete containing Potsdam sandstone was used in the construction of a number of highway structures, bridges over the seaway[31] and in dams[32]. The sandstone consists of angular to subrounded quartz grains with either calcite or silica cement; only the rock with the silica cement is reactive. In the reactive rock, many of the quartz grains are surrounded by secondary quartz overgrowths[33] which are in optical continuity with the grains. The interface between the quartz grains and the overgrowths is marked by a layer of inclusions (Figure 8.12). When slices of the sandstone are immersed in alkali the quartz overgrowths are preferentially dissolved, indicating that they have a higher free energy than the centre of the grains; this may be the result of some lattice misfit between the grains and the overgrowths. A similar potentially reactive quartz sandstone occurs at Mount Wilson near Calgary, Alberta (Figure 8.2), but mortar bar test results[30] show the rock to be only marginally expansive; there is no record of its use in concrete.

8.3 Canadian Standards relating to alkali–silica reactivity

8.3.1 *Introduction*

The 1986 supplement to the portions of the standard relating to alkali-aggregate reactivity incorporates the results of recent research. The Canadian

(CSA) standard is in two parts[3], the first, CAN3-A23.1, *Concrete Materials and Methods of Concrete Construction*, specifies the types of reactivity that are found in Canada and the methods to be used in the evaluation of concrete aggregates;, the second, CAN3-A23.2, *Methods of Test for Concrete*, specifies the test procedures to be used. The major change incorporated into the 1986 supplement to the Canadian Standard is the inclusion of flow diagrams which outline the procedure to be followed in the evaluation of an aggregate. There are three flow diagrams, B1, B2 and B3. In diagram B1, the general procedures to be followed starting with field investigation and sampling of the aggregate source are outlined (Figure 8.13). Aggregates may be accepted based on field performance alone, provided that the proposed concrete mix design is not significantly different from that used successfully in the past. If an aggregate has not been used previously, and provided it satisfies the physical requirements, e.g. frost resistance, the next step in its evaluation is a petrographic investigation to determine to which type of potentially reactive aggregate the rock belongs, so that appropriate tests can be selected. The importance of the petrographic investigation cannot be overemphasised because if inappropriate tests are made an incorrect diagnosis may result. The appropriate test protocols are outlined in flow diagrams B2 and B3. The protocol for investigation of alkali–carbonate reactive aggregates, which lies outside the scope of this chapter, is outlined in diagram B2. The procedure to be followed in assessing the potential reactivity of alkali–silica and slow/late-expanding alkali–silicate/silica aggregates is outlined in flow diagram B3 (Figure 8.14). Table 8.1 shows expansion limits for mortar bar and concrete prism tests as specified in the supplement to the Canadian Standard A23.1, Appendix B.

8.3.2 CSA standard test methods

8.3.2.1 *A23.2-15A, petrographic examination of aggregates for concrete.* Petrographic examinations are made to determine the physical and chemical properties of an aggregate that can be deduced by petrographic means, to classify the constituents into non-reactive or one of the three categories of potentially reactive aggregates and to determine the relative amounts of potentially reactive components, e.g. chert, in the aggregate. Petrographic examination as used in this context may include, in addition to the use of the petrographic microscope, the use of X-ray diffraction, electron microprobe, SEM and thermoanalytical techniques. The standard specifies that the procedures in ASTM C295[34] should be followed.

8.3.2.2. *CSA A23.2-14A, potential expansivity of cement aggregate combinations (concrete prism method)*[3]. The accelerated version of this test, in which the concrete prisms are stored at 38°C and 100% RH, is specified for determining the potential reactivity of slow/late-expanding alkali–silicate/silica reactive aggregates. The test specifies the procedure to be used in making and measuring the length change of the concrete prisms. A standard mix design is specified in which the alkali content of the cement is adjusted to 1.25% Na_2O eq. by the addition of NaOH to the mix water. The cement used must have an alkali content of $1.0\pm0.2\%$ Na_2O eq. The standard also permits the testing of job mixes, the effect of the use of low-alkali cement or of the addition of supplementary cementing materials (pozzolans) or of curing at elevated temperature on the expansion of concrete, but in such cases two tests are prescribed, one with the standard mix, the other with the modified

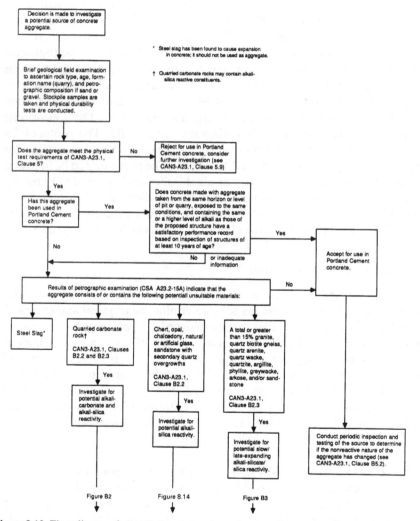

Figure 8.13 Flow diagram for preliminary stage in evaluation of aggregates for alkali–aggregate reactivity. From CAN3-A23.1 and CAN3-A23.2-M88.

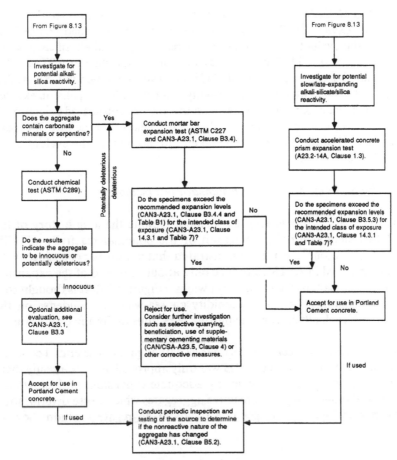

Figure 8.14 Flow diagram showing procedure to be followed in evaluating the potential reactivity of alkali–silica and slow/late-expanding alkali–silicate/silica reactive aggregates. From CAN3-A23.1 and CAN3A23.2-M88.

mix or elevated curing conditions. The 1990 revision of the standards recommends the inclusion, in the test series, of concrete prisms made with a known natural reactive aggregate to act as an internal standard.

8.3.2.3 *ASTM C289, the chemical method*[35]. The chemical method is not recommended as a final acceptance test for potentially alkali–silica reactive aggregates without some confirmatory testing because of its low reliability[36]. It was included in the CSA standards because it could be a useful method of monitoring the quality of an aggregate source containing some potentially alkali–silica reactive material.

8.3.2.4 *ASTM C227, potential reactivity of cement aggregate combinations (mortar bar method*[37]). The mortar bar method is the specified test in CSA A23.1 for the evaluation of the potential reactivity of alkali–silica reactive aggregates (Figure 8.14). The test is carried out as specified in ASTM C227 and evaluated according to ASTM C33[5], however it is noted in CSA A23.1, Appendix B.3.4.4, that 'cracking is usually observed when expansion exceeds 0.05% and hence expansions greater than this may be considered deleterious'; furthermore the Canadian standard also recommends that the rate of expansion at the end of the test should be taken into account in the evaluation of the aggregate.

8.3.2.5 *Problems encountered with standard test methods*

(1) *ASTM C227, the mortar bar method.* Despite the long history of the use of this method for the evaluation of the potential reactivity of ASR aggregates, Hooton[38] recently demonstrated that mortar bar containers with wicks, as specified in ASTM, produce only about one-third of the expansion observed in similar containers without wicks, (Figure 8.15). It is thought that in wicked containers excessive humidity causes water to condense on the samples and leach alkali from the mortar bars, thus inhibiting expansion.

(2) *CSA A23.2-14A, concrete prism test.* Although the accelerated concrete prism test run at 38°C and 100% RH was only approved in 1986, already there appear to be problems with achieving adequate expansions with slow/late-expanding alkali–silicate/silica reactive aggregates which cause deterioration of concrete in highway bridges due to alkali–aggregate reaction. Several

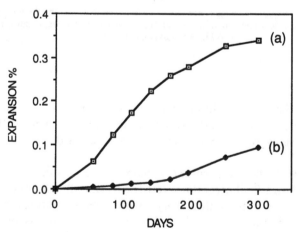

Figure 8.15 The effect of wicks on expansion of mortar bars in ASTM C227 made with silicious limestone from Ottawa. (a) Containers with wicks. (b) Containers without wicks.

factors may contribute to the reduced expansion levels which have been observed with the concrete prism test:

(a) The problem of excess humidity which affects the mortar bars may also affect the expansion of concrete prisms which are stored over water in sealed containers similar to those specified in ASTM C227.

(b) The CSA standard A23.2-14A specifies a slump of 80 ± 10 mm for the concrete to be used in making the prisms; this may be too high and could result in a dilution of the alkalis in the pore solution. Earlier successful tests by the author[19] were made using zero slump concrete. Tests are in progress to optimise the mix design and the specifications for the storage containers for the concrete prism test.

(3) *ASTM C289, chemical method.* This test is considered to give rather unreliable results, particularly with slow/late-expanding alkali–silicate/silica reactive aggregates such as some argillites, quartz arenites and greywackes[36]. The reason why this test gives erroneous results is not clear, but it may be related to the mechanism of reaction of the slow/late-expanding aggregates in which very little gel is produced indicating, possibly, that significant expansion can occur without much silica being dissolved. It has also been shown that limits for the permissible amount of dissolved silica need to be established for different types of aggregates and even for different occurrences of the same type of aggregate[36].

(4) *CSA A23.2-15A, petrographic examination of aggregates for concrete.* It is sometimes possible to determine the potential reactivity of an aggregate by petrographic examination, but the reliability of the determination is likely to be poor and ancillary testing is generally needed to confirm it. For example, in the case of the silicious limestones from the Trenton Group, the silica is often too fine to be observed in the petrographic microscope, although it can be seen in acid-insoluble residue examined in an SEM. The reactivity of some slow/late-expanding alkali–silicate/silica reactive aggregates has been related to the microcrystalline quartz content of the aggregate (Figure 8.10); however, difficulties with accurately determining the microcrystalline quartz content of the aggregate limit the usefulness of the method and, furthermore, its wider applicability remains to be demonstrated.

Attempts have been made to relate the reactivity of quartz-bearing aggregates to undulatory extinction in quartz grains. The procedure to be followed in determining the undulatory extinction angle is given by Dolar-Mantuani[39]. Recent research results have shown that there is little if any correlation between undulatory extinction angles and the expansivity of quartz-bearing aggregates[25,40]. The author[25] has suggested that the apparent correlation of undulatory extinction with reactivity may be due to the occurrence of microcrystalline quartz in aggregates. Undulatory extinction in quartz results from the rock being subject to metamorphic conditions under which silica is frequently mobilised and may recrystallise as microcrystalline quartz. More research is needed to develop accurate petrographic procedures for determining the potential reactivity of aggregates.

8.4 Effect of de-icing salt application on deterioration of concrete due to alkali–aggregate reaction

The addition of NaCl has been shown to increase the concentration of OH^- in the pore solution of cement paste[41]. It has also been shown that immersion of concrete test prisms containing alkali–silica reactive aggregate in salt solution enhances expansion and deterioration of the concrete[42]. It would therefore not be surprising if the addition of de-icing salts, commonly NaCl and $CaCl_2$, to structures exacerbated deterioration of the concrete due to AAR. In parts of Nova Scotia, where the local aggregates, quartzites and quartzarenites, are potentially reactive, most highway structures over 5 years old exhibit cracking due to SLASSR; the highways are heavily salted during the winter months. No evidence of concrete deterioration was found in high-rise or residential construction in which the same aggregates that caused deterioration in the bridges were used. However, it is not certain to what extent the use of de-icing salts contributed to the deterioration of the highway structures because the concrete in them usually has a higher cement content than that used in residential or in most high-rise construction. Increasing the cement content of concrete increases the alkali content of the pore solution and hence increases expansion and deterioration caused by alkali–aggregate reaction. The lack of deterioration in high-rise structures may in part be because of the rapid drying of smooth vertical surfaces after wetting by fog or rain with the result that the mean moisture level in the concrete may be below that required to sustain slow/late-expanding alkali–silicate/silica reaction. The type of reactive aggregate used and its degree of reactivity also affect the deterioration of concrete due to alkali–aggregate reaction. For example, in Montreal and Quebec City where silicious limestones are commonly used as concrete aggregates, cracking and deterioration of structures not subject to de-icing salt application is common[43,44], although not as widespread as in highway structures.

8.5 New test methods

The two test methods in most common use in Canada, the concrete prism method and the mortar bar method, may both take up to a year or more to complete; this is an unacceptably long time. The increased awareness of the potential reactivity of aggregates in many parts of the country has created an

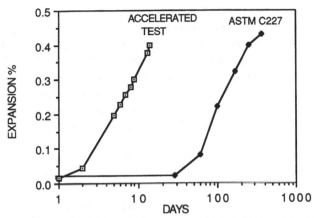

Figure 8.16 Comparison of mortar bar expansions in ASTM C227 and in the accelerated method. The mortar bars were made with silicious limestone from Ottawa. From Ref. 43.

urgent need to develop rapid, reliable test methods which would permit the evaluation of an aggregate in a few weeks.

A number of rapid mortar bar test methods have been proposed; they may be divided into two categories:

(1) Tests in which the mortar bars are autoclaved[45,46]
(2) Tests in which the mortar bars are stored in NaOH solution at elevated temperature and at 1 atmosphere[47,48].

The Oberholster[47] test in which the samples are stored for 14 days in 1 M NaOH at 80°C was selected for development in a number of laboratories because of its simplicity and also because it is conducted at 1 atmosphere, thereby reducing the risk that some of the observed expansion might be due to hydrothermal reactions which could occur under autoclave conditions. Expansions of mortar bars made with silicious limestone in the accelerated test after 2 weeks compare well with those obtained with the C227 test (Figure 8.16).

A wide variety of aggregates have been evaluated with the rapid test. On average, the accelerated test gave three times the expansion found with the concrete prism test. On this basis, if the expansion limit for the concrete prism test is 0.04% then that for the accelerated mortar bar test should be about 0.12%. However, some results indicate that expansions up to 0.15% may be acceptable with certain aggregates.

8.6 Methods of counteracting ASR

Up until recently, the lack of information concerning the potential reactivity of aggregates resulted in few organisations taking precautions to avoid concrete deterioration due to alkali–aggregate reaction. The exceptions are provincial hydroelectric companies, which have usually evaluated aggregates for use in dams, and the Ministry of Transportation of Ontario, which approves aggregate sources for use in highway structures. Recently the railways have begun specifying low-alkali cement for all concrete construction in an effort to avoid future concrete deterioration due to alkali–aggregate reaction. In the 1960s, heavy media separation was used to remove a potentially reactive opaline shale from a fine gravel used as concrete aggregate in a dam in Saskatchewan[7].

Research on the effect of supplementary cementing materials (pozzolans) on expansion due to alkali–aggregate reaction has only been pursued intensively for the past few years[49–52].

Soles *et al.*[50,52] showed that a slag (A1) which failed to reduce expansion by 75% in C441[53] (Figure 8.17), as specified in ASTM C618[54], was effective in reducing the expansion of mortar bars made with a reactive argillite for up to 2 years. A fly ash (L1) containing 2.6% alkali (Na_2O eq.), which also failed to reduce expansion by 75% in C441 (Figure 8.17), was more effective in reducing expansion due to alkali–aggregate reactivity in laboratory tests with slow/late-expanding alkali–silicate/silica reactive aggregate than fly ash (T), which meets the specifications in ASTM C618 but which contains 6.8% alkali (Na_2O eq.).

There is evidence in the literature[55,56] that some pozzolans only postpone expansion rather than prevent it. Oberholster[55] showed that replacement of 5% of cement by silica fume postponed expansion of concrete cubes made with a reactive aggregate for 3 years, but subsequently expansion proceeded at the same rate as in the sample containing 100% cement. Replacement of

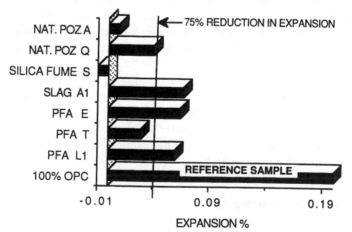

Figure 8.17 Effectiveness of a number of supplementary cementing materials in reducing expansion of mortar bars in ASTM test C441. Data from Ref. 50.

10% of the cement by silica fume prevented expansion for 4 years, but only time will tell how effective it will be in the long term.

8.7 Concluding remarks

Research on alkali–aggregate reactivity is now being actively pursued in a number of Canadian laboratories. Much of the research is being conducted in Ontario, Quebec and the eastern Maritime provinces, where the use of cements with relatively high alkali contents has resulted in a considerable number of cases of concrete deterioration due to alkali–aggregate reaction. Problems have been identified in the two main test methods used to evaluate the potential reactivity of aggregates, the mortar bar test ASTM C227 and the concrete prism test CSA CAN3-A23.2-14A, which could result in incorrect diagnosis. Research is currently in progress to improve the concrete prism and mortar bar test methods. Development of improved petrographic procedures for the evaluation of aggregates and of the accelerated mortar bar test, in which length change of samples stored in 1 M NaOH at 80°C for 2 weeks is monitored, are being actively pursued. Stockpiles of documented reactive aggregates have been established for use as reference samples in test programmes and for research purposes.

Supplementary cementing materials (pozzolans) have not been widely used in the past in Canada to counteract the deleterious effects of alkali–aggregate reaction in concrete; however, it is expected that the use of these materials will increase with increased awareness of the problems of alkali aggregate reaction in the construction industry. Results of recent research show that the effectiveness of some supplementary cementing materials in reducing expansion in concrete varies with the physical and chemical properties of the materials and with the type of aggregate used in the concrete.

References

1. Stanton, T.E. (1940) Expansion of concrete through reaction between cement and aggregate. *Proc. ASCE* **66**, 1781–1811.
2. Swenson, E.G. (1957) Cement aggregate reaction in concrete of a Canadian bridge. *ASTM Proc.* **57**, 1043–1056.
3. CSA CAN3-A23.1, Appendix B, Supplement No. 2 (1986) to CSA Standards CAN3-A23.1-M88 and CAN3-A23.2-M88. *Concrete Materials and Methods of Concrete Construction, Methods of Test for Concrete.* Canadian Standards Association, Rexdale, Ontario, Canada.
4. Chauret, E. (1958) Les réactions agrégat–ciment dans le Beton. *L'Ingenieur (Montreal)* **44**, 174 (Summer), 12–19.
5. ASTM C33-54 (1954) *Standard Specification for Concrete Aggregates.* American Society for Testing and Materials, Philadelphia, PA, pp. 11–20.
6. Fournier, B., Berube, M.A. and Vezina, D. (1987) *Proceedings of the Seventh International Conference on Alkali–Aggregate Reaction*, pp. 23–29.
7. Price, G.C. (1961) Investigation of concrete materials for South Saskatchewan River Dam. *ASTM Proc.* **61**, 1155–1179.

8. Swenson, E.G. (1957) A reactive aggregate undetected by ASTM tests. *ASTM Bull.* **226** (Dec), 48–51.
9. Rogers, C. and Worton, S. (1988) *Alkali Aggregate Reactions in Canada. A Bibliography*, Report EM-73. Ministry of Transportation, Engineering Materials Office, Ontario, p. 26.
10. Gillott, J.E. and Swenson, E.G. Mechanism of the alkali–carbonate rock reaction. *Q. J. Engng. Geol.* **2**, 7–23.
11. Swenson, E.G. (1970). *Alkali-Expansivity of Some Concrete Aggregates in Nova Scotia.* Atlantic Industrial Research Institute, Nova Scotia Technical College, Halifax, unpublished report.
12. Duncan, M.A.G., Swenson, E.G., Gillott, J.E. and Foran, M.R. (1973) Alkali–aggregate reaction in Nova Scotia. I. Summary of a 5 year study. *Cement Concr. Res.* **3**, 55–69.
13. Duncan, M.A.G., Gillott, J.E. and Swenson, E.G. (1973) Alkali–aggregate reaction in Nova Scotia. II. Field and petrographic studies. *Cement Concr. Res.* **3**, 119–128.
14. Duncan, M.A.G., Swenson, E.G. and Gillot, J.E. (1973) Alkali–aggregate reaction in Nova Scotia. III: Laboratory studies of volume change. *Cement Concr. Res.* **3**, 233–245.
15. Gillott, J.E. and Swenson E.G. (1973) Some unusual alkali-expansive aggregates. *Engng. Geol.* **7**, 181–195.
16. Hobbs, D.W. (1980) *Influence of Mix Proportions and Cement Alkali Content Upon Expansion Due to the Alkali–Silica Reaction*. Report 534. Cement and Concrete Association, UK.
17. Rogers, C.A. (1983) *Alkali Aggregate Reactions, Concrete Aggregate Testing and Problem Aggregates in Ontario: A Review*. Report EM-31. Ministry of Transportation and Communications, Engineering Materials Office, Ontario, Canada, p. 38.
18. Gillott, J.E., Duncan, M.A.G. and Swenson, E.G. (1973) Alkali–aggregate reaction in Nova Scotia. IV. Character of the reaction. *Cement Concr. Res.* **3**, 521–535.
19. Grattan-Bellew, P.E. (1981) A review of test methods for alkali-expansivity of concrete aggregates. *Proceedings of the Fifth International Conference on Alkali–Aggregate Reaction in Concrete*, S252/9.
20. Grattan-Bellew, P.E. (1987). Unpublished report, NRC, Ottawa.
21. Leaming, S.F. (1968) *Sand and Gravel in the Straight of Georgia Area.* Geological Survey of Canada Paper 66–60. Geological Survey of Canada, Department of Energy Mines and Resources, Ottawa, Canada, p. 149.
22. Bérubé, M. A., Fournier, B., Vezina, D. and Frenette, J. (1987) Alkali–aggregate reactivity in concrete structures from the Quebec City area. *Field Trip Guide Book*, Rapport GGL 87-19. Dept. de Géologie, Université Laval, p. 30.
23. Grattan-Bellew, P.E. (1978) Study of expansivity of a suite of quartzwackes, argillites and quartz arenites. *Proceedings of the Fourth International Conference on the Effects of Alkalies in Cement and Concrete*, pp. 113–140.
24. Magn, E.R., Rogers, C.A. and Grattan-Bellew, P.E. (1987) Influence of alkali–silicate reaction on structures in the vicinity of Sudbury, Ontario. *Proceedings of the Seventh International Conference on Alkali–Aggregate Reaction*, pp. 17–22.
25. Grattan-Gellew, P.E. (1987). Is high undulatory extinction in quartz indicative of alkali-expansivity of granite aggregates? *Proceedings of the Seventh International Conference on Alkali–Aggregate Reaction*, pp. 434–439.
26. Dolar-Mantuani, L.M. (1972) Harmful constituents in natural concrete aggregates in Ontario. *Proceedings of the 24th International Geological Congress, Montreal*, Section 13, pp. 227–234.
27. Kneller, W.A., Kriege, H.F., Saxer, E.L., Wilbamd, J.T. and Rohrbacher, T.J. (1968) *The Properties and Recognition of Deleterious Cherts which Occur in Aggregate used by Ohio Concrete Producers*. University of Toledo, Report 1014: 201
28. Gillott, J.E. (1975) Alkali–aggregate reactions in concrete. *Engng. Geol.* **9**, 303–326.
29. Wilson A.E. (1964) *Geology of the Ottawa–St Lawrence Lowland, Ontario and Quebec.* Geological Survey of Canada Memoir 241, Department of Mines and Technical Surveys, Ottawa, Canada, p. 66.
30. Grattan-Bellew, P.E. and Gillott, J.E. (1987) Three decades of studying alkali reactivity of Canadian aggregates. *Katharine and Bryant Mather International Conference, American Concrete*. Inst. Special Publication, SP 100-70 Vol. 2, pp. 1365–1384.
31. Houde, J., Lacroix, P. and Morneau, M. (1987) Rehabilitation of railway bridge piers

heavily damaged by alkali–aggregate reaction. *Proceedings of the Seventh International Conference on Alkali–Aggregate Reaction*, pp. 163–167.

32. Albert, P. and Raphael, S. (1987) Alkali–silica reactivity in the Beauharnois Powerhouses, Beauharnois. *Proceedings of the Seventh International Conference on Alkali–Aggregate Reaction*, pp. 10–16.

33. Bérard, J. and Lapierre, N. (1977) Réactivitiés aux alcalis du grès de Potsdam dans les bétons. *Can. J. Civ. Engng.* **4**, 332–344.

34. ASTM C295 (1985) Standard practice for petrographic examination of aggregates for concrete. *Annual Book of ASTM Standards, 04.02 Concrete and Mineral Aggregates.* American Society for Testing and Materials, Philadelphia, PA, pp. 220–230.

35. ASTM C289 (1985) Standard test method for potential reactivity of aggregates (chemical method). *Annual Book of ASTM Standards, 04.02 Concrete and Mineral Aggregates.* American Society for Testing and Materials, Philadelphia, PA, pp. 200–207.

36. Grattan-Bellew, P.E. (1983) Evaluation of test methods for alkali–aggregate reactivity, *Proceedings of the Sixth International Conference on Alkalis in Concrete*, pp. 303–314.

37. ASTM C227-81 (1985) Standard test method for potential reactivity of cement–aggregate combinations (mortar bar method). *Annual Book of ASTM Standards, 04.02 Concrete and Mineral Aggregates.* American Society for Testing and Materials, Philadelphia, PA, pp. 156–161.

38. Hooton, R.D. (1987) *Effect of Storage Containers on Expansion of Mortar Bars made with Spratt Aggregate (Siliceous Limestone)*. Minutes of Meeting of CSA Subcommittee A5, Cement Aggregate Reactivity. Laval University, Quebec City, October 6, 1987.

39. Dolar-Mantuani, L.M. (1983) *Handbook of Concrete Aggregates, A Petrographic and Technological Evaluation.* Noyes Publications, Park Ridge, NJ, p. 345.

40. Mullick, A.K., Wason, S.K., Sinha, S.K. and Rao, L.H. (1987) Evaluation of quartzite and granite aggregates containing strained quartz. *Proceedings of the Seventh International Conference on Alkali–Aggregate Reaction*, pp. 428–433.

41. Nixon, P.J., Canham, I., Page, C.L. and Bollinghaus, R. (1987) Sodium chloride and alkali–aggregate reaction. *Proceedings of the Seventh International Conference on Alkali–Aggregate Reaction*, pp. 110–114.

42. Swamy, R.N. and Al-Asali, M.M. (1987) New test methods for alkali–silica reaction. *Proceedings of the Seventh International Conference on Alkali–Aggregate Reaction*, pp. 324–328.

43. Bérard, J. and Roux, R. (1986) La viabilité des bétons du Québec: le role des granulats. *Can. J. Civ. Engng.* **13**, 12–24.

44. Bérubé, MA. and Fournier, B. (1987) Le barage Sartigan dans la Beauce (Québec) Canada: un cas-type de déterioration du béton par les réactions alcalis-granulats. *Can. J. Civ. Engng.* **14**, 372–404.

45. Tang, M-S., Han, S-F. and Zhen, S-H. (1983) A rapid method for identification of alkali–reactivity of aggregate. *Cement Concr. Res.* **13**, 417–422.

46. Tamura, H. (1987) A test method on rapid identification of alkali reactivity aggregate (GBRC rapid method). *Proceedings of the Seventh International Conference on Alkali–Aggregate Reaction*, pp. 304–308.

47. Oberholster, R.E. and Davies, G. (1986) An accelerated method for testing the potential alkali reactivity of siliceous aggregates. *Cement Concr. Res.* **16**, 181–189.

48. Yoshioka. Y., Kasami, H., Ohno, S. and Shinozake, Y. (1987) Study on a rapid test method for evaluating the reactivity of aggregates. *Proceedings of the Seventh International Conference on Alkali–Aggregate Reaction*, pp. 314–318.

49. Durant, B., Bérard, J. and Soles, J.A. (1987) Comparison of the effectiveness of four mineral admixtures to counteract alkali–aggregate reaction. *Proceedings of the Seventh International Conference on Alkali–Aggregate Reaction*, pp. 30–35.

50. Soles, J.A., Malhotra, V.M. and Suderman, R.W. (1987) The role of supplementary cementing materials in reducing the effects of alkali–aggregate reaction. *Proceedings of the Seventh International Conference on Alkali–Aggregate Reaction*, pp. 79–84.

51. Perry, C., Day, R.L., Joshi, R.C., Langan, B.W. and Gillott, J.E. (1987) The effectiveness of twelve Canadian fly ashes in suppressing expansion due to alkali–silica reaction. *Proceedings of the Seventh International Conference on Alkali–Aggregate Reaction*, pp. 93–97.

52. Soles, J.A., Malhotra, V.M. and Chen, H. (1987) Canmet investigations of supplementary

248   THE ALKALI–SILICA REACTION IN CONCRETE

cementing materials for reducing alkali–aggregate reactions. I. Granulated/pelletized blast furnace slags. *International Workshop on Granulated Blast-Furnace Slag in Concrete, Toronto, Canada,* 1987.

53. ASTM C441-81 (1985) Standard test method for effectiveness of mineral admixtures in preventing excessive expansion of concrete due to the alkali–aggregate reaction. *Annual Book of ASTM Standards,* 04.02 *Concrete and Mineral Aggregates.* American Society for Testing and Materials, Philadelphia, PA, pp. 283–286.

54. ASTM C618-85 (1985) Standard specification for fly ash and raw or calcined natural pozzolan for use as a mineral admixture in Portland cement concrete. *Annual Book of ASTM Standards,* 04.02 *Concrete and Mineral Aggregates.* American Society for Testing and Materials, Philadelphia, PA, pp. 382–385.

55. Oberholster, R.E. and Davies G. (1987) The effect of mineral admixtures on the alkali–silica expansion of concrete under outdoor exposure conditions. *Procceedings of the Seventh International Conference on Alkali–Aggregate Reaction,* pp. 60–65.

56. Xu, HuaYong and Chen, M. (1987) AAR in Chinese engineering practices. *Proceedings of the Seventh International Conference on Alkali–Aggregate Reaction,* pp. 253–257.

9 Alkali–aggregate reaction— New Zealand experience

D.A. ST. JOHN

Abstract

New Zealand has a potentially serious alkali–aggregate problem because reactive materials are present in 30% of the aggregates used in concrete. Although the widespread use of low-alkali cement has reduced this problem, a limited use of high-alkali cement has resulted in 60 cases of AAR. ASTM C289 and C227 test results confirm the necessity for the use of preventive methods against alkali–aggregate reaction (AAR) and have provided the assessment of the potential risk of using these reactive materials in concrete. Field inspections of structures supported by petrographic examination of concrete have identified AAR, validated laboratory test results and revealed previously unknown combinations of reactive aggregates. It is concluded that the control of AAR in New Zealand requires the application of preventive measures such as the use of low-alkali or Portland pozzolan cements.

9.1 Introduction

New Zealand lies along the circumpacific belt of volcanic rocks which cover an area of 20 000 km^2 in the North Island (see Figure 9.1.). As this 20 000 km^2 area includes half the population of New Zealand, the aggregates derived from these volcanic rocks are present in a substantial percentage of the concrete used in the country. Many of these volcanic aggregates are potentially reactive with cement alkalis, and damage to concrete structures can occur if they are used without adequate controls. This chapter gives a brief description of the approach taken in New Zealand to allow the safe use of these volcanic aggregates in concrete, the tests carried out and work planned for the future.

9.2 The factors controlling AAR in New Zealand

New Zealand, surrounded as it is by a very large oceanic area, has a marine climate which varies from temperate to near-subtropical, so that exposure is essentially one of wetting and drying. Persistent westerly winds deposit large amounts of sea salts onto coastal areas. However, chloride attack on reinforcing steel is not serious as an ample and well-distributed rainfall washes the salt from exposed surfaces of structures. Severe secondary deterioration in AAR caused by frost and salt attack, which are often evident overseas, is either absent or insignificant in New Zealand.

9.2.1 *Cement manufacture and alkali levels*

Cement is produced in New Zealand in modest-sized plants which currently produce up to a million tonnes per year. Since 1945, when records of alkali analyses first became available, much of the cement produced or imported contained less than 0.7% alkalis[1]. The exceptions are one of the original works, which until 1968 often produced cement with alkalis in excess of 1.0% (see Figure 9.2), and two smaller, now defunct, works which operated for about a decade. The high-alkali cements produced by these two small plants were not used in any large-scale construction or public works. Until the recent installation of a suspension preheater-type kiln system, one of the larger works consistently produced cement with an average of 0.3% alkalis. Thus low-alkali cements have always been readily available in New Zealand, and since 1968 all cement produced has contained less than 0.6% alkalis. This ready availability of low-alkali cement has been an important factor in limiting AAR in New Zealand.

9.2.2 *The role of the Ministry of Works and Development (MWD)*

In 1940 New Zealand had a population of 1.5 million people. By 1970 this population had doubled. During this period of considerable growth, which called for major road-building, extensive construction of hydro and geothermal projects and other major public works, MWD played a leading role both in design work and construction. At times, the ministry was purchasing up to 50% of all the cement produced and was able to exert pressure through specifications and purchasing contracts to keep the alkali contents of most cements below 0.6% alkalis. However, it was not until 1968 that this economic pressure finally eliminated all high-alkali cement production (see Figure 9.2). While low-alkali cements were used in all major projects this did not always apply to smaller public works. As a result AAR is present in some of these smaller structures.

As early as 1943 the MWD recognised that a potential alkali–aggregate problem existed in New Zealand. Since volcanic aggregates were to be used in many of the hydro stations to be built immediately after World War II, MWD initiated investigations into AAR by the Department of Scientific and Industrial Research[2]. They also took active steps to prevent AAR by using low-alkali cements and pozzolan. It is this early application of Stanton's work[3,4], learnt by New Zealand engineers training at the US Bureau of Reclamation, that has been an important contributing factor to minimising AAR in New Zealand.

9.2.3 *The role of the New Zealand Department of Scientific and Industrial Research* (*DSIR*)

Two divisions of DSIR have been involved in the investigation of AAR. The New Zealand Geological Survey has carried out the geological investigation of rocks to be used as aggregates, while the Chemistry Division (previously the Dominion Laboratory) has investigatd aspects of the chemistry and performed the tests on cement and concrete. Initially the Chemistry Division's work was primarily concerned with specific materials intended for use in hydro dams. Since 1960 it has carried out active research into many aspects of the wider problem. The work undertaken by DSIR forms the main body of extant knowledge about AAR in New Zealand.

9.2.4 *The distribution and geology of New Zealand aggregates*

There is an extensive area of potentially reactive volcanic rocks in the North Island[5]. The large area of acid-intermediate volcanic rocks shown in Figure 9.1 contains some areas of andesite, particularly in the Coromandel Peninsula and Bay of Plenty. However, most of the area between Lake Taupo and the Bay of Plenty is underlain by acid rocks, including widespread ignimbrite sheets. The extent of volcanic rocks in the South Island is limited, and only basalt and phonolite, which are probably non-reactive, are used as aggregates. It was decided that reliable data were required on the quantities of each rock type being used for concrete aggregate and how they varied from area to area. In 1980 a Quarries and Minerals Survey (QUAM) was undertaken in which all the mineral production was located and geologically identified[6]. The results extracted from QUAM, for the use of concrete aggregates by rock type, are summarised in Table 9.1.

The main concrete material used in New Zealand is greywacke. The term 'greywacke' covers a broad group of rocks which are indurated sand/silt/ claystones available in many parts of the country. Providing it is not heavily weathered and the silt/claystone fractions are excluded, greywacke makes an excellent concrete aggregate which is non-reactive with high-alkali cement.

It is evident that 75% of all concrete aggregates are extracted from alluvial deposits and that 95% of aggregates containing volcanics are also alluvial materials. In Taranaki, where the only rock type around Mt Egmont is andesite, all local aggregates are extracted from rivers and terraces and are not quarried from hard rock. In most other areas containing volcanics the alluvial aggregates consist of varying combinations of rhyolitic varieties, dacite, andesite and lesser basalt with greywacke. In Auckland, 50% of the

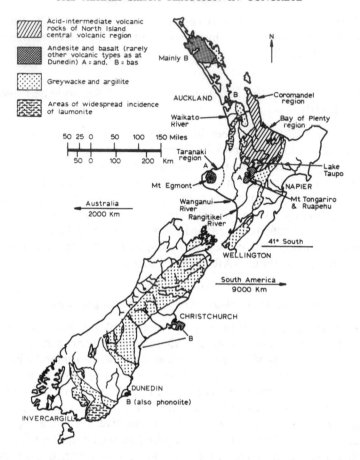

Acid-intermediate volcanic rocks of North Island central volcanic region

Andesite and basalt (rarely other volcanic types as at Dunedin) A = and. B = bas

Greywacke and argillite

Areas of widespread incidence of laumonite

Mainly B

AUCKLAND

Waikato River

50 25 0   50   100   150 Miles

50 0   100   200 Km

Coromandel region

Bay of Plenty region

Taranaki region

Mt Egmont

Lake Taupo

NAPIER

Australia
2000 Km

Wanganui River

Rangitikei River

Mt Tongariro & Ruapehu

41° South

WELLINGTON

South America
9000 Km

CHRISTCHURCH

B

DUNEDIN

B (also phonolite)

INVERCARGILL

Figure 9.1 The distribution of main rock types used as concrete aggregates in New Zealand. Locations of regions and geographical features referred to in text are shown. From Ref. 5.

sand used in concrete is taken fom the lower Waikato River and consists of 30–45% rhyolite, dacite and andesite grains mixed with non-reactive materials.

Most of the volcanic aggregates in current use are geologically recent and fresh. Glassy matrices are often present and in the more silicious types cristobalite and tridymite may be present. Rhyolitic varieties include obsidian, pitchstone, pumice, ignimbrite and lithoidal rhyolite. Dacites are less common in occurrence but a wide range of andesites are present. The Egmont andesites have about 30% of glassy matrix, while in the Tongariro and eastern andesites it is variable from 10 to 75%, as is the degree of devitrification[7]. It has been considered that this glassy matrix may be the principal reactive constituent in these volcanic rocks. Where available, basalt and in one small area phono-

Table 9.1 Building aggregate volumes from QUAM (m³) (averaged for 1982–84).

	m³	Percentage of total	Reactivity
Quarried materials			
Greywacke	799 000	12.4	
Limestone	68 000	1.1	
Basalt	793 000	12.3	
Miscellaneous	34 000	0.5	
Rhyolite/dacite	17 000	0.3	Reactive
Alluvial materials			
Greywacke	2,548 000	39.6	
Greywacke + other	654 000	10.2	
Quartz sands + other	345 000	5.4	
Quartz sand + volcanics	3 000	0.1	Reactive
Mixed volcanics	178 000	2.8	Reactive
Greywacke + volcanics	499 000	7.8	Reactive
Reactive North Island materials		30	
Reactive South Island materials		0	

lite are used. These rocks do not usually have glassy texture and up to now have not caused any problems with AAR. Apart from limited use of quartz sands, schist and granite and minor use of limestone there are no other rock types that are used in concrete.

9.2.5 The extent of the potential alkali–aggregate problem

Potentially reactive rock types used as concrete aggregates are limited to the acid-intermediate volcanic rocks, namely rhyolite to andesite. So far basalts have been assumed to be non-reactive. Some basalts can have glassy texture that may be reactive, and adequate petrological examination and some caution is necessary when making decisions about the use of basaltic rocks. New Zealand does not have a problem with amorphous silica minerals such as opal or chalcedony, although this could occur in the future if some of the volcanic rocks in the Coromandel area were used in concrete[5]. As yet the problems experienced overseas with the alkali–silicate type of rocks[8], such as greywacke, argillite, sandstone, granite and quartzite, have not been found in New Zealand. The New Zealand greywacke is non-reactive and argillites are not used.

Thus the problem of AAR in New Zealand is limited for the present to the range of volcanic rock types described above. The distribution of these rocks in the North Island is so extensive that without adequate investigation and control New Zealand could have a major and costly problem with concrete structures damaged by AAR.

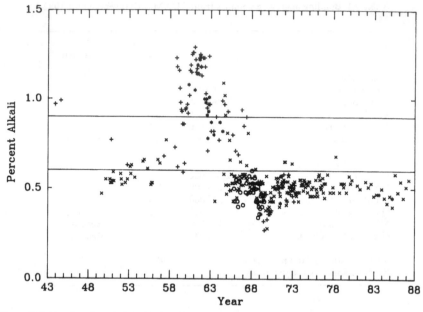

Figure 9.2 Results of available alkali analyses of the production from one of the original cement works for the period 1943–1988 showing the marked production of high-alkali cement between 1958 and 1968. +, ordinary; ×, low-alkali; *, rapid-hardening; O, special. From Ref. 1.

9.3 Investigations of materials and structures

9.3.1 *Field surveys of structures*

The first known case of AAR in a New Zealand structure was identified in 1970. By 1980 the number of known cases had increased to three, and by 1988 60 had been identified. This illustrates that, in New Zealand, AAR has been a problem where the frequency is directly related to the extent and efficiency of field surveys. Field evidence is of great importance in the investigation of AAR, and information obtained from surveys takes precedence over all other test data. It should validate laboratory tests and can indicate hitherto unknown areas that require investigation. MWD has specifically designed the inspection techniques to suit the requirements of New Zealand conditions[9]. An area is targeted for survey, and structures such as road bridges are inspected. AAR is only one of several deteriorative processes that are identified. The field survey procedure used where AAR is suspected is as follows:

(a) Any unexplained cracking or exudation is noted during the inspections. These inspections must be carried out by staff who are familiar with all aspects of AAR.

(b) Any structure with unexplained cracking is revisited with an experienced concrete petrographer to decide on locations for sampling. This inspection gives the petrographer the opportunity to examine the structure and its deterioration.

(c) The structure is sampled by coring, the only acceptable method of sampling.

(d) A full petrographic examination is made of the cores to determine the presence and extent of AAR, the types of aggregate used and those involved in the reaction, the volumes of the components, the quality of the concrete and other faults present. Wherever necessary the cores are tested in water and sodium hydroxide solutions for residual expansion.

(e) The report should include not only the results of the petrographic examination but also any available details of materials and mix design or other data. The results need to be placed in the context of previous known cases of AAR and the extant body of test data.

9.3.2 Results of field surveys and inspection of structures

MWD has completed a number of field surveys, primarily on bridges and dams. No hydro dam has as yet been found with any signs of AAR. This justifies the actions taken by MWD in specifying low-alkali cement and the use of pozzolans in these structures. In the Taranaki area[1] where the reactive Egmont andesite is present, 10 bridges have been identified undergoing AAR (see Table 9.2). Apart from one bridge built in 1978, these bridges were built either before 1939, and hence before Stanton's work on AAR[3], or between 1958 and 1968, when one of the larger cement works was producing high-alkali cement. While six of these bridges contain Egmont andesite, four contain aggregates that are mixtures of volcanics with other materials obviously brought in from outside the area. Records show that aggregates in two of these bridges came from the Rangitikei River, some 50–80 km to the south. This predominantly greywacke aggregate was found to contain up to 5% volcanics in the fine aggregate, which is apparently sufficient to cause mild AAR[10]. The presence of these volcanics in the Rangitikei River had not been previously identified, and further confirmation that other cases of AAR had occurred with these aggregates was received recently. Two rail bridges, both built in the early 1960s and both also containing aggregate from the Rangitikei River, are reported to be undergoing unexplained, expansive-type cracking.

In the Waikato and Tongariro regions, field survey of bridges has located 35 structures suspected of undergoing AAR. These have yet to be confirmed. This is not unexpected, as the Waikato region contains a range of volcanic aggregates. As yet no survey has been carried out in the Auckland metropolitan area or in the Bay of Plenty and Coromandel. In the case of Auckland the supply of reactive Waikato River sands will have resulted in cases of AAR. Recently, DSIR inspected four bridges for MWD which will undoubt-

Table 9.2 Structures undergoing AAR.

	Age	Cement	Aggregates	Reactant	Source	Severity
Structures identified by MWD field surveys						
Patea	1959	HA	Grey/volc	Rhyolite	Rangitikei	Moderate
Cobham	1962	HA/LA	Grey/volc	Rhyolite	Rangitikei	Moderate
Mangaongaonga	1967	NK	Grey/volc	Rhyolite	Wanganui	Moderate
Heimama	1928	NK	Egmont	Andesite	Heimama	Moderate
Reikohua	1930	NK	Egmont	Andesite	N Taranaki	Severe
Uruti	1930	NK	Egmont	Andesite	N Taranaki	Moderate
Mangaoraka	1938	NK	Egmont	Andesite	N Taranaki	Moderate
Kai-iwi	1938	NK	Grey/volc	Andesite	NK	Mod/severe
Leperton	1962	NK	Egmont	Andesite	N Taranaki	Mod/severe
Waiwakaiho	1978	LA	Egmont	Andesite	Waiwakaiho	Minimal
Waikato and Tongariro regions—35 bridges not yet investigated						
Other structures						
Whenuapai	1963	HA	Bas/grey/volc	Rhyolite	Waikato	Severe
Tuakau	1930	NK	Bas/grey/volc	Rhyolite	Waikato	Severe
Fairfield	1937	NK	Bas/grey/volc	Rhyolite	Waikato	Moderate
Te Henui	1962	HA	Egmont	Andesite	N Taranaki	Severe
Whangamomona	1962	HA	Egmont	Andesite	N Taranaki	Moderate
Whangamomona	1962	HA	Egmont	Andesite	N Taranaki	Moderate
NP reservoir	1962	HA	Egmont	Andesite	N Taranaki	Severe
Waiwakaiho railbridge	1936	NK	Egmont	Andesite	N Taranaki	Moderate
Te Whaiau	1961		No details known			Moderate
Wanganui	1961		No details known			Moderate
Whakapapa	1960		No details known			Severe
Onepoto	1968		No details known			Moderate
Te Atatu	1974		No details known			Severe
Pahurehure	1965		No details known			Severe
Slippery Creek	1965		No details known			Minimal
Rail Bridge	1965		No details known			Moderate
Rail Bridge	1965		No details known			Moderate

HA, high alkali; LA, low alkali; bas, basalt; grey, greywacke; volc, volcanic; NK, not known

edly prove to be undergoing AAR. One of these is serious enough to justify consideration being given to replacement of the deck beams.

Other cases of AAR are reported in Table 9.2. Once again most of these structures were built before 1939 or between 1958 and 1968. There are two cases of considerable concern. These are the Waiwakaiho River bridge built in 1978 and one of the Auckland bridges built in 1974. Both of these were probably built using low-alkali cement, which raises some interesting questions about the use of low-alkali cement as a primary prevention against AAR. The AAR in the Waiwakaiho River bridge is only minimal, but in the case of the Auckland bridge the reaction is severe. The concrete in the Auckland bridge probably contains basalt coarse aggregate and Waikato River sand, but this is still to be confirmed.

Because New Zealand's climate is mild and both salt and frost attack are limited or absent, the manifestation of AAR is truly representative of the reaction itself. Expansive cracking is generally limited and subtle, and tends to be more severe in areas of extreme wetting and drying. Gel formation and pop-outs are not common. However, petrographic examination shows that the cracking and gel is often more extensive internally, especially in those cases where it is known that considerable expansion has taken place.

9.3.3 *Petrological examination of aggregates*

Aggregates being tested in the laboratory are always examined petrographically to determine the types and amounts of volcanic materials present together with descriptions of the rock textures[5]. In the past this has also been carried out on aggregates for major construction. The present local practice is for engineers and contractors to depend largely on existing geological and laboratory test data in assessing the potential reactivity of aggregates. This is a dangerous practice, as is evidenced by the current problems associated with the Rangitikei and Waikato River aggregates. Considering the modest cost and speed of petrographic examination there appears to be little justification for omitting this type of testing.

9.3.4 *The use of ASTM C289*[7]

ASTM C289, *Test Method for Potential Reactivity of Aggregates* (*Chemical Method*), has proved to be most useful when applied to screening New Zealand volcanic rocks. The problems encountered by others in applying this test to sedimentary and metamorphic materials[8] have not been encountered in New Zealand because the sedimentary materials all appear to be non-reactive. From the test results in Figure 9.3 it will be evident that an individual result from testing an aggregate has little meaning if there is not a body of test data with which it can be compared. This especially applies to results that fall close to the division line drawn in Figure 2 of ASTM C289. For instance, the results from testing Egmont andesite, Figure 9.3B, show some of the aggregates to be in the 'innocuous' area of the graph. Given the way the results are grouped, this is a questionable interpretation of the data. The accumulated results have been subdivided into groups for ease of comparison in Figure 9.3.

9.3.5 *The use of ASTM C227*[7]

The more important results from using ASTM C227, *Potential Alkali Reactivity of Cement–Aggregate Combinations* (*Mortar Bar Method*), are given in Figure 9.4a–g. These illustrate the need to test volcanic aggregates for 'pessimum proportions' which vary not only with composition of the aggregate

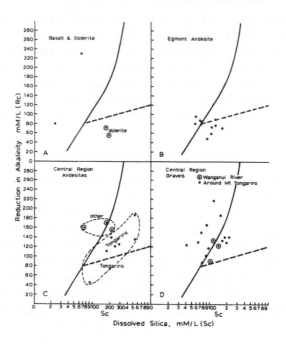

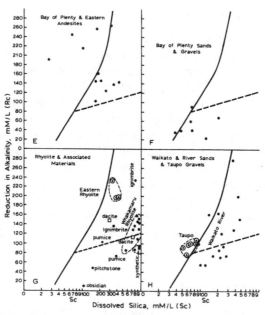

Figure 9.3 Results of ASTM C289 testing applied to some New Zealand aggregates. From Ref. 7.

but to some extent with alkali content. Gravels and sands are far more dangerous than the parent volcanic rocks because their reactivity varies with location and depends on the amount of non-reactive material present. There is no alternative to systematic laboratory testing of these materials.

Generally the use of ASTM C227 for testing New Zealand materials has been satisfactory. The problems encountered elsewhere when the test has been used with sedimentary and some metamorphic materials has not been applicable to New Zealand volcanics[8]. Rather the greatest problem has been the inability to carry out a sufficiently wide range of testing because of limitations of space and finance.

9.3.6 The results from using ASTM C289 and ASTM C227[7]

9.3.6.1 *Basalt.* Only three test results exist for basalt (Figure 9.3A). One represents a major aggregate used in Auckland and the other is from a basalt well to the south. Both are geologically fresh yet show large differences. The dolerite sample originally classed as a basalt when it was used in some hydro dams appears to be 'reactive'. This is confirmed by mortar bar tests. The two results obtained for the dolerite show the difference in results obtained when this material is tested in two laboratories.

9.3.6.2 *Egmont andesite.* The results from testing Egmont andesite from the Taranaki region fall into a well-defined group (Figure 9.3B). They also illustrate the care that is needed in interpretation. Mt Egmont is a recently extinct, isolated volcanic cone with a ring plain built up from numerous lava flows, so that there is variation of material around the mountain. However, there is no correlation between geographic location of samples and position of the results on the graph. The results indicate that Egmont andesite does not have any 'pessimum proportions'. This has been confirmed by extensive mortar bar and concrete testing (Figure 9.4a).

9.3.6.3 *Central region andesites.* The central region andesites form two groups (Figure 9.3C), one group from around Mt Tongariro and some results from two isolated andesitic cones to the north of Lake Taupo. The results from testing the Tongariro materials are scattered with indications that the material should show 'pessimum proportions'. This has been confirmed by mortar bar tests (Figure 9.4b). There are no mortar bar test data for the andesitic cones north of Lake Taupo.

9.3.6.4 *Central region gravels.* The central region gravels (Figure 9.3D) are from rivers and streams immediately around Mt Tongariro and from one large river draining part of the area to the south. A common diluent is greywacke, and this appears to be making some of the materials less reactive, probably because the 'pessimum proportion' is avoided. The results from the

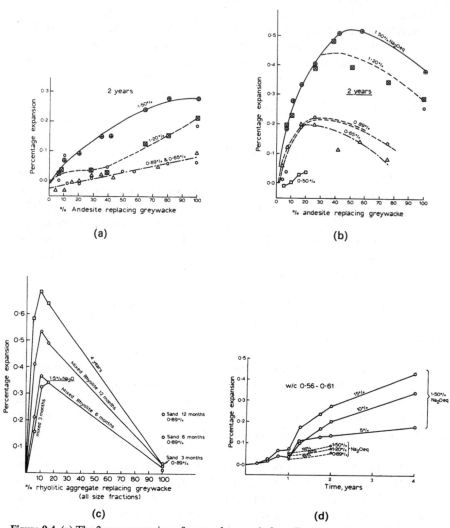

Figure 9.4 (a) The 2-year expansion of mortar bars made from Egmont andesite and greywacke. Percentage values on curves (1.50%, 1.20%, 0.89% and 0.65%) refer to Na$_2$O eq. (b) The 2-year expansion of mortar bars made from Tongariro andesite and greywacke. (c) The expansion at various ages of mortar bars made from Whakamaru rhyolite and greywacke. (d) The expansion of mortar bars made from Taupo obsidian and greywacke. Percentage values on the upper three curves (5%, 10%, 15%) refer to percentage of obsidian.

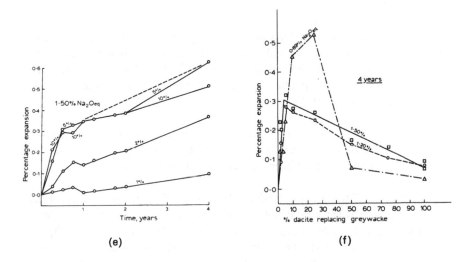

(e)   (f)

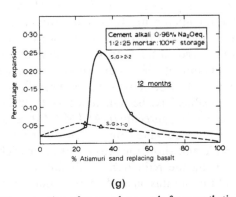

(g)

Figure 9.4 *cont'd* (e) The expansion of mortar bars made from synthetic cristobalite and grey-wacke. Percentage values on curves (10%, 5%, 2%, 1%) refer to percentage of cristobalite. (f) The expansion of mortar bars made from Tauhara dacite and greywacke at designated alkali contents. (g) The 12-month expansion of mortar bars made from Atiamuri sand from the Waikato River and Ongaroto dolerite. The upper curve represents materials from which pumice has been removed. The lower curves represent material primarily consisting of pumice. From Ref. 7.

Wanganui River are not strictly comparable as rhyolite is also present in these materials derived from some tributaries rising in the acid volcanics of the central region.

9.3.6.5 *Eastern and Bay of Plenty andesites* (Figure 9.3E). It has been shown petrographically that there is a wide variation in the texture of these andesites, and this is reflected in the results from ASTM C289. Unfortunately, insufficient mortar bar testing has been carried out on these samples to allow interpretation of these results.

9.3.6.6 *The Bay of Plenty gravels.* These materials contain not only andesite but also some rhyolite (Figure 9.3F). This appears to make them all 'reactive' but no mortar bar tests have been carried out to confirm this.

9.3.6.7 *Rhyolite.* Rhyolite is considered to be one of the most reactive of the New Zealand volcanic rocks, often showing marked 'pessimum proportions'. ASTM C289 results show these materials to be 'potentially reactive' (Figure 9.3G). The Whakamaru rhyolites form a compact group which have marked 'pessimum proportions' confirmed by mortar bar tests (Figure 9.4c). The eastern rhyolites from the Bay of Plenty form another group and also appear to be 'potentially reactive'. There are no test data or field evidence to confirm this. Both pumice and ignimbrite are often present as minor constituents in sands. Pumice, which tests 'reactive', does not result in expansion because its vesicular nature can accommodate any gel formed. There are no mortar bar test results available for ignimbrite. Both pitchstone and obsidian can be considered as glassy varieties of rhyolite. Both test as 'reactive' and mortar bar tests confirm that obsidian tends to show 'pessimum proportion' with alkalis such that small amounts of obsidian may cause expansion with low-alkali cement (Figure 9.4d). For comparison the synthetic materials, borosilicate glass and cristobalite, have been included. Both of these test 'reactive'. The glass is used as standard aggregate in ASTM C441, and mortar bar tests confirm the reactivity of cristobalite (Figure 9.4e). The two results for dacite are on the same sample and show the effect that extensively rewashing this material has in changing the reduction in alkalinity and dissolved silica. Mortar bar tests confirm that this dacite has 'pessimum proportions' (Figure 9.4f).

9.3.6.8 *Waikato River sands.* The sands from the Waikato River (Figure 9.3H), consisting of mixtures of rhyolites and andesites with non-reactive materials, are all 'reactive' or 'potentially reactive'. Both mortar bar tests (Figure 9.4g) and field experience with concrete structures confirm this reactivity. These sands are probably the most damaging reactive materials currently being used in New Zealand. The gravels and sands from some rivers flowing into Lake Taupo, the source of the Waikato River, contain from 5 to 30% of volcanic glassy materials, together with greywacke. The materials are marginally 'reactive' but there are no other data to confirm this.

9.3.7 Concrete tests[11]

There has only been one major investigation of AAR using concrete as the test material. Egmont andesite was chosen as the test aggregate since it is the only aggregate extracted in the Taranaki region. It is also an ideal material as this particular andesite from the Stony River seems relatively uniform and has no 'pessimum proportion'. The testing involved casting thousands of cylinders and beams using 4 m^3 of concrete. The extensive work involved is the reason why concrete testing has not been repeated on other New Zealand materials. The results shown in Figure 9.5a and 9.5b correlate well with the mortar bar tests as well as giving detailed information for the use of concrete containing Egmont andesite. The results from outdoor exposure, while showing considerable variation, are still consistent with those obtained from mortar bar testing in Figure 9.4a and will be more representative of the behaviour of concrete in structures.

9.3.8 The use of pozzolans[12]

A diatomaceous pumicite was used on some hydro dams as an on-site admixture to the concrete, initially to help minimise possible AAR but later because it gave desirable plastic properties to the concrete. Since low-alkali cement was used, the pozzolan was only an additional safeguard against AAR. Apart from this use, pozzolan has not been used commercially, and blended Portland pozzolan cements have not found favour, probably because of the ready availability of low-alkali cement. This situation may change as imported cements with higher alkali contents become available, and some additional protection against AAR may be necessary.

9.3.9 Petrographic examination of concrete

The use of large-area thin sections for the petrographic examination of concrete has been in routine use at DSIR since 1975[13]. This technique has been of critical importance in the evaluation of AAR in New Zealand. It has enabled the unequivocal identification of the presence and extent of AAR in structures and of the rock types undergoing reaction. In addition, other faults in the concrete are detected, quality is assessed, and the volumes of components estimated. It has been possible to recognise cases of AAR in which the outward appearance of the damage is subtle and insignificant. Because the visual observations made during inspection can be confirmed and interpreted, the field surveys have been made more effective for the detection of AAR.

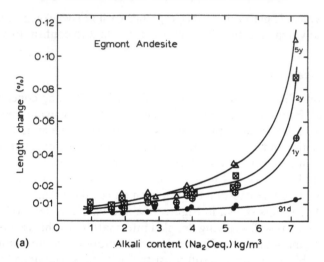

(a)

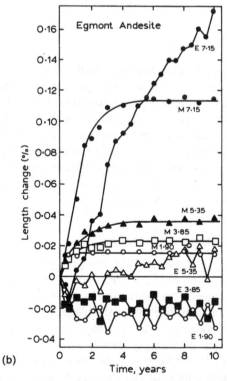

(b)

Figure 9.5 (a) The expansion of concrete prisms, moist-cured at 21°C, made from Egmont andesite versus the alkali content of the concrete. (b) The length change versus time for concretes with individual alkali contents (kg/m³) and subjected to moist curing conditions (M) and to outdoor exposure (E). Cement content 600 kg/m³. From Ref. 11.

For comparison purposes a number of concretes affected by AAR from Canada, USA, South Africa and the UK have also been examined. In some cases the structures involved have also been inspected. Discussion of detailed petrographic texture is inappropriate, but some general conclusions are of interest. While there has been a tendency for highly visual and spectacular cases to be featured in the literature, these are not the norm. It has been found that many cases of AAR, some quite serious, do not show spectacular external symptoms. The petrographic textures of reacting flint, chert and opaline silicas differ from most other reactive aggregates. Since many of the early petrographic studies were concentrated on these aggregates, this has tended to give a distorted view of the petrographic detail of AAR[14].

9.3.10 *Engineering investigation and remedial measures*

As few structures in New Zealand have been seriously affected by AAR little work has been needed in this field. The most serious case was an airbase pavement in which early remedial measures included reinstatement of slot drains and replacement of some adjacent areas of heaved bitumen runways. Testing of cores taken from the concrete[15] gave a mean tensile splitting strength of 3.3 MPa (CV = 18%), indicating a 30% reduction in strength due to the expansive cracking. However, it was concluded that sufficient flexural strength remained for normal use of the concrete. New concrete work was laid abutting the affected area, effectively locking the pavement. This resulted in a spectacular case of heaving. Figure 9.6 illustrates the results of

Figure 9.6 This spectacular case of damage occurred in an airbase pavement when 20 years after construction some new concrete was laid abutting one end. The results of restraining the expansion due to AAR were released rapidly during one night.

the active movement of this 20-year-old concrete. Recently, it was felt that small spalls of concrete might be sucked into the jet engines of aircraft at the airbase, and as a result the entire affected area was replaced.

A 50-year-old bridge affected by AAR is showing spalling of concrete over beam reinforcement and reasonably severe cracking of arches. Epoxy grouting of the arches was attempted, but petrographic examination indicated that penetration down the approximately 100-μm-wide cracks was minimal. Spalling over reinforcement has been successfully remedied by removing faulty concrete and replacing the cover. The original problem seems to have been poor cover over the reinforcement by porous concrete, possibly exacerbated by AAR.

In all other cases the extent of AAR is not great enough to warrant any action apart from inspection, although this may change with time. In many cases the elements affected are massive piers and abutments which can withstand considerable damage and still retain structural integrity. Also, because of seismic design requirements in New Zealand, public structures tend to have high concentrations of reinforcing steel, expecially in areas of shear stress. Design requirements in this country for spacing of stirrups in columns and beams are also conservative.

9.3.11 *Cement specifications and Codes of Practice for the control of AAR*

There are currently no codes of practice dealing with AAR, but NZS 3122 (1974) *Specification for Portland Cement (Ordinary, Rapid Hardening, and Modified)* provides an option for the specifying authority to specify low-alkali cement where potentially reactive aggregates are to be used. Before 1974 the requirement for low-alkali cement had to be specifically written into individual contracts. However, MWD engineers have always specified the use of low-alkali cement for major structures. Therefore the question remains as to why, with probably less than 10% of high-alkali cement being available for a decade, the number of publicly owned structures affected by AAR has risen to 60.

The answer lies in the presence of a very large area of highly reactive rocks in the North Island. Even the slightest relaxation in specification and control is likely to result in cases of AAR. For example, on one bridge the cast *in-situ* piers and abutments are pattern cracked, while the precast beams, probably containing different materials, are unaffected. On another bridge this situation is reversed and the bridge beams are seriously affected and may require replacement. An airbase pavement, recently partly replaced at cost of NZ$1.5 M was laid using two contractors. One half is sound and the other has expanded due to AAR. The specifications for another badly affected bridge clearly required the use of low alkali cement, but failure to check the cement being used resulted in the inadvertent use of high-alkali cement.

These cases highlight the problem that exists in New Zealand. On large projects careful specification and checking of materials is carried out. On smaller jobs the degree of knowledge and selection of specifications and monitoring are limited, and it is not always easy to control subcontractors who carry out work away from the site. Thus preventive measures against AAR such as the obligatory use of low-alkali or Portland pozzolan cements in areas where reactive aggregates are present appear to be the only practicable solution. The areas where reactive aggregates are present are not clearly defined and include only about 20% of the country. However this area contains half the population. Whether the answer lies in requiring all concrete to be mixed with either low-alkali or Portland pozzolan cements, or merely placing this requirement on the affected areas, is in the final analysis a matter of politics and economics. Either way there may be an economic penalty. In the first case it is possible that more expensive cements will be required in many structures where they are not necessary, and in the second there will be inevitable misuse of high-alkali cement, resulting in costly maintenance and possible early replacement of structures.

9.3.12 *The economics of cement production in New Zealand*

The economics of producing low-alkali cement is often raised when discussing the problem of AAR. It has been suggested that the production of low-alkali cement has resulted in higher prices for cements in New Zealand. However, enquiries show that one of the cement works uses naturally occurring low-alkali raw materials and two of the other works 'burn hard' to produce the necessary strength characteristics, and as a result can also comply with the necessary alkali limits. Only one cement plant suffers a disadvantage, and this is mainly because of increased wastage of raw materials at the quarry. The high price of New Zealand cement is primarily the result of high transport costs, currently about 60% of their final cost, and to a lesser extent the modest size of plant, and is only in small part the result of the production of low-alkali cements.

9.3.13 *Current investigations and future research*

Investigation of affected structures identified from field surveys and for individual clients continues. One of the problems in this area has been to determine the alkali levels present in the concrete. For most structures the details of the mix designs are not available, and, although cement contents can be estimated from measured volumes of hardened cement paste, this does not allow the alkali content to be determined. Past attempts at estimating alkali contents in concretes, especially those containing basalt as part of the aggregate, indicate that large errors are possible. A research project, funded by the National Roads Board, is in progress to determine and overcome the types of errors likely in estimating alkalis directly on concretes containing the range of New Zealand aggregates currently in use. In the past it has been considered that the glass matrix of volcanic aggregates is the principal reactive component in New Zealand rocks. Recent Japanese work[16] has claimed that some correlation exists between reactivity and both cristobalite and tridymite often present in the rock matrix. A cooperative programme to compare Japanese and New Zealand rocks is in progress. Some limited commercial investigations into the formulation of Portland pozzolan cements are also being undertaken.

There has been no systematic testing of aggregates in New Zealand over the last 20 years. Investigations of affected structures have clearly indicated a number of areas where the knowledge required for engineers to evaluate aggregates is lacking. It is hoped that aggregate testing will continue to be an area of major investigation during the next decade. The renewed testing of aggregates will start with mortar bar tests on materials from the Rangitikei and lower Waikato Rivers almost immediately.

9.4 Concluding comments

The above description of the New Zealand experience with AAR is of a general nature. Detailed discussion of the results from testing specific reactive materials tends to be unique to the country concerned, and the results cannot necessarily be translated to materials from other countries unless there is a close correlation between the rock types involved. The nature of the alkali–aggregate reaction is such that the only effective method of dealing with this problem is to prevent its occurrence. This requires a thorough knowledge of the aggregates being used and their potential reactivity as measured by laboratory testing. Field studies on structures are needed to confirm laboratory testing, and effective control measures must be implemented. For those countries that have ignored the possibility of alkali–aggregate reaction until recently there is the problem that testing can take up to a decade or longer to produce a large enough body of data to be of use. However, providing effective field surveys and adequate petrographic examination procedures are used these countries will have numerous cases of reaction in structures which will give a wide range of useful evidence. The New Zealand experience shows that even when the problem is recognised and reasonably close control is exercised on the alkali content of cements, alkali–aggregate reaction can only be minimised and not eliminated.

References

1. St John, D.A. (1988) Alkali–aggregate reaction and synopsis of data on AAR. *N.Z. Concrete Construction*, April and May.

2. Hutton, C.O. (1945) The problem of reaction between aggregate materials and high alkali cements. *N.Z. J. Sci. Tech.* **26B**, 191–200.
3. Stanton, T.E. (1940) The influence of cement and aggregate on concrete expansion. *Engineering News Record*, **1**(Feb), 59.
4. Stanton, T.E. (1943) Studies to develop an accelerated test procedure for the detection of adversely reactive cement–aggregate combinations. *Proc. ASTM* **43**, 875–904.
5. Watters, W.A. (1969) Petrological examination of concrete aggregates. *Proceedings of the National Conference on Concrete Aggregates*, Hamilton, New Zealand, pp. 48–54.
6. St John, D.A., Singers, W.E. and Aldous, K. (1988) *Quarries and Minerals Database: Users' Manual.* Chemistry Division Report No. C.D. 2388.
7. Kennerley, R.A. and St John, D.A. (1969) Reactivity of aggregates with cement alkalis. *Proceedings of the National Conference on Concrete Aggregates, Hamilton, New Zealand*, pp. 35–47 and 199–204.
8. Duncan, M.A.G., Swenson, E.G., Gillot, J.E. and Foran, M.R. (1973) Alkali–aggregate reaction in Nova Scotia. Pts I to IV. *Cement Concr. Res.* **3**, 55–69, 119–128, 233–244, 521–535.
9. Rowe, G.H., Freitag, S.A. and Stewart, P.F. (1985) *Concrete Quality in Bridges: Gisborne, Hawkes Bay, Wairarapa, Manawatu, Taranaki and Wellington.* MWD Central Laboratories Report 4-84/4.
10. Doyle, R.B. (1988) *The Occurrence of Alkali Reactive Materials in the Lower Waikato River Sands and Rangitikei River Aggregates.* MWD Central Laboratories Report 4-88/1M.
11. Kennerley, R.A., St John, D.A. and Smith, L.M. (1981) A review of thirty years of investigation of the alkali-aggregate reaction in New Zealand. *Proceedings of the Fifth International Conference on Alkali–Aggregate Reaction in Concrete*, S252/12.
12. Kennerley, R.A. and Clelland, J. (1959) An investigation of New Zealand pozzolans. *N.Z. DSIR Bulletin* **133**, 66.
13. Abbott, J.A. and St John, D.A. (1981) *The Preparation of Large Area Thin Sections in Concrete.* Chemistry Division Report C.D. 2311.
14. Idorn, G.M. (undated) *Studies of Disintegrated Concrete.* Danish National Institute of Building Research, Progress Reports Series N, Nos 2–6.
15. McNamara, G. (1985) *RNZAF Whenuapai Hardstanding Taxiway Pavement Cracking Concrete Strength Evaluation.* MWD Auckland Engineering Laboratory, Report No. AEL 85/80.
16. Katayama, T. and Kaneshige, Y. (1986) Diagenetic changes in potential alkali–aggregate reactivity of volcanic rocks in Japan—a geological interpretation. *Proceedings of the Seventh International Conference on Concrete Alkali–Aggregate Reactions*, pp. 489–495.

10 Alkali–silica reaction—Japanese experience

S. NISHIBAYASHI, K. OKADA, M. KAWAMURA,
K. KOBAYASHI, T. KOJIMA, T. MIYAGAWA, K. NAKANO
and K. ONO

10.1 Outline of basic studies in Japan

10.1.1 *Mechanisms of alkali–aggregate reactions*

Some research concerning the mechanisms of alkali–silica reactions (ASR) had been conducted in Japan before concrete structures were found to have been damaged by ASR. Studies in the early days, before cracking due to ASR was confirmed in Japan in 1983, were mainly concerned with elucidation of the ASR mechanisms through a combination of EDXA analysis and microhardness measurements. The following findings were reported in a series of papers between 1979 and 1983 by Kawamura *et al.*[1-3]

(1) The alkali concentration within the peripheral layer of opal grains embedded in cement paste rapidly increases with time. The microhardness in the region decreases with time. The thickness of the soft region increases as the ASR occurring within the opal grain proceed. The hardness decreases as water gradually penetrates into the reacting region (Figures 10.1 and 10.2).

(2) The solidification of the softened region within about 50 μm of the interface after 14 days that was found in coarse opal grain embedded in cement paste appears to relate to the intrusion of a relatively large amount of Ca^{2+} (Figure 10.3).

(3) The fly ash used in these studies did not inhibit the ASR at all, but facilitated the mobilisation of Ca^{2+} into reacting aggregate grains, as shown in Figure 10.4.

The third finding led to the proposal of a concept of the mechanisms responsible for the preventive effects of fly ash on the ASR.

Since the increase in alkali–aggregate reaction problems in Japan, the interest of workers has been focused on the elucidation of the mechanisms of the inhibition of the ASR by fly ash, blast furnace slag and silica fume. The major results obtained between 1983 and 1988 in Japan are as follows:

(1) The enhancement of expansion in Beltane opal mortars containing silica fume is attributable to the delay in softening of the gels formed in the mortars[4].

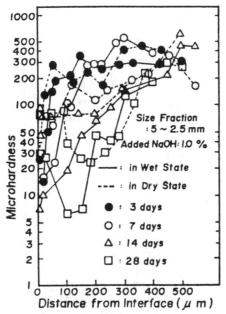

Figure 10.1 Microhardness within opal particles embedded in the cement paste with 1% added NaOH. From Ref. 3.

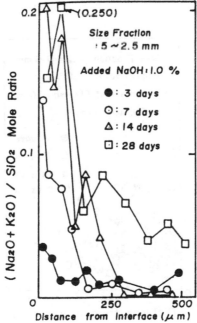

Figure 10.2 Distribution of alkalis concentration within opal particles of 5–2.5 mm. From Ref. 3.

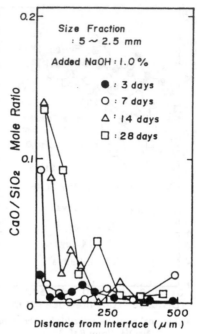

Figure 10.3 Distributions of calcium concentration within opal particles. From Ref. 3.

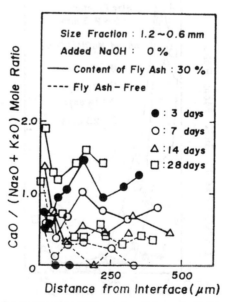

Figure 10.4 CaO/(Na$_2$O + K$_2$O) ratios within opal particles embedded in the cement pastes with and without fly ash. From Ref. 3.

(2) Some factors other than the reduction in alkalinity in pore solutions are related to the prevention of expansion due to the ASR caused by the incorporation of pozzolans and slag[5,6].

(3) Silica fumes differ widely in their preventive effects on the alkali–silica expansion of mortars with an opal aggregate[7].

(4) The effectiveness of fly ashes in inhibiting the expansion of mortar with a silicious aggregate which reacts very quickly varies widely, depending on their alkali content and pozzolanic activity[8].

10.1.2 Microstructure of reaction products and petrographic studies

Andesitic rocks are the aggregates which were considered to be responsible for deterioration due to alkali–aggregate reaction in Japan[9]. The reactive components in the rocks were cristobalite, tridymite and volcanic glass[9]. Since then, ASR-affected concrete structures, containing cherty rocks as well, have been found in several regions in Japan[10]. Several workers who conducted petrographic examinations in the cherty rocks in the ASR-affected concretes identified cryptocrystalline quartz, chalcedony and strained coarse quartz as potential alkali-reactive components[10–12]. They also investigated the microstructure of the reaction products by means of scanning electron microscopy (SEM), polarising microscopy and X-ray diffractometry. Tatematsu et al.[10] examined reaction products on the fracture surfaces of aggregate grains in concrete cores drilled from a damaged concrete structure.

The result of their investigation led to the classification of reaction products into three types: (1) $Na_2O–K_2O–SiO_2$, (2) $Na_2O–K_2O–CaO–SiO_2$ and (3) $Na_2O–K_2O–CaO–SiO_2$. They reported that the reaction products of types (1) and (2) were jelly-like. They also confirmed that the appearance of

Figure 10.5 Scanning electron micrograph of rosette-like products. From Ref. 10.

Figure 10.6 Scanning electron micrograph of rosette-like products. From Ref. 11.

type (3) reaction product was rosette-like (Figure 10.5). Such rosette-like reaction products were also found in other affected concrete (Figure 10.6). As shown in Figure 10.7, Morino *et al.* found a reaction product among the interstices of microcrystalline quartz crystals in the cherty rock contained in the concrete cores drilled from a 10-year-old concrete structure[11].

10.1.3 *Effect of mineral admixtures on alkali–aggregate reactions*

The effect of mineral admixtures such as blast furnace slag and fly ash in counteracting alkali–aggregate reactions was specially investigated because

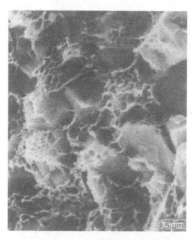

Figure 10.7 Scanning electron micrograph of microcrystalline quartz and reaction products in a 10-year-old concrete structure. From Ref. 11.

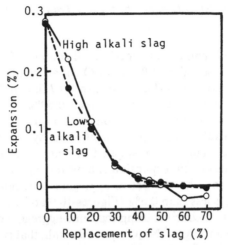

Figure 10.8 Relation between replacement of slag and expansion.

these materials are produced as byproducts in very large quantities in Japan. Quality data on slags and fly ashes indicate that the total alkali content ranges from 0.4 to 0.7% for slag and from 0.8 to 3.5% for fly ash. Silica fumes are almost all imported and are only rarely produced in Japan.

Tests to determine the effectiveness of these mineral admixtures in counteracting alkali–aggregate reactions show that slag and fly ash are able to control expansive reactions[13-15] when Portland cement is replaced by 30% or more of slag and 20% or more of fly ash. Typical results are shown in Figures 10.8 and 10.9.

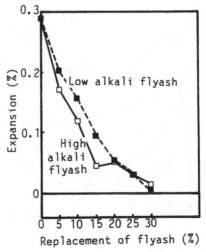

Figure 10.9 Relation between replacement of fly ash and expansion.

10.1.4 *Influence of cements and chemical admixtures on alkali–aggregate reactions*

10.1.4.1 *Influence of cements.* The total alkali content of ordinary Portland cement in Japan was less than 1.07% Na_2O eq. at maximum, and has been gradually reduced between 1983 and 1987. In 1987 the total alkali content of ordinary Portland cement in Japan averaged 0.60% with a maximum of 0.69%, according to statistics collected on cement quality.

The degree of alkali–silica expansion effected by the alkali content in ordinary Portland cement was investigated by many researchers. The results obtained[15,16] indicate that the expansion of mortar bars using typical andesitic reactive aggregate and cherty reactive aggregate does not occur when the total alkali content is less than 0.6% (Figures 10.9 and 10.10). On the other hand, the expansion of concrete using the same aggregate does not occur in concrete with less than 3 kg/m³ Na_2O eq. total alkali, that is 3 kg/m³ is judged to be the critical content for alkali–silica expansion (Figures 10.11 and 10.12).

10.1.4.2 *Influence of chemical admixtures.* The influence of chemical admixtures such as air-entraining agents, water-reducing agents and super-plasticisers differs with the kind of chemical admixtures. It is considered that the characteristics of chemical admixtures which influence alkali–silica expansion are the alkali content of the admixtures and the dispersing power of the cement particles. For this reason, chemical admixtures having high alkali content and high dispersing power sometimes show a high capacity to promote ASR (Figure 10.13)[17].

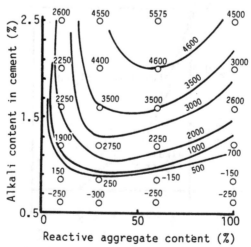

Figure 10.10 Equiexpansion contour map.

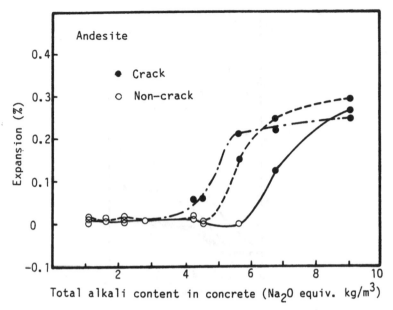

Figure 10.11 Relation between total alkali content and expansion.

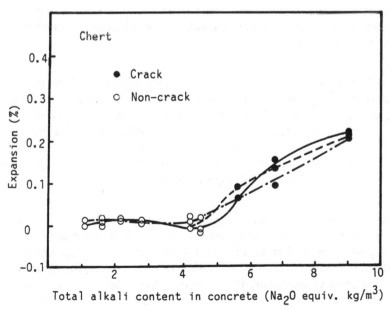

Figure 10.12 Relation between total alkali content and expansion.

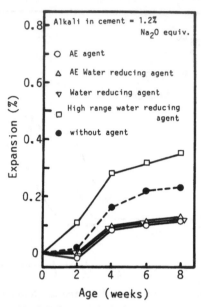

Figure 10.13 Influence of admixtures on expansion.

10.1.5 *Influence of environment conditions on alkali–aggregate reactions*

Expansion caused by alkali–aggregate reactions is affected by temperature, humidity and externally supplied alkali. Typical results obtained in recent investigations are as follows.

10.1.5.1 *Influence of temperature.* Alkali–silica expansion depends on temperature, and a pessimum temperature exists in the relation between temperature and expansion. Results[18] of mortar bar tests with typical andesitic reactive aggregate indicate that maximum expansion occurs at a temperature of 40°C. At 60°C, the expansion is small and saturates at an early stage. At 20°C, however, although expansion is delayed, it increases rapidly at a later stage and exceeds that at 60°C (Figure 10.14).

10.1.5.2 *Influence of externally supplied alkali.* In general, alkali–silica expansion is said to be accelerated by seawater or by alkali sulphates in soil, but the degree of expansion varies over a very wide range, depending on the conditions. From the results[15] of testing of concrete specimens made with andesitic reactive aggregate, it was found that, when the specimens were exposed in a half-immersed condition in seawater, the expansion was extremely accelerated (Figure 10.15).

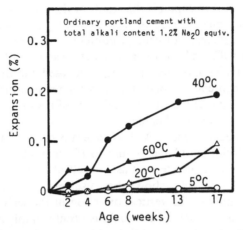

Figure 10.14 Influence of curing temperature on expansion.

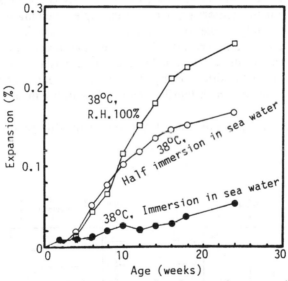

Figure 10.15 Influence of seawater on the expansion of concrete (concrete made with 500 kg/m^3 ordinary Portland cement and reactive andesite coarse aggregate).

10.2 Concrete structures damaged by ASR and their repair

10.2.1 *Examples of concrete structures damaged by ASR*

Causes of cracking in concrete structures have generally been considered to be thermal stress, drying shrinkage, settlement of placed concrete, settlement of formwork, freezing and thawing, expansion arising from the corrosion of reinforcing bar, excess of load and differential settlement.

Appearance of ASR in Japan is the result of the use of reactive mountain gravel or crushed stone because of lack of river gravel, increase of alkali content in cement, increase of cement content in concrete, and use of sea sand. The damage suffered by a concrete structure as a result of ASR depends upon the characteristics of the reactive aggregate used, the amount of reactive aggregate, alkali content of concrete, ambient temperature and moisture content, size and shape of the structure, degree of restraint of the structural member and existence of initial cracks.

The pattern of cracking due to ASR is generally different from that due to thermal stress or drying shrinkage. Figure 10.16a shows a schematic cracking pattern in a reinforced concrete wall as a result of thermal stress or drying shrinkage and Figure 10.16(b) shows that resulting from ASR. The former cracking occurs mainly in the vertical direction of the wall and at an early age. On the other hand, cracking due to ASR occurs mainly in the horizontal direction and develops with age. ASR sometimes yields cracking in the axial direction at the top of the wall.

Cracking in reinforced concrete as a result of ASR is apt to occur in the direction of the main reinforcing bar. For example, cracking in a beam or column occurs mainly in the axial direction. The reason for fewer cracks occurring in the direction perpendicular to the axial direction is considered to be the compressive stress developed in the concrete as a result of restraint of the expansion in the direction by the reinforcing bar. Therefore, it is inferred that a relatively high tensile stress is developing in the reinforcing bar.

On the other hand, at portions of concrete with little reinforcement or

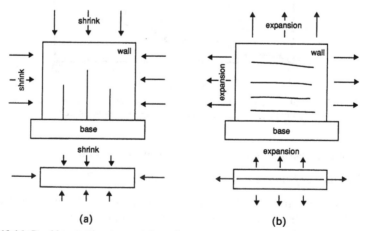

Figure 10.16 Cracking pattern in a reinforced concrete wall. (a) Cracking due to temperature drop or drying shrinkage. (b) Cracking due to ASR.

without reinforcement, as for example at the free end of a beam or abutment, on the surface of a sea defence structure or in a concrete block, etc., the crack pattern due to ASR is irregular or map-like.

A part exposed to direct sunshine and rain generally has more cracks than the sheltered part of the same structure.

Figures 10.17 to 10.32 are examples of concrete structures in Japan damaged by ASR.

Figure 10.17 is a typical example of ASR cracks in reinforced concrete beams in which the main cracks are developing in the horizontal direction. Cracking at the top of a culvert box is shown in Figure 10.18. The cracks are seen to be developing mainly in the horizontal direction, which is the axial direction.

Figure 10.19 shows cracking in a bridge column in which vertical cracks are developing. No cracks were found developing in most of the inside columns of this bridge, which were sheltered from direct sunshine and rain.

Figure 10.20 shows cracking in an abutment. These cracks also include those on the coating material used for the repair earlier. The main cracks in the abutment are in the horizontal direction.

Figure 10.21 shows cracking in a retaining wall at the seaside, where horizontal cracking is again dominant. A typical case of cracking in a staircase is shown in Figure 10.22 where the axial cracking is again dominant. An example of cracking at the top of a wall type of structure is illustrated in Figure 10.23.

Figure 10.24 shows an example of the effect of temperature or water on ASR. More cracks were found to be developing on the left-hand side of the pier than on the right-hand side, which was partially sheltered by the superstructure.

Figures 10.25 and 10.26 are examples of irregular cracking occurring at the free end of structures, where restraint by the reinforcing bar is relatively weak. Cracks in Figure 10.27 are developing at the corner of the abutment, where restraint by reinforcing bar is also weak. Figures 10.28 and 10.29 are examples of irregular cracking in concrete with less reinforcement or without reinforcement.

Figures 10.30 and 10.31 are examples of cracking in plain concrete. In these examples, fewer but wider cracks are developing.

Figure 10.32 is an example of deformations arising from ASR. The retaining wall on the left-hand side is pushing up the adjacent retaining wall and causing damage at the joint.

10.2.2 *Field investigation*

Various investigations of the ASR-damaged concrete structure were carried out, generally by cores drilled from the structures. The investigations included measurement of core expansion at 40°C and 100% relative humidity (RH),

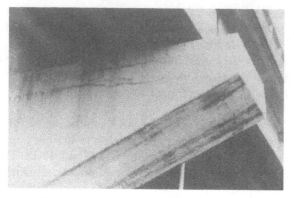

Type of structure	T-type pier
Year of construction	1970
Reactive aggregate	Bronzite andesite
Compressive strength of core	230 kgf/cm^2
Corrosion of reinforcing bar	Partially
Crack width	0.6 mm
depth	6 cm
Ambient condition	Partially sheltered at the seaside once repaired

Figure 10.17 Cracking in a beam.

Type of structure	Box culvert
Year of construction	1972–1974
Reactive aggregate	Bronzite andesite
Compressive strength of core	150 kgf/cm^2
Corrosion of reinforcing bar	Partially
Crack width	10–12 mm
depth	30 cm
Ambient condition	Unsheltered

Figure 10.18 Cracking in a box culvert.

Type of structure	Column
Year of construction	1972–1976
Reactive aggregate	Bronzite andesite
Compressive strength of core	230 kgf/cm^2
Corrosion of reinforcing bar	Partially
Crack width	3 mm
depth	8 cm
Ambient condition	Partially sheltered (no cracking in columns where not exposed to direct sunshine and rainfall)

Figure 10.19 Cracking in a pier—1.

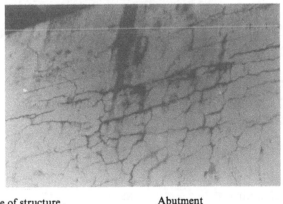

Type of structure	Abutment
Year of construction	1970
Reactive aggregate	Bronzite andesite
Compressive strength of core	230 kgf/cm^2
Corrosion of reinforcing bar	Partially
Crack width	2–4 mm
depth	9–12 cm
Ambient condition	Partially sheltered at the seaside once repaired

Figure 10.20 Cracking in an abutment.

Type of structure	Retaining wall
Year of construction	NI
Reactive aggregate	Chert, slate
Compressive strength of core	NI
Corrosion of reinforcing bar	NI
Crack width	10 mm
depth	NI
Ambient condition	Unsheltered at the seaside

NI: Not investigated

Figure 10.21 Cracking in a retaining wall.

Type of structure	Staircase
Year of construction	NI
Reactive aggregate	Bronzite andesite
Compressive strength of core	NI
Corrosion of reinforcing bar	NI
Crack width	5–7mm
depth	NI
Ambient condition	Unsheltered

NI: Not investigated

Figure 10.22 Cracking in a staircase.

Type of structure	River defence
Year of construction	1971
Reactive aggregate	Bronzite andesite
Compressive strength of core	NI
Corrosion of reinforcing bar	NI
Crack width	10–15 mm
depth	NI
Ambient condition	Unsheltered

NI: Not investigated

Figure 10.23 Cracking in a river defence.

Type of structure	Pier
Year of construction	1969–1970
Reactive aggregate	Tuff
Compressive strength of core	NI
Corrosion of reinforcing bar	No corrosion
Crack width	0.1 mm
depth	3 cm
Ambient condition	Unsheltered

NI: Not investigated

Figure 10.24 Cracking in a pier—2.

Type of structure	T-type pier
Year of construction	NI
Reactive aggregate	Bronzite andesite
Compressive strength of core	NI
Corrosion of reinforcing bar	NI
Crack width	9 mm
depth	14 cm
Ambient condition	Unsheltered

NI: Not investigated

Figure 10.25 Cracking at the end of a beam.

Type of structure	Retaining wall
Year of construction	NI
Reactive aggregate	NI
Compressive strength of core	NI
Corrosion of reinforcing bar	NI
Crack width	5–7 mm
depth	NI
Ambient condition	Unsheltered

NI: Not investigated

Figure 10.26 Cracking at the end of a retaining wall.

Type of structure	Abutment
Year of construction	1973–1975
Reactive aggregate	Chert, slate
Compressive strength of core	270 kgf/cm^2
Corrosion of reinforcing bar	Partially
Crack width	3 mm
depth	12 cm
Ambient condition	Unsheltered at the seaside

NI: Not investigated

Figure 10.27 Cracking at the corner of an abutment.

Type of structure	Wing of abutment
Year of construction	NI
Reactive aggregate	NI
Compressive strength of core	NI
Corrosion of reinforcing bar	NI
Crack width	5–10 mm
depth	NI
Ambient condition	Unsheltered

NI: Not investigated

Figure 10.28 Cracking in the wing of an abutment.

Type of structure	Sea defence
	(plain concrete)
Year of construction	NI
Reactive aggregate	Chert, slate
Compressive strength of core	140 kgf/cm^2
Corrosion of reinforcing bar	
Crack width	8–12 mm
depth	33 cm
Ambient condition	Unsheltered at the seaside
NI: Not investigated	

Figure 10.29 Cracking in a sea defence wall of plain concrete.

Type of structure	Stopper (plain concrete)
Year of construction	NI
Reactive aggregate	Bronzite andesite
Compressive strength of core	NI
Corrosion of reinforcing bar	
Crack width	10 mm
depth	NI
Ambient condition	Unsheltered
NI: Not investigated	

Figure 10.30 Cracking in a stopper of plain concrete.

Type of structure	Precast concrete block (plain concrete)
Year of construction	1969–1970
Reactive aggregate	Chert, slate
Compressive strength of core	220 kgf/cm^2
Corrosion of reinforcing bar	
Crack width	10–15 mm
depth	NI
Ambient condition	Unsheltered in the sea

NI: Not investigated

Figure 10.31 Cracking in a precast concrete block.

Type of structure	Retaining wall
Year of construction	NI
Reactive aggregate	Bronzite andesite
Compressive strength of core	175 kgf/cm^2
Corrosion of reinforcing bar	NI
Crack width	4 mm
depth	14 cm
Ambient condition	Unsheltered

NI: Not investigated

Figure 10.32 Deformation and cracking of a retaining wall.

rock identificiation by mineralogical inspection, measurement of alkali content, measurement of crack width and depth, checking of corrosion of reinforcing bars, compressive strength and Young's modulus of cores and ultrasonic pulse velocity of the structures.

The reactive aggregates identified as causing distress in the structures shown in Figures 10.17 to 10.32 are bronzite andesite, chert and slate. Bronzite andesite is the most widely occurring reactive aggregate in Japan. Figures 10.33 and 10.34 show the reaction rim and exuded gel of bronzite andesite at the cut surface of a drilled core.

Table 10.1 gives an example of the comprehensive documentation carried out in such field studies.

Figure 10.35 relates the degree of cracking to core expansion of various concrete structures in which bronzite andesite was used at different mix proportions. These results indicate that degree of damage or cracking due to ASR depends upon the mix proportion of bronzite andesite in total coarse aggregate and the alkali content of concrete. The amount of expansion of the cores from concrete structures with extensive cracking exceeded 500×10^{-6}.

Figure 10.36 shows the relation between the width and depth of crack due to ASR. According to these results, there is a tendency for wider cracks to be deeper. However, the crack depth in most of the reinforced concrete structures investigated remained within the range of the concrete cover.

Figure 10.37 relates the compressive strength and Young's modulus of cores. These results indicate that the compressive strength of concrete drops as a result of ASR. These results also indicate that Young's modulus of the reacted concrete is very low. However, Young's modulus back-analysed from the deflection of the beam measured at the loading test was not so low (Figure 10.38)[19]. The low Young's modulus measured by the core may be the result of release of restraints existing in the structure. However, the difference is not clearly understood.

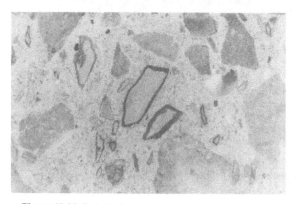

Figure 10.33 Reaction rim on the cut surface of a core.

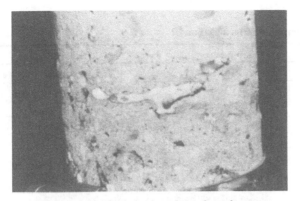

Figure 10.34 Exuded gel on the surface of a core.

Figure 10.39 shows the variation of ultrasonic pulse velocity in a concrete structure damaged by ASR and the compressive strength of the drilled cores. These results indicate that there is an almost linear relationship between them.

10.2.3 *Repairs by coating*

An efficient method to inhibit ASR has not been established yet. Since ASR requires water for the reaction, it is inferred that waterproofing might be effective in inhibiting ASR.

Various types of coating materials have been tested in Japan. Some of the materials have been applied to bridge substructures damaged by ASR. Polyurethane resin and epoxy resin were selected as waterproofing coating materials. Polybutadiene resin, silane resin and polymer cement were selected as aeration type of coating. Cracks wider than 0.3 mm were normally injected before coating. Before the coating repair, crack width and depth, corrosion of steel, reactive aggregate used, expansion of core drilled from the structure, etc. were investigated. After coating, strain measurement was carried out periodically at the surface of the structure.

Tables 10.2 to 10.6 show several examples of structures repaired by coating. It should be noted that the measured strains include also strain due to temperature changes. The residual expansion of core at the time of coating as well as the total expansion should be taken into account when judging the efficiency of the coating by the degree of the strain after coating.

The following statements can be made from these field tests:

(1) Waterproofing type of coating is not so effective since the piers in Table 10.2 and 10.3 expanded greatly after coating and cracked again.
(2) The effect of polybutadiene cannot be concluded at this stage since the

Table 10.1 Example of the investigation

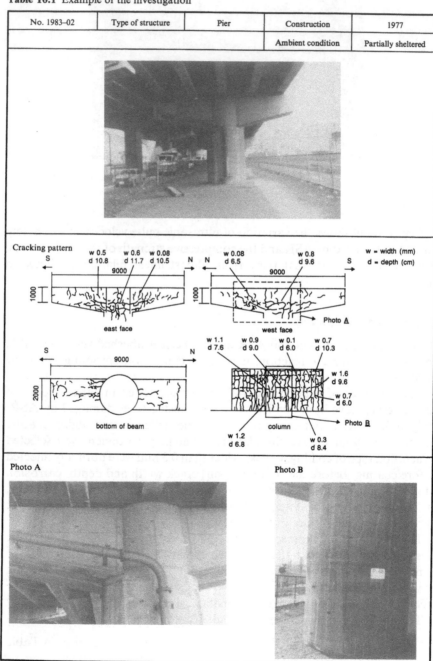

Mix proportion of concrete	Maximum size of aggregate (mm)	Slump (cm)	Air content (%)	W/C (%)	S/a (%)	C	W	S	G
	25	8	4	55	41	345	191	745	1070

	Design strength σck = 240 kgf/cm^2
Source	Coarse aggregate $\cdots$ T. Island, Fine aggregate $\cdots$ Y. River
Cracking	Maximum width = 1.5 mm, Maximum depth = 12 cm
Abnormal change	Nil
Time of inspection	October, 1983
Age of concrete when investigated	6 years old

Test results of core	Compressive strength (kgf/cm^2)	Young's modulus (kgf/cm^2)	Poisson's ratio	Ultrasonic wave velocity (m/s)	Depth of neutralisation (cm)
	230	10.3×10^4		3740	1.1

Ultrasonic wave velocity in structure	3700 m/s
Maximum spontaneous potential	300 mV
Corrosion of steel	Nil
Reactive aggregate	Bronzite andesite
Reactive coarse aggregate/ Total coarse aggregate	40%
Expansion of core ($\times 10^{-6}$)	Total expansion = 1000, Release expansion = 350, Residual expansion = 650

Alkali content,	Alkali content			Chlorine content
Chlorine content (% of concrete by weight)	Equivalent Na$_2$O	Na$_2$O	K$_2$O	
	0.28	0.20	0.12	0.11

ACI chemical method	Dissolved silica (Sc)	Reduction in alkalinity (Rc)	Judgment
	576 mM/l	64 mM/l	deleterious

Mortar bar method ($\times 10^{-6}$)	3 months	6 months	Judgment
	2568	2827	potentially reactive

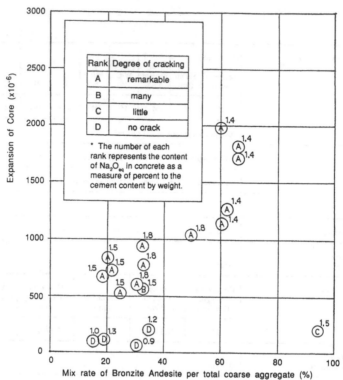

Figure 10.35 Degree of cracking and expansion of the drilled core.

residual expansion of the abutment in Table 10.4 is very little, although the abutment has not recracked after coating.

(3) Silane resin looks effective for the pier in Table 10.5. However, the coating was applied at a low stage of the expansion. Therefore, more information is required to conclude the efficiency of silane resin.

(4) Polymer cement appears to be working well for the pier in Table 10.6. The strain occurring in the pier seems to be the result only of the temperature change although the residual expansion of the core is very large. However, this result is limited to 1 year after coating. Therefore, more observation is required before judging the efficiency of the repair.

10.3 Test methods—standard and new rapid tests and preventive measures in Japan for alkali–aggregate reaction

This section deals with trends and problems of testing methods, preventive measures for alkali–aggregate reactions and the present situation of research in Japan on the rapid testing method.

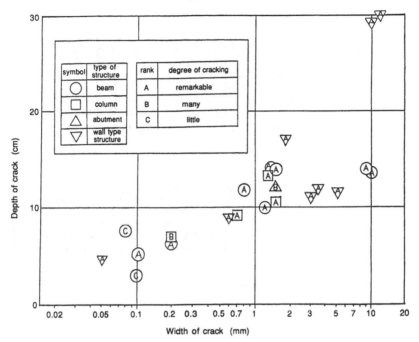

Figure 10.36 Relation between crack width and depth.

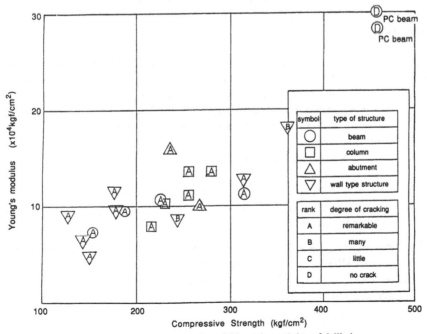

Figure 10.37 Compressive strength and Young's modulus of drilled cores.

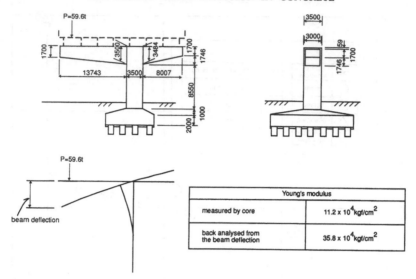

Figure 10.38 Loading test of a beam.

10.3.1 *Standard test method*

The standard test method which has been tentatively established by the Japan Concrete Institute (JCI)[20] comprises chemical, mortar bar and concrete specimens tests.

10.3.1.1 *Chemical test.* Besides adding a more detailed commentary to the testing methods, the JCI decided that the accuracy of testing results would be improved by changing the method of drying used in the determination of dissolved silica content. Evaporation has been replaced by dehydration using perchloric acid, as the former method was subject to comparately large errors. In addition the combined use of the gravimetric method and atomic absorption spectroscopy was recommended.

10.3.1.2 *Mortar bar method.* With several test results as references, and broadly following the ASTM specifications, the following test conditions were provided for the mortar bar testing method:

(1) Specimen size $= 4 \times 4 \times 16$ cm, as specified in the Japanese Industrial Standard (JIS) for the cement mortar test.
(2) Total alkali content $= 1.2\%$ Na_2O eq. (adjusted by NaOH).
(3) Water–cement ratio $= 0.50$.

Otherwise ASTM testing conditions and expansion limits were adopted. However, it was recommended that the minimum period of measurement

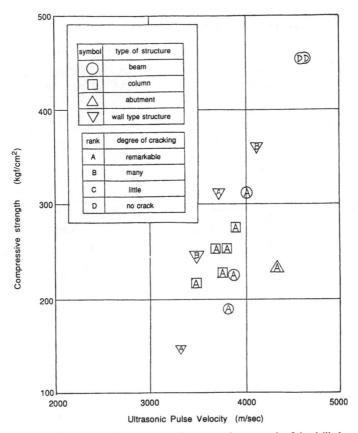

Figure 10.39 Ultrasonic pulse velocity and compressive strength of the drilled core.

should be 1 year, because some aggregates showed large expansion after longer periods of curing.

10.3.1.3 *Concrete specimens method.* Tests using concrete specimens have as their objective the evaluation of the reactivity of the aggregates and the estimation of the cracking damage due to alkali–aggregate reactions under conditions of actual use. At present, studies are mainly being made to test the five items below:

(1) Effects of shape and dimensions of the test specimen.
(2) Effects of storage conditions.
(3) Effects of the unit cement content and the total alkali content.
(4) Comparisons of the test results obtained from concrete specimens and mortar bars.

Table 10.2 Repair by coating—1.

Structure	Pier	Before repair (1983)
Construction	1977	
Investigation	1983	
Repair	1985	
Crack width	1.5 mm	
depth	12 cm	
Corrosion of steel	Nil	
Reactive aggregate	Bronzite andesite	
Expansion of core		
($\times 10^{-6}$)		
total	1000	
release	350	After repair (1985)
residual	650	
Ambient condition	Partially sheltered	
Crack injection	Nil	
Coating primer, putty	Epoxy resin (0.4 kg/m^2)	
main	Polyurethane resin (0.4 kg/m^2)	

Expansion after coating

July 1987

March 1988

July 1988

Table 10.3 Repair by coating—2.

Structure	Pier	
Construction	1978	
Investigation	1984	
Repair	1985	
Crack width	1 mm	
depth	13 cm	
Corrosion of steel	Not investigated	
Reactive aggregate	Bronzite andesite	
Expansion of core ($\times 10^{-6}$)		
total	800	
release	500	
residual	300	
Ambient condition	Partially sheltered	
Crack injection	Flexible epoxy resin	
Coating primer, putty	Epoxy resin (0.4 kg/m^2)	
main	Flexible epoxy resin (0.4 kg/m^2)	

Before repair (1984)

After repair (1985)

Expansion after coating

February 1986

November 1986

March 1988

Table 10.4 Repair by coating—3.

Structure	Abutment	Before repair (1984)
Construction	1969	
Investigation	1984	
Repair	1986	
Crack width	4 mm	
depth	15 cm	
Corrosion of steel	Partially	
Reactive aggregate	Bronzite andesite	
Expansion of core		
($\times 10^{-6}$)		
total	350	
release	250	
residual	100	
Ambient condition	Partially sheltered	After repair (1986)
	at the seaside	
	once repaired	
Crack injection	Flexible epoxy resin	
Coating primer, putty	Epoxy resin (1.0 kg/m^2)	
main	Polybutadiene resin	
	(1.5 kg/m^2)	

August 1987

Table 10.5 Repair by coating—4.

Structure	Pier	After repair (1985)
Construction	1969	
Investigation	NI	
Repair	1985	
Crack width	NI	
depth	NI	
Corrosion of steel	NI	
Reactive aggregate	NI	
Expansion of core ($\times 10^{-6}$)		
total	NI	
release	NI	
residual	NI	
Ambient condition	Partially sheltered	
Crack injection	Flexible epoxy resin	
Coating main	Silane resin (0.4 kg/m^2)	

NI: not investigated

Expansion after coating

Table 10.6 Repair by coating—5.

Structure	Pier
Construction	1971
Investigation	1985
Repair	1986
Crack width	2 mm
depth	16 cm
Corrosion of steel	Partially
Reactive aggregate	Bronzite andesite
Expansion of core ($\times 10^{-6}$)	
total	2000
release	600
residual	1400
Ambient condition	Partially sheltered, once repaired
Crack injection	Flexible epoxy resin
Coating primer, putty	Flexible polymer cement (0.6 kg/m^2)
main	Flexible polymer cement (0.9 kg/m^2)

Before repair (1985)

After repair (1986)

July 1988

Expansion after coating

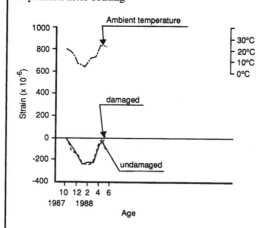

(5) Investigation of the interrelationship between length change, total alkali content (or cement content) and strength.

The items measured are length change, dynamic modulus of elasticity, ultrasonic pulse velocity and crack growth pattern.

10.3.2 Prevention of alkali–aggregate reactions in new construction

One of the following measures should be taken for the prevention of alkali–aggregate reactions in new concrete structures, according to the tentative recommendations[21] proposed by the Japanese Ministry of Construction:

(a) Use of aggregates considered innocuous by means of either the chemical method or the mortar bar method.
(b) Use of low-alkali Portland cement, i.e. less than 0.6% Na_2O eq.
(c) Use of an appropriate blended cement, such as blast furnace slag cement of type B or C with a slag content of 30–60% or 60–70% respectively, or admixture such as blast furnace slag and fly ash.
(d) Control of the total alkali content in concrete. The total alkali content of the concrete mix shall not exceed 3 kg/m^3 of Na_2O eq., when Portland cement is used.

10.3.3 Maintenance of concrete structures damaged by alkali–aggregate reactions

The maintenance, inspection and evaluation of structures affected by alkali–aggregate reactions have been extensively attempted by the Hanshin Expressway Public Cooperation[15]. The main items of inspection are as follows:

(1) The accumulated total length of the cracks and wider expansion cracks.
(2) Inspections of alkali–aggregate reactions are conducted by measuring the expansion and observing the exudation of gel from the drilled core, and identifying the aggregate mineral.
(3) Evaluation of the inspection results is performed on the total expansion (more than 100×10^{-6}) and on the accumulated total crack length (more than 100 m) together with the gel.
(4) Repair, such as the injection of epoxy resin into cracks and synthetic resin coating or impregnation, shall be done in case of (3).
(5) Long-term monitoring after repair or strengthening is to be carried out.

10.3.4 Some new rapid testing methods for determining alkali–aggregate reactivity in concrete in Japan

10.3.4.1 Takenaka Komuten Co. Ltd.[22]. The test method of immersing specimens in NaOH solution at an early age and then keeping them under high temperature and high humidity is effective in accelerating the reaction and the expansion of mortar bar. The temperature required is 80°C, the immersion period is 24 hours and the concentration of the NaOH solution is 1 N. The results of the rapid test correlate well with those of the mortar bar method, and the reactivity of the aggregate can be evaluated at 7 days.

10.3.4.2 *General Building Research Co. Ltd.*[23]. The mortar specimens [dimensions = $4 \times 4 \times 16$ cm; S:C = 1:1, water–cement ratio (w/c) = 0.50, alkali content = 2.5% Na_2O eq.], after 3 days of curing, are placed in boiling water in the pressure vessel (0.5 kgf/cm^2, 110°C) for 2 hours. Cracking, changes in the ultrasonic pulse velocity and dynamic Young's modulus of the specimens are inspected just before and after boiling. The aggregate is evaluated as innocuous if there is no cracking or if the reduction ratios of the ultrasonic pulse velocity or dynamic Young's modulus are less than 5% or 15% respectively.

10.3.4.3 *Tottori University—1*[24]. The specimens [dimensions = $4 \times 4 \times 16$ cm, C/S = 1:2.25, w/c = 0.45, total alkali content = 2.0% Na_2O eq. (by NaOH)] are subjected to the autoclave treatment. The test conditions based on the results of several experiments are as follows: pressure in the autoclave kiln = 0.1 MPa; duration of the autoclave treatment = 4 hours; age of the specimen = 24 hours after casting; measurement of the length change = 24 hours after autoclave treatment. The expansion that occurs during the autoclave treatment corresponds to about 30% of that under normal storage conditions. The expansion limit of 0.02% is proposed as the standard for evaluation of the reactivity of aggregate.

10.3.4.4 *Tottori University—2*[25]. The concrete specimens (dimensions = $10 \times 10 \times 40$ cm) are subjected to repeated cycles of immersion in a testing bath (20°C) and drying in an oven (60°C). The length change, the reduction in the dynamic Young's modulus and the ultrasonic pulse velocity were measured until 50 cycles in order to determine the degree of deterioration due to alkali. When the expansion is greater than 0.1% and the reduction in the dynamic Young's modulus is less than 85%, it must be assumed that injurious cracking damage has occurred in the concrete.

10.3.4.5 *Kyoto University*[26]. The concrete specimens (dimensions = $10 \times 10 \times 40$ cm) are subjected to repeated cycles of 40°C, 100% RH for 12 hours, and 20°C, 60% RH for 12 hours. This wetting and drying exposure simulates the most severe exposure conditions anticipated in Japan. Under these test conditions, the effects of various repair systems (both the waterproofing type and the water-repellent type) were examined to determine how they would perform in the actual exposure conditions occurring in Japan. The repair systems include bisphenol-amine epoxy coating, polybutadiene urethane coating, methyl methacrylate impregnation, sodium silicate impregnation, silane monomer impregnation, acrylic polymer cement mortar lining and so on.

References

1. Kawamura, M., Takemoto, K. and Hasaba, S. (1982) Elucidation of alkali–silica reaction mechanisms by the combination of EDXA and microhardness measurements. *Transactions of the Japan Concrete Institute* **4**, 1–8.
2. Kawamura, M., Takemoto, K. and Hasaba, S. (1983) The mechanisms of prevention due to alkali–silica reaction by pozzolanic additives. *Transactions of the Japan Concrete Institute* **5**, 91–96.
3. Kawamura, M., Takemoto, K. and Hasaba, S. (1983) Application of quantitative EDXA analyses and microhardness measurements to the study of alkali–silica reaction mechanisms. *Proceedings of the Sixth International Conference on Alkalis in Concrete*, pp. 167–174.
4. Kawamura, M., Takemoto, K. and Hasaba, S. (1986) Effect of silica fume on alkali–silica expansion in mortars. *Proceedings of the Second International Conference on Fly Ash, Silica Fume, Slag and Natural Pozzolans in Concrete*, pp. 999–1012.
5. Kawamura, M., Takemoto, K. and Hasaba, S. (1984) Effect of various pozzolanic additives on alkali–silica expansion in mortars made with two types of opaline reactive aggregates. *Review of the 38th General Meeting*, The Cement Association of Japan, pp. 92–95.
6. Kawamura, M., Takemoto, K. and Hasaba, S. (1986) Effectiveness of various flyash and blast furnace slags in preventing alkali–silica expansion. *Review of the 40th General Meeting*, The Cement Association of Japan, pp. 276–279.
7. Kawamura, M., Takemoto, K. and Hasaba, S. (1987) Effectiveness of various silica fumes in preventing alkali–silica expansion. *Proceedings of the Katharine and Bryant Mather International Conference on Concrete Durability*, pp. 1809–1819.
8. Kawamura, M., Takemoto, K. and Hasaba, S. (1988) Correlation between pore solution composition and alkali–silica expansion in mortars containing various flyashes and slags. *International Journal of Cement Composites and Lightweight Aggregate Concrete* **10**, 215–223.
9. Kawamura, M., Takemoto, K. and Hasaba, S. (1983) Case studies of concrete structures damaged by the alkali–silica reaction in Japan. *Review of the 37th General Meeting*, The Cement Association of Japan, pp. 88–89.
10. Tatematsu, H., Takada, J. and Tanigawa, S. (1986) Characterization of alkali aggregate reaction products. *Clay Science* **26**, 143–150 (in Japanese).
11. Morino, K., Shibata, K. and Iwatsuki, E. (1987) Alkali aggregate reactivity of cherty rock. *Clay Science* **27**, 199–210 (in Japanese).
12. Morino, K., Shibata, K. and Iwatsuki, E. (1987) Alkali aggregate reactivity of andesite containing smectite. *Clay Science* **27**, 170–179 (in Japanese).
13. Okada, K., Adachi, T. and Nagao, Y. (1986) Effect of the properties of blast-furnace slag on expansion caused by alkali–silica reaction. *Review of the 40th General Meeting*, The Cement Association of Japan, pp. 266–267.
14. Kobayashi, S., Kawano, H., Numata, S. and Chikada, T. (1986) Some consideration on the mechanism of effectiveness of ground granulated blast furnace slag in preventing alkali–silica reaction. *Review of the 40th General Meeting*, The Cement Association of Japan, pp. 268–271.
15. Committee on ASR of Hanshin Expressway Public Corporation (1986) *Committee Report on Alkali–Aggregate Reaction*. Hanshin Expressway Public Corporation, Osaka (in Japanese).
16. Committee on Concrete of the Cement Association of Japan (1988) *Study on Expansion Properties of Concrete made with Alkali-Reactive Aggregate*. The Cement Association of Japan, Tokyo (in Japanese).
17. Nakano, K., Kobayashi, S. and Arimoto, Y. (1984) Influence of reactive aggregate and alkali compounds on expansion of alkali–silica reaction. *Review of the 38th General Meeting*, The Cement Association of Japan, pp. 96–99.
18. Nakano, K., Kobayashi, S., Nakaue, A. and Ishibashi, H. (1986) Influence of alkali contents

and curing conditions on expansion of mortar bars due to alkali–silica reaction. *Review of the 40th General Meeting*, The Cement Association of Japan, pp. 254–257.

19. Hanshin Expressway Public Corporation and Ohtori Consultant (1981) A loading test on the beam of Hanshin Expressway, Hanshin Expressway Public Cooperation and Ohtori Consultant, Osaka (in Japanese).

20. Kishitani, K., Nishibayashi, S. and Morinaga, S. (1987) Response of JCI to alkali–aggregate reaction problem—guideline for determining potential alkali reactivity. *Proceedings of the Seventh International Conference on Alkali–Aggregate Reaction in Concrete*, pp. 264–268.

21. Makita, M., Wakisaka, Y., Moriya, S. and Kawano, H. (1986) Methods of testing reactive aggregate. *Concrete Journal* **24**(11), 33–39 (in Japanese).

22. Yoshioka, Y., Kasami, H., Ohno, S. and Shinozuka, Y. (1987) Study on a rapid test method for evaluating the reactivity of aggregates. *Proceedings of the Seventh International Conference on Alkali–Aggregate Reaction in Concrete*, pp. 314–318.

23. Tamura, H. (1987) A test method on rapid identification of alkali reactivity aggregate (GBRC rapid method). *Proceedings of the Seventh International Conference on Alkali–Aggregate Reaction in Concrete*, pp. 304–308.

24. Nishibayashi, S., Yamura, K. and Matsushita, H. (1987) A rapid method of determining the alkali–aggregate reaction in concrete by autoclave. *Proceedings of the Seventh International Conference on Alkali–Aggregate Reaction in Concrete*, pp. 299–303.

25. Nishibayashi, S., Yamura, K., Hayashi, A. and Imaoka, S. (1988) Effects on the alkali–aggregate reaction of the cycles of wet and dry. *Proceedings of the Japan Concrete Institute* **10**, 789–794 (in Japanese).

26. Miyagawa, T., Sugashima, A., Kobayashi, K. and Okada, K. (1987) Repair of concrete structures damaged by alkali–aggregate expansion. *Transactions of the Japan Concrete Institute* **9**, 219–226.

11 Alkali–silica reaction—Indian experience

A. K. MULLICK

Abstract

Multidisciplinary investigations carried out to identify the causes of distress in a concrete dam and a concrete spillway have established the occurrence of alkali–silica reaction (ASR) in concrete in India for the first time; these cases are described. The evaluation of aggregates for use in new constructions shows that the reactivity of quartzite and granitic aggregates results from the presence of strained quartz, even in the absence of secondary silica minerals. Revised methods of testing and modified norms for evaluation of such aggregates are described. The characteristics of the cements currently produced, including blended cements, and of additives like pozzolana and slag, are discussed.

11.1 Introduction

Commencing in the early 1950s, post-independence India witnessed a spurt in developmental activities which involved construction of various concrete structures. As a result, there are in India a large number of hydraulic structures such as concrete dams and bridges, which are presently more than 30–40 years old, the age by which problems due to any ASR in concrete should become apparent. A majority of these structures are in satisfactory condition, which bears testimony to the care and attention paid to design, workmanship and quality control in the projects.

However, confidence that the problem of distress due to ASR in concrete may not exist in India has been somewhat thwarted by the report of two concrete irrigation structures in which distress during a service life of nearly 30 years has now been attributed to ASR[1]. These, coupled with similar cases in some other concrete dams or bridge substructures (which have not been fully investigated) and reports of occurrence of ASR in neighbouring Pakistan, have focused considerable attention on the problem. As a result, aggregates and cementing materials for most of the new dam projects are being exhaustively evaluated, and research has begun on the choice of appropriate cementing systems and rehabilitation techniques. This chapter describes the contemporary experience of these aspects in India.

11.2 Manifestation of the problem

Some of the earliest references to the occurrence of ASR in concrete structures in India were made in 1962[2]; however, few details were documented. The first reported cases of distress in a concrete spillway and a concrete gravity dam and the powerhouse structure can be found in Refs. 3 and 4 along with details of comprehensive investigations carried out, which are summarised below.

11.2.1 *Hirakud Dam spillway*

The first investigation relates to a concrete spillway in one of the longest earth dams in the world. The structure at the time of investigation was nearly 27 years old. It had suffered extensive cracking, mostly in the walls of openings like galleries, shafts and adits. Typical 'map' cracking was superimposed with longitudinal horizontal cracks. The extent of cracking had been increasing with time. In addition, malfunctioning of radial crest gates, snapping of bolts which fix the sluice gate roller tracks and guide-rails to the concrete, and deflection of side walls of the adit gallery were noticed[3].

Samples of concrete from such locations where typical signs of ASR were present showed the unmistakeable presence of ASR, as evidenced by off-white, translucent to opaque agglomeration of fluffy gel-type deposits in voids bordering the aggregates, or on the aggregates; the aggregates had a dark reaction rim around their edges, visible to the naked eye (Figure 11.1).

Scanning electron microscopy (SEM) of representative samples obtained from concrete cores showed the aggregate boundaries to contain a reaction rim altering their edges, sometimes with microfractures in the aggregate pieces, with a considerable amount of white reaction products (Figure 11.2). In many instances, such cracks in the aggregates were apparently caused by the formation of the reaction products inside the aggregates[5].

The reaction products thus formed were found to be essentially non-crystalline, gel-type (Figure 11.3). An EDAX point scan on the aggregate and the surrounding rim showed the presence of a considerable amount of silicon, less calcium and small amounts of potassium (Figure 11.4). The presence of sodium could not be detected by EDAX. In view of the gel-type nature of the reaction products and the elemental composition, these can be described as lime–alkali–silica gel, identified and described by Regourd[6] as well as Thaulow and Knudsen[7]. In addition, the reaction products were occasionally found to be crystalline in nature (Figure 11.5) and their com-

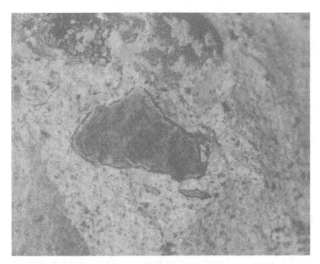

Figure 11.1 Concrete core sample showing dark reaction rim around the aggregates and typical ASR gel.

position as detected by EDAX was similar to that in Figure 11.4. Away from the reaction zone, the mortar phase was generally found to be cracked and gel formation was associated with such cracking[5].

11.2.1.1 *Aggregate types*. On petrographic examination the following three types of reactive coarse aggregates were identified[8]:

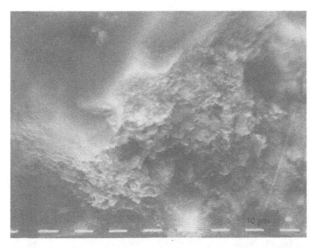

Figure 11.2 Scanning electron micrograph of the gel formation around an aggregate, altering its edges.

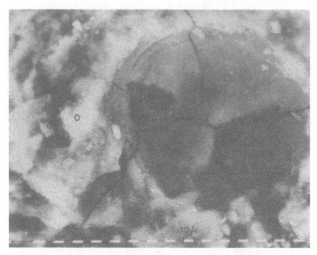

Figure 11.3 Typical gel formation and microcracking in the mortar phase.

(a) *Quartzite river shingles*, consisting predominantly of crystalline quartz (β form) with grains of varying dimensions, cemented by either crystalline silica or ferruginous matter. In a number of cases, the quartz grains were cemented by near-opaque or semi-opaque to translucent cryptocrystalline silica. The refractive index of around 1.53 indicated the material to be chert or chalcedony[9]. Quartz grains very often showed wavy (undulatory) extinction; more than 20% of the grains showed wavy extinction, the undulatory extinc-

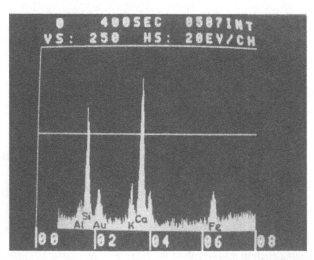

Figure 11.4 EDAX spectrum corresponding to the point marked 'O' in Figure 11.3.

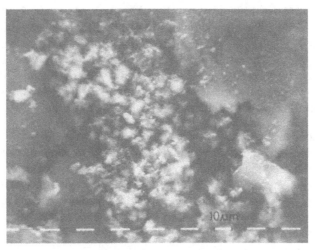

Figure 11.5 Crystalline nature of products of ASR as seen under SEM.

tion (UE) angle ranging from 19° to 28°, determined by the procedure suggested by Dolar-Mantuani[9]. Occasionally the river shingle consisted of dark
fine-grained plagioclase, horn-biotite and hornblende which showed alteration to chlorite.

(b) *Granitic rocks.* These comprised granite, granodiorite or granite
porphyry. The rock was porphyroblastic in texture and showed evidence of
action of direct pressure, manifested by wavy or undulatory extinction of
quartz. Large subidioblastic to xenoblastic plates of feldspar constituted
both orthoclase and perthitic microcline as well as plagioclase, in which the
polysynthetic twin lamellae were often deformed. The mineral analysis showed
60% alkali feldspar, 10% plagioclase feldspar, 22–25% quartz, 5–7% hornblende and biotite and 3–5% accessory minerals. Fracturing and wavy extinction were common in quartz grains. The extinction of individual grains ranged
from 12° to 25°. In the patchy (mottled) variety the orthoclase, feldspar,
occurred as large plates or laths (up to 8 mm), and this rock can be termed
granite porphyry. A third type consisted of a smaller amount of quartz (15%),
predominantly feldspar, and a comparatively large amount of biotite and
hornblende. This type is termed granodiorite. The gradation between the
three types was slight, such that they should not be considered separately.
The rocks showed common and uniform alteration to chlorite (from biotite
and hornblende) and sericite (from feldspar).

(c) *Diorites.* The rock showed a predominance of plagioclase and perthitic
microcline. Biotite, hornblende and quartz occurred in varying amounts. The
accessories were sphene, epidote and titaniferous magnetite. A modal analysis
of typical rock showed 20–30% plagioclase, 11–14% hornblende, 6–12%

biotite, 10–18% perthitic or pure microcline/orthoclase and 6–10% quartz. The quartz grains very often showed wavy extinction and cracking, the feldspars showed bending of twin lamellae and cracks along cleavage planes. More than 25% of quartz grains showed wavy extinction with a UE angle 15–30°. The rock showed conspicuous secondary alteration, which was manifested by conversion of hornblende to biotite and chlorite, feldspars to sericite and of perthitic feldspar to kaolin and other clay minerals[8].

Laboratory evaluation had earlier revealed the river shingles to be potentially reactive, and these were inadvertently used in some locations during the peak construction period. However, the other two types comprising crushed aggregates were considered to be 'innocuous' according to the criterion prevalent in the 1950s. As a result, no effort was made to obtain 'low-alkali' cement, and cement used from two sources probably contained 0.8–1.0 total alkalis (Na_2O eq.).

11.2.2 Rihand Dam and powerhouse structure

The concrete gravity dam and adjacent powerhouse of this hydroelectric project, 25 years after their construction, showed extensive distress, which was attributed to ASR[4]. External manifestations included cracking of concrete, misalignment of machinery and difficulties in the operation of gates, cranes and passenger lifts as a result of movements in concrete. In the powerhouse, the rotor assembly had sunk in relation to the stator, leading to fouling of rotor blades, and high spill current resulting in frequency tripping of the machines was reported. The rotor runner assembly had risen in relation to the speed ring. The horizontal labyrinth clearance at both top and bottom was progressively reduced in a longitudinal direction and increased in the transverse direction. There was horizontal displacement of about 3 cm between the powerhouse crane girders in different bays, and intake gates did not seal properly. The powerhouse seemed to have tilted upstream[4].

Examination of the concrete samples with a hand magnifying glass and visual examination on the broken surface of the concrete revealed typical white deposits associated with ASR in the voids in concrete and on aggregates, but they were not frequent. On the other hand, reaction rims around aggregates, another manifestation of ASR[10], were quite conspicuous. In some cases, the broken surface of the concrete showed the formation of a thin white rim around the aggregates. In most cases, however, a uniform dark band in the peripheral zone of the aggregate was observed[11].

A complete scan of the sample morphology with SEM showed that the aggregates contained a reaction rim altering their borders, sometimes with microcracks either in the aggregate or in the mortar phase, similar to Figures 11.2 and 11.3[11]. The presence of a fluffy gel-type formation with occasional crystalline deposits was observed, in which potassium was predominant. The fact that the needle-like crystal structures were not ettringite was verified with the help of EDAX, which showed the absence of sulphur. The aggregates were observed to form white fluffy gels, giving an impression that these were oozing out from the main aggregate[12]. The aggregate sample itself contained alkalis, originating from alkali feldspars. However, EDAX analysis of the reaction products showed a much larger amount of potassium.

11.2.2.1 *Aggregate types*. Petrographic examination of aggregates extracted from concrete samples indicated these to be mainly biotite granite, muscovite granite and mica granite[12]. In each of these rocks, the quartz content varied from 32 to 45% and alkali (sodium–potassium or sodium–calcium) feldspars such as orthoclase, microcline and plagioclase from 35 to 45%; varying amounts of biotite, muscovite and other accessories, including iron ore, chlorite and apatite, were present. The average grain size of quartz varied from 0.10 to 0.20 mm, with small grains up to 0.03 mm in some cases and large grains up to 0.45 mm in others. Nearly 50–80% of the quartz exhibited strain effect with a UE angle varying from 25 to 30° (Figure 11.6).

Potash feldspar in most of the cases was orthoclase which was found to have altered to sericite. Plagioclase feldspars were found to have altered to clay minerals (Figure 11.7). Biotite occurred in the form of laths and sometimes as sheath-like structures. These and muscovite showed bending effect. The normal granitic texture of the aggregate was considerably disrupted because of the high degree of alteration of minerals. Laboratory evaluation according to the criteria existing at the commencement of the project had indicated the aggregates to be 'innocuous' and there was felt to be no

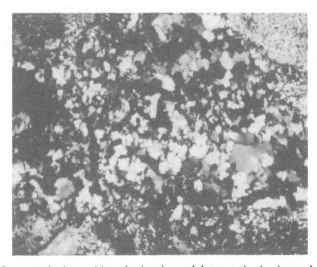

Figure 11.6 Quartz grains in granitic rocks showing undulatory extinction (crossed nicols, × 50).

Figure 11.7 Alteration of feldspar to clay minerals in granitic rock aggregates (crossed nicols, × 50).

need to obtain low-alkali cement or even regularly to monitor the alkali content of the cement used[4].

11.2.2.2 *Structural interaction*. The penstock gallery structure located on the toe of the concrete gravity dam comprised six blocks, corresponding to the six overflow bays. Each block had four reinforced concrete frames constructed integrally with the dam body interconnected with reinforced concrete beams running parallel to the axis of the dam (Figure 11.8). The major structural distress noticed in the penstock gallery and in the adjacent powerhouse structure after nearly 25 years of operation included multiple horizontal cracks on the face of columns marked 9–10–11 in Figure 11.8 (close to penstock) and extending to a height of 1 m or so from the foot of the columns, wide horizontal cracks near mid-height and snapping of main longitudinal reinforcement in column 9–10–11, spalling of concrete at the end section of beam 5–6, closing of 25-mm expansion joints between gallery and scroll concrete, and relative horizontal shifts in the beams supporting the generator floor[13].

For such distress in concrete structures, temperature effects, deleterious chemical reactions and relative settlement of foundation merit *prima facie* consideration as probable causes. From the records, it could be ascertained that the concrete was precooled and the temperatures recorded were within the limits[4]. This, along with the fact that the movements were still continuing, eliminated temperature effects as a cause. The dam was founded on granite rock which was considered as nearly ideal. Finite element analysis of the dam

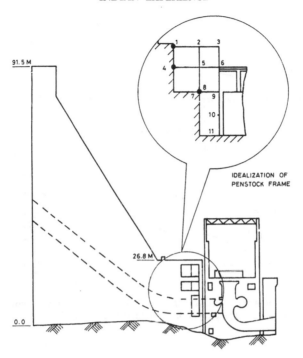

Figure 11.8 Concrete gravity dam and powerhouse structure—typical intake section.

section, including a portion of the foundation rock, and allowing differential modulus of elasticity of the rock to result in differential settlement revealed that this could not have caused the distress noticed in the frames[13].

Further analysis was carried out by imposing various levels of horizontal and vertical displacements of nodes 1, 4, 7 and 8 (Figure 11.8) representing the expansion in the dam body due to ASR. Because the deformation characteristics of all the frames were identical and all the frames had the same configuration and mechanical properties, a two-dimensional analysis was considered adequate. A systematic search was made to correlate the distress observed with the possible combination of vertical and horizontal movements transmitted to the joints of the frames due to volume change in the main dam.

From the analysis, it was concluded that the relative displacements due to ASR expansion could cause distress in the members of the penstock gallery frame, as observed[13]. In addition, the reaction in the turbine block, which could be either passive or active, could aggravate the distress in column 9–10–11. Horizontal and vertical displacements of 12.5 mm and 3 mm respectively, imposed relative to node 13 in Figure 11.8, produced the bending moment diagram shown in Figure 11.9. Visual observations of the shifts in the beams,

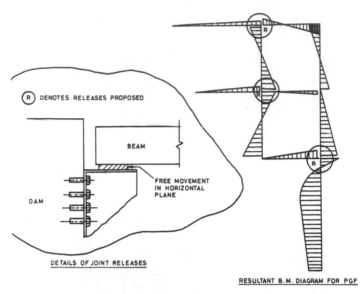

Figure 11.9 Bending moment in penstock gallery frame corresponding to induced horizontal and vertical displacements and reaction from the turbine block.

closing of 25-mm expansion joints and further long-term observations[14] established that displacements of such magnitude were entirely feasible.

11.2.3 *Characterisation of the reaction products*

The general description of the microstructure of reaction products as observed by SEM has been mentioned already. Since these were the first reported cases of ASR in India, detailed examination of the reaction products was undertaken in order to compare them with the features reported in the literature[6,7,15].

A composite gel sample was made in both cases by carefully scooping out the gel from various locations; this was used for the chemical analyses, X-ray diffraction and optical microscopy. The chemical analysis of the gels is presented in Table 11.1. The alkali contents were determined by flame photometry. The compositions were similar to the ranges indicated by others as representative of alkali–silica gel[15].

11.2.3.1 *Petrography.* The composite gel was petrographically examined under a polarising microscope in immersion liquids. In the case of quartzite aggregates containing secondary silica minerals, the material showed the following distinct composition: (i) amorphous gel-type matter of irregular shape with a refractive index of 1.48–1.50 which compared favourably with

Table 11.1 Chemical composition of ASR gel.

SL no.	Constituents	Quantity (%)	
		Quartzite aggregates	Granite aggregates
1	Loss on ignition	14.04	16.70
2	SiO_2	43.31	49.36
3	CaO	21.76	15.94
4	Al_2O_3	2.78	1.77
5	Fe_2O_3	0.66	0.49
6	MgO	0.83	0.49
7	Alkalis		
	(a) Na_2O	3.74	3.88
	(b) K_2O	12.88	11.71

1.455–1.502 reported by Mather[15]; (ii) distinct small grains of chert or chalcedony with a refractive index of 1.50–1.52 (Figure 11.10); (iii) crystalline material with patches of opaque mineral; and (iv) crystals with slight anisotropy and no birefringence, presumably of the crystalline white deposits with a refractive index of 1.42–1.48[16]. The white material also showed occasional grains of aragonite and calcite with a refractive index of 1.65–1.66. In the case of granite aggregates containing strained quartz, similar features were noted, except for the presence of secondary silica such as chert or chalcedony.

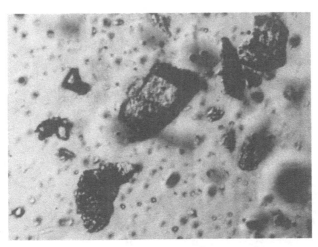

Figure 11.10 Grains of partially crystalline opaline silica in the reaction products (polarised light, × 25).

11.2.3.2 *X-ray diffraction analysis.* Typical X-ray diffractograms of the alkali–silica gel obtained in both cases are given in Figure 11.11. The nature of the gel was predominantly amorphous, and in addition to the typical cement hydration products, some new crystalline products, believed to be due to ASR, were identified[16]. In the case of quartzite pebbles, the peaks at 2θ degrees = 5.5, 16.7 and 25.7 (Cu–Kα) were assigned to a crystalline alkali–silicate hydrate of composition $NaSi_{17}O_{13}(OH)_3 \cdot 3H_2O$, and other prominent peaks at 2θ degrees = 31, 29.8 and 29.4 to a composition $K_2Ca(SO_4)_2 \cdot H_2O$. In contrast, in the case of granitic aggregates, the prominent peaks at 2θ degrees = 6.9, 13.6, 30.5 and 53.3 were ascribed to a composition of crystalline potassium–sodium–calcium–silicate hydrate $(K_2Na_2Ca)_{16}Si_{32}O_{80} \cdot 2H_2O$. These compositions are different from these reported earlier[17].

11.2.3.3 *Infra-red spectroscopy (IR).* The use of IR to study ASR has been reported before, when certain absorption bands in the 650–1600 wave number region were taken as characteristic of silica gel and calcium carbonates[18]. Composite samples of the reaction products in both the cases discussed in this chapter were studied by recording the IR spectra in the range 4000–200/cm, and compared with a synthetic silica gel. The samples were prepared by grinding and passing through a 45-μm sieve before drying in an oven at about 110°C for a few hours. The fine powder was made into a pellet with KBr.

The results are presented in Figure 11.12. In the case of the reaction products (alkali–silica gel) obtained in the two cases, the broad band observed in the region 3000–3600/cm is due to the O–H stretching vibrations. The

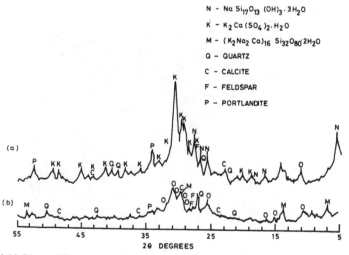

N – $Na\,Si_{17}O_{13}\,(OH)_3 \cdot 3H_2O$

K – $K_2\,Ca\,(SO_4)_2 \cdot H_2O$

M – $(K_2Na_2\,Ca)_{16}\,Si_{32}O_{80}\,2H_2O$

Q – QUARTZ

C – CALCITE

F – FELDSPAR

P – PORTLANDITE

Figure 11.11 X-ray diffractogram of products of ASR. (a) Quartzites, (b) granite aggregates.

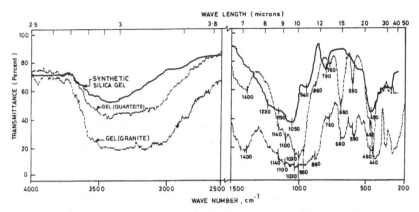

Figure 11.12 IR spectra of products of ASR and synthetic silica gels.

bands at 1400 and 1140/cm are assigned to carbonates and feldspars respectively. While the band at 1100/cm is due to monosulphates, the bands at 680 and 590/cm are assigned to SO_4^{2-} group. The band at 1000/cm indicates the presence of hydrated calcium silicates. Lastly, the peaks at 1030, 960, 860, 760, 460 and 440/cm are assigned to different modes of SiO_4 vibrations.

Similarly, the IR spectrum from the synthetic silica gel contains a broad band in the range 3000–3600/cm. Most of the bands characteristic of different modes of SiO_4 vibrations, such as 1230, 1150, 1050, 940, 790 and 450/cm are observed. The occurrence of a 'shoulder' at 590/cm, which is assigned to SO_4^{2-}, is also observed. The absence of bands at 1400, 1100, 1000 and 680 described earlier is quite understandable as the alkali–silica gel was extracted from the concrete cores. In all other respects, the IR spectra of the two materials were similar.

In summary, the microstructure of the ASR products was predominantly of the amorphous gel type, with occasional crystals. In the case of metastable silica minerals, distinct reaction products seemed to be formed, whereas in the case of strained quartz the result of ASR was alteration in and of the aggregates[16].

11.2.4 *Repairs*

Long-term observations on concrete core samples immersed in KOH solution at different temperatures indicated that the expansion potential of the concretes was not yet exhausted[14]. The repair techniques, therefore, had to take into account the present state of distress as well as possible aggravation in future. Analyses of stability against overturning as well as sliding failure, assuming cracked sections, indicated the margins of safety to be satisfactory[3,4]. The remedial measures were accordingly aimed at minimising the rate of reaction, restoring structural integrity and accommodating possible future expansion of concrete.

Cracks at the upstream faces were sealed with epoxy grout, followed by epoxy painting to minimise ingress of water. It was recommended that chemical grouting be carried out in the mass of the structures effectively to fill up the cracks and restore their monolithic behaviour. In the case of the spillway, at locations where cracking was extensive, stitching across the cracks by providing anchor rods to act as shear pins was suggested[3]. The drainage holes were cleaned to make them function effectively to reduce the uplift pressure.

In the reinforced concrete columns of the powerhouse in which reinforcement had snapped, additional reinforcement was provided and the columns were than jacketed with steel plates. To accommodate future expansion, the fixed joints in the penstock gallery frame were released (Figure 11.9) and the expansion joint at the toe of the dam was made functional. For operation of gates, gantries, cranes, etc., portions of concrete were chipped off to provide easy movement. Extensive instrumentation for periodic monitoring of the structures was recommended[3,4].

11.3 Types of reactive aggregates in India

In view of the vast size of the country and the wide variations in the geomorphological characteristics of natural rocks from one location to another, any generalisation as to the reactivity of Indian aggregates could be misleading. A broad classification of the common rock types used as natural aggregates for concrete in different parts of India is shown in Figure 11.13. The common rock types so identified have been found to be reactive to varying degrees in different parts of the world, depending upon their modal composition, essentially the presence of secondary silica minerals such as cherts, chalcedony, opal, etc.[9]

11.3.1 Previous assessments

The presence of secondary silica minerals had been known to be responsible for the reactivity of common aggregates for 40 years or more, and the first comprehensive assessment of certain Indian aggregates was made from this chemical point of view[19]. Accordingly, the following common rock types were identified as potentially reactive on the basis of their composition as well as results of the rapid chemical test (ASTM C289) and mortar bar expansion test (ASTM C227):

(1) Fine-grained, glassy to microcrystalline basalts containing more acidic glassy phases and occurring in the Deccan Plateau, west coast, Maharashtra, Madhya Pradesh, Gujarat, Andhra Pradesh, Jammu and Kashmir, West Bengal and Bihar.

Figure 11.13 Types of natural aggregates commonly used in India (schematic map—not to scale).

(2) Sandstones containing secondary silica minerals such as chalcedony, crypto- to microcrystalline quartz, opal and quartzites having a reactive binding matrix and occurring in Madhya Pradesh, West Bengal, Bihar and Delhi.

(3) Granites and pigmetites containing opal, rhyolites and glasses and occurring in south India, notably Tamil Nadu and Karnataka.

(4) Trap aggregate containing reactive constituents occurring in Jammu and Kashmir and Deccan Plateau.

Since reactivity of aggregates due to the presence of secondary silica minerals was known at the time some of the major concrete dams were built in India in the early 1950s, such aggregates were used with due caution. Bhakhra Dam in Punjab, which was the highest concrete gravity dam in the world when completed, used river gravels composed predominantly of quartzites, metasandstones, greywackes and sandstones, which contained cherts, chalcedonic sandstones and glassy andesites—2.3% of the total mass on an average—as well as limestone and dolomites. Similarly, the natural sand contained less than 2% cherts. These aggregates were proven to be innocuous after exhaustive laboratory evaluation and were used without any detrimental effect till now[20]. In the case of Hirakud Dam spillway, the quartzite river shingles containing cherts and chalcedony, which proved to be potentially reactive in laboratory evaluation, were used inadvertently.

Gogte in 1973 postulated that the reactivity of common aggregates could also be due to mineralogical and textural features of the crystal rocks, in which the commonly accepted susceptible forms of silica, such as opal and chalcedony, were absent[2]. He ascribed the reactivity of such aggregates to the presence of strained quartz. As a modification of the then existing test procedures of the Indian Standards (IS) and ASTM, Gogte recommended that mortar bar tests be carried out at a temperature of 50°C and suggested a criterion of mortar bar expansion above 0.05% in 6 months as indicative of reactivity. Accordingly, Gogte identified a number of granites, charnockites and quartzites and schistose rocks, mostly from Andhra Pradesh, Karnataka and Tamil Nadu in south India, as well as basalts from Maharashtra and Gujarat in western India, as being potentially reactive because of the presence of strained quartz. In most of these rocks, samples showed strongly undulatory, fractured and granulated quartz, nearly 35–40% of quartz grains showing a UE angle between 18 and 20°. On the other hand, rocks which contained less than 20% strained quartz, or in which most of the quartz showed uniform or faint undulatory extinction, were considered 'innocuous'. Sandstones from Andhra Pradesh, Rajasthan, Himachal Pradesh and Madhya Pradesh owed their reactivity to presence of cherts as a detrital constituent and sometimes as binding matrix. Sandstones devoid of cherts but containing a few grains of strongly undulatory quartz showed expansion in mortar bar tests within tolerable limits[2].

11.3.2 *Aggregates for new constructions*

From the foregoing, it would transpire that most of the reactive aggregates in India are those containing strained quartz as the reactive component. In addition, reactivity of granitic rock aggregates is also partly due to the presence of alkali feldspar, which can undergo alterations as a result of the action of hydrothermic solution and the normal process of weathering[12]. Laboratory experiments by Van Aardt and Visser had shown that alkali feldspars can release alkali in the presence of calcium hydroxide and water[21], in which case they can supplement the alkali derived from cement. While detection of potential alkali reactivity of aggregates containing such secondary silica minerals is quite straightforward, e.g. by rapid chemical test or mortar bar test, detection of reactivity due to the presence of strained quartz poses some problems. In such cases, rapid chemical test may not detect the potential reactivity, and the threshold values for mortar bar expansion may have to be somewhat lowered[1,22]. As a result of this, attention is being focused on the effects of strained quartz in evaluating concrete aggregates for use in a number of concrete dams to be constructed in India. Indeed a large volume of data on aggregates from different sources has now become available and is summarised below.

A large number of such aggregate samples were of the quartzite type, while others were granitic rocks containing feldspars and mica-bearing phases in substantial quantities as well as other varieties[23]. In addition, composite samples containing rocks of more than one type were also involved. A summary of the petrographic details of the aggregates is shown in Table 11.2. The strain effect in quartz grains was measured in terms of UE angle[9]. Alkali feldspars in the granitic aggregates were found to have altered to clay minerals or sericite. Metastable secondary silica minerals were not detected in any of these aggregate samples. As such, any potential alkali–silica reactivity of these aggregates could be ascribed mainly to the presence of strained quartz.

11.3.2.1 *Test procedure.* Knowing that the existing standardised test procedures ASTM C227, C289, C586 and IS:2386:Part VII may not detect the potential reactivity of such aggregates, some improvisation in the testing regimes was necessary. Accordingly, the aggregate samples were subjected to rapid chemical test for a total duration of 24 hours, 3 days and 7 days. Similarly, the mortar bar expansion test was carried out at 38°C as per IS:2386:Part VII, as well as at 60°C, as suggested by Buck[24]. Three samples of ordinary Portland cement containing a total of 1.00, 0.57 and 0.25% alkalis (Na_2O eq.) were used in mortar bar tests. The composition of the mortar bars for both 38 and 60°C regimes was identical, which permitted direct comparison of the results. Exploratory tests were also carried out on aggregate samples immersed in KOH solution (1 N) at 60°C and changes observed by SEM and IR spectroscopy[25].

11.3.2.2 *Test results and discussions.* As anticipated, the rapid chemical test ASTM C289 for 24 hours showed most of the aggregate samples to be innocuous. When tested after prolonged storage for 3 and 7 days, the results shifted somewhat to the 'right' of the demarcation line in the case of quartzite aggregates (Figure 11.14). Many of the other aggregates continued to be in the 'innocuous' zone. It is not surprising that the limiting curve of Figure 11.14, developed in relation to mortar bar tests on aggregates containing secondary silica minerals, should not be valid for aggregates containing strained quartz, which are relatively slowly reactive.

A summary of the mortar bar expansion tests is given in Table 11.3. All the aggregate samples exhibited increased expansion with increasing alkali content in the cements, when tested in either of the temperature regimes. A

Table 11.2 Description of aggregates.

Sl no.	Type	Quartz grain size (mm)	Strain effect UE (degrees)	UE (%)	Modal composition (%)					
					Qz	*Ir*	*Bio*	*Chl*		
1	Quartzite	0.025–0.12*	31–40	90	94	1	2	3		
		0.025*	20–30	90						
2	Quartzite	0.075–0.575*	32–40	90	91	2	2	3		
		0.075*	22–33	90						
3	Quartzite	0.040–0.0525*	33–39	90	93	2	2	3		
		0.0125*	23–28	92						
4	Quartzite	0.030–0.125*	31–42	85	89	3	8	—		
		0.005–0.30*	21–28	90						
5	Quartzite	0.025–1.375	33–35	25	79	2	5	2(12mica)		
6	Quartzite	0.1–0.3	30–40	65	78	7	5	10		
7	Quartzite	1.0–3.0	25–40	70	88	4	8			
					Qz	*Fels*	*Gm*	*Ir*		
8	Greywacke	0.025–0.40	15–20	25	50	7	40	3		
9	Orthoquartzite	0.2–1.5	15–20	75	94					
					Aug	*Fels*	*Ir*	*Br*	*Qz*	*Bio*
10	Dolerite	0.175–1.0	22–28	15	44	26	14	10	4	2
					Qz	*Or*	*Pl*	*M*	*Bio*	*Acc*
11	Granite	0.006–1.5	11–15	15	40	12	6	30	7	5
12	Biotite granite	0.075—1.25*	30–40	75	45	31	8	—	12	4
		0.075*	23–28	23						
13	Biotite gneiss	0.050–0.550	25–30	20	40	22	13	9	14	2
14	Augen gneiss	0.050–0.625	26–32	30	42	28	11	5	9	5
15	Granite gneiss	0.1–0.8	20–25	50	40	37	9	—	8	6
16	Phyllite (very fine grained)		20–25	70	37	26	—	—	13	24
17	Sand		30–45	80	70	5	—	2	7	16
					Qz	*Ir*	*Fels*	*Bio*	*Chl*	
18	Sand (medium–fine)		17–19	20	60	4	9	5	7	

*Denotes two generations of quartz.

Qz, quartz; Ir, iron oxide; Bio, biotite; Chl, chlorites; .Fels, feldspar; Aug, augite; Gm, groundmass; Or, orthoclase; Pl, plagioclase, M, muscovite; Acc, accessory minerals.

few results are shown in Figure 11.15. Such an alkali-dependent expansion of the aggregates would, *prima facie*, classify them as 'alkali-reactive'.

For quantifying the effects of 'strained quartz', various parameters, namely grain size, proportion of quartz in the modal composition, percentage of quartz showing strain effect and UE angle, were considered. Except for very fine-grained (smaller than 0.1 mm) rocks, average grain size did not exhibit any discernible influence on the resultant expansion. The amount of expansion was more dependent upon the precentage of quartz grains showing strain effect[23].

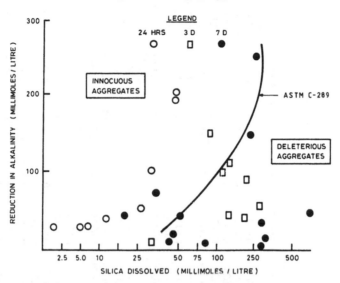

Figure 11.14 Results of rapid chemical tests on aggregates containing strained quartz.

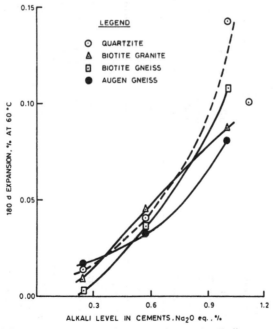

Figure 11.15 Dependence of mortar bar expansion on the alkali content in cements.

Table 11.3 Summary of mortar bar expansion tests at 38°C and 60°C.

Sl no.	Type of aggregate*	0.25%				0.57%				1.00%			
		38°C		60°C		38°C		60°C		38°C		60°C	
		3 months	6 months	3 months	6 months	3 months	6 months	3 months	6 months	3 months	6 months	3 months	6 months
1	Quartzite	—	—	0.01	0.014	0.011	0.014	0.036	0.040	0.076	0.103	0.165	0.204
2	Quartzite	—	—	0.004	—	0.021	0.0026	0.029	0.041	0.038	0.046	0.157	0.196
3	Quartzite	—	—	0.004	—	0.021	0.0026	0.029	0.041	0.038	0.046	0.157	0.201
4	Quartzite	—	—	0.009	—	0.018	0.022	0.038	0.047	0.064	0.082	0.0178	0.230
5	Quartzite	—	—	0.0148	0.0148	0.0208	0.0188	0.0304	0.0396	0.0248	0.0220	0.0608	0.0844
6	Quartzite	—	—	—	—	0.0374	0.0374	—	—	0.020	0.029	0.060	0.075
7	Quartzite	—	—	—	—	0.0192	0.0192	0.0311	0.0311	0.037	0.041	0.072	0.092
8	Greywacke	—	—	—	—	—	—	—	0.0096	—	—	—	0.0480
9	Orthoquartzite	—	—	—	—	0.016	0.016	0.0356	0.0364	0.0224	0.0416	0.0504	0.0720
10	Dolerite	—	—	—	—	—	—	—	—	—	—	0.048	0.0488
11	Granite	—	—	—	—	—	—	—	—	—	—	0.0512	—
12	Biolite granite	—	—	0.10	—	0.014	0.018	0.033	0.045	0.035	0.035	0.080	0.087
13	Biolite gneiss	—	—	0.0036	0.0036	0.0176	0.0204	0.0316	0.0356	0.0312	0.0364	0.0706	0.103
14	Augen gneiss	—	—	0.016	0.016	0.014	0.0116	0.00296	0.0328	0.0246	0.0222	0.0642	0.0732
15	Granite gneiss	—	—	—	—	—	—	—	0.0288	—	—	—	0.0560
16	Phyllite	—	—	—	—	—	—	—	0.0380	—	—	—	0.090
17	Sand	—	—	—	—	0.014	0.0164	0.0152	0.0172	0.0348	0.0348	0.0600	0.0764
18	Sand	—	—	—	—	0.0212	0.0252	0.0280	0.0344	0.0288	0.0448	0.0576	0.0960
19	Composite 1	—	—	—	—	0.0148	0.0148	0.0212	0.0204	—	—	0.0352	0.0584
20	Composite 2	—	—	—	—	0.0112	0.0112	0.0348	0.0348	0.0380	0.0396	0.0488	0.0640
21	Composite 3	—	—	—	—	0.0128	0.0152	0.0232	0.0368	0.0144	0.020	0.0416	0.0608
22	Composite 4	—	—	—	—	0.0104	0.0104	0.0264	0.0264	0.0128	0.0144	0.0392	0.0416

Expansion, (%) with cements having total alkalis

*Details as for Table 11.2

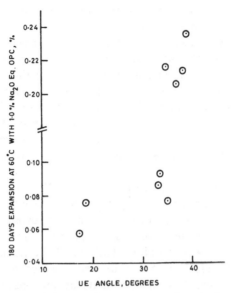

Figure 11.16 Average UE angle in quartzite aggregates and mortar bar expansion with high-alkali cements.

Since the amount of quartz grains was relatively greater in quartzite rocks than in other aggregates, these were considered separately. For both the sets of aggregates, the amount of expansion increased with the UE angle in the quartz grains (Figure 11.16).

After continuous immersion in 1 N KOH solution at 60°C, quartzite aggregates revealed differences in IR spectra in the region above 850 cm^{-1}. A broad and intense peak in the 1600 cm^{-1} region, due possibly to OH bending mode, was identified. Under SEM, typical gel formation and microcrystalline growth on the aggregates could be seen. However, the trend was not very clear. Fuller details are given in Ref. 25.

11.3.3 Revised criteria and standardisation

From Table 11.3, it can be seen that, except for aggregates 1 and 4 (Table 11.2), the maximum expansion with 1.00% alkali cement at 38°C was generally within the limit of 0.05% at 3 months that is stipulated for reactive aggregates. The rapid chemical tests also failed to indicate the potential reactivity of these aggregates (Figure 11.14). Yet the dependence of expansion in mortar bars on the level of alkalis in cements would suggest that the aggregates were indeed responsive to the alkali in the system (Figure 11.15). Field experiences with similar aggregates, as reported in 11.3, would confirm the potential reactivity of such aggregates. As such, revised criteria for these types of aggregates containing strained quartz are necessary.

A summary of the recommendations made in relation to reactive aggregates, including those containing strained quartz, is reproduced in Table 11.4. Recognising the relatively slower kinetics of expansion reaction of aggregates containing strained quartz, Buck suggested that either the limits of expansion at 38°C (ASTM C227) be applied at later ages too or that different limits be fixed for the 60°C regime[24]. Gogte had suggested a lower limit of expansion of 0.05% at 6 months when tested at 50°C[2]. The Canadian (CSA) specifications CAN3-A23.1 and A23.2-M77 (1986) have also recognised the relatively slow expansion in concrete prism tests with aggregates whose reactivity is due to strained quartz, and have suggested a limit of 0.04% expansion at 38°C after 1 year.

Presently Indian Standard specification IS:383 (1970) does not specify any limit of expansion in mortar bar tests, perhaps because until now little concern has been expressed. In view of recent experiences, a revised criterion is being proposed, for which the reasoning is as follows[23]. In the current investigations, expansion at 60°C with low-alkali (0.57%) cements seldom exceeds 0.05% at 90 days (Table 11.3). Since the well-accepted remedy for use with reactive aggregates is such low-alkali cements, expansions recorded with such cements can be considered as acceptable. On the other hand, aggregates similar to those used in hydraulic structures which have exhibited distress due to ASR (e.g. 11, 12, 15 of Table 11.2) resulted in expansion exceeding 0.05% at 3 months with 1.00% alkali cement at 60°C. In many cases, formation of ASR gel was noticed on the broken surfaces of the mortar bars with high-alkali cement (Figure 11.17). A limit of acceptable expansion of the order of 0.05% at 3 months or 0.06% at 6 months, when tested with higher alkali (about 1.00%) cement at 60°C is, therefore, proposed[23].

Attempts have been made to set 'prescription'-type limits of UE angle or percentage of quartz grains exhibiting strain effect[24]. 'Performance'-type specifications of the nature of limiting expansion in mortar-bar tests is to be

Table 11.4 Suggested criteria for reactive aggregates.

Sl no.	Authority	Temperature	Percentage expansion after			UE angle	Percentage quartz showing strain effect
			90 days	6 months	1 year		
1	ASTM C227	100°F (37.8°C)	0.05	0.10	—	—	—
2	Ref. 24	38°C	—	0.05	0.10	—	—
		60°C		0.025	0.04	15°	20
3	Ref. 2	50°C	—	0.05	—	Strongly undulatory	40
4	Ref. 23	60°C	0.05	0.06	—	25°	25 for quartzite 15 for granitic aggregates

Figure 11.17 White ASR gel formed in mortar bars made with high-alkali cement and aggregates containing strained quartz.

preferred. Note that the suggested limits were reached by quartzite aggregates, in which a minimum of 25% quartz grains showed strain effect with UE angle of 25° or above (Table 11.1). The limits in the case of granitic aggregates would be somewhat lower, because of the role of alkali feldspars and mica-bearing phases.

It is proposed that these limits are incorporated in the national specification (IS:383:1970). The relevant test procedures (IS:2386:Part VII) would stipulate mortar bar tests at 60°C for aggregates containing strained quartz and petrographic examination to quantify strain effect in terms of UE angle would also be recommended in IS:2386:Part VIII.

11.4 Cement systems

Whenever alkali-reactive aggregates are encountered, it has been customary to use cement of low alkali content—the limit being 0.6% Na_2O eq. As elsewhere in the world, availability of low-alkali ordinary Portlant cement in India has been somewhat limited. Figure 11.18 indicates that the average level of alkalis in cements in India has increased during the last two decades[26]. This, to a large extent, is due to the manufacturing technologies adopted for conservation of energy and requirements of environmental protection, in that modern dry-process cement plants require hot exit gases containing the volatiles as well as the kiln dust to be recirculated in the process stream and not vented to the atmosphere. Use of blended cements like Portland pozzolana cement and Portland slag cements would also tend to increase the level of total alkalis in the cement, because pozzolanas and slags in general contain alkalis higher than in the cement clinker. In India, the proportion of blended cements is 70% of the total production and dry-process cement plants now constitute 72% of the total installed capacity.

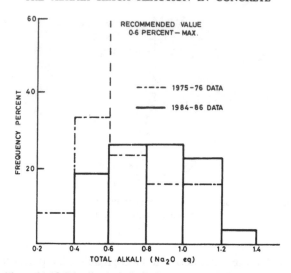

Figure 11.18 Distribution of alkali contents in Indian cements.

A comprehensive investigation has been carried out on the role of indigenous blended cements as well as pozzolana and slags used in commercial production of such blended cements in India in alleviating ASR[27]. The results show that, in general, blended cements are helpful, and optimum results are obtained when the amount of substitution of pozzolana or slag is relatively high, i.e. of the order of 25–30% in the case of pozzolana and more than 50% in case of slags. Although Indian Standard specifications permit pozzolana contents in blended cements to vary between 10 and 25%, the average in practice is of the order of 11–15%. Similarly, the slag content in Portland slag cement is less than 50% whereas the maximum permitted is 65%. In the manufacture of blended cements with a prefixed quantity of pozzolana or with slags interground in the cement to meet the other requirements of specifications, the flexibility to add larger doses of cement substitutes becomes somewhat restricted.

One question that has often worried engineers is the safe limit of total alkalis in blended cements when the additives (pozzolana or slag) are not separately available for analysis. Depending upon the hydraulic activity of the slag or pozzolana, part of the alkalis contributed by them becomes available in the pore solutions[28]. In certain specifications, a limit of 0.9% total alkalis in the case of Portland slag cements in which the slag content is greater than 60% has been suggested[10]. No such limit has, however, been established for commercially produced Portland pozzolana cements. The investigation cited showed that a limit of 0.6% total alkalis in ordinary Portland cements corresponded to a limit in the case of Portland pozzolana cements of the order of 0.8–0.9% (Figure 11.19). Nevertheless, the safe values depend upon a host of factors such as the chemical composition of the cement clinker and the pozzolana, the reactivity of the pozzolana to lime–water systems and the reactivity of the aggregates, thereby making any generalisation hazardous. It is prudent, therefore, to establish a safe aggregate–cement–pozzolana (or slag) combination by prior trials. For many of the new constructions reported in 11.3, use of active pozzolana as part replacement of cement has been envisaged.

In addition to lowering the total soluble alkali content in the concrete to the extent that cement is replaced by active silicious pozzolana and hydraulic slags, they also combine with the CH liberated during the hydration of cement. It has been reported that if the CH liberated can be fully consumed by large proportions of slags, ASR would not occur[29]. In the context of the foregoing, modified cement compositions having no C_3S phase or a lower C_3S phase merit consideration[12]. In these cement systems, the amounts of CH liberated upon hydration are considerably lower. Use of such cements with known reactive aggregates is presently under investigation[30].

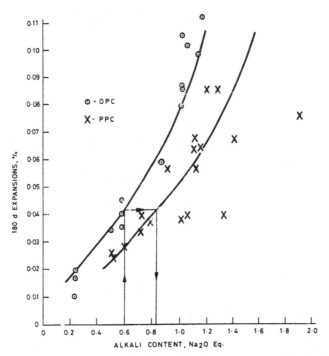

Figure 11.19 Relative performance of ordinary Portland cements and Portland pozzolana cements in mortar bar expansion tests with reactive aggregates.

11.5 Concluding comments

Instances of ASR in concrete structures in India have mainly been due to the presence of silicious aggregates such as quartzites, granites, granodiorites, granite porphyry and diorites etc containing strained quartz. The potential reactivity of these slowly reactive aggregates could not be detected by the test procedures and evaluation norms existing at the time of construction. This has led to modifications in the test methods and adoption of revised threshold values for mortar bar expansion tests, according to which aggregates proposed for many new constructions are now judged as being potentially reactive. The use of low-alkali cements, along with relatively large dosages of active pozzolana, is contemplated in such situations. Although the availability of low-alkali cements is somewhat restricted as yet, it has been possible to meet the demand through indigenous sources.

Acknowledgements

This contribution is based on the research carried out at the Construction Development Institute (CDI) of the National Council for Cement and Building Materials, New Delhi, and is presented with the permission of Dr H. C. Visvesvaraya, Chairman of the Council. The author is grateful to him for the guidance and suggestions at various stages of the investigations. The help of C. Rajkumar, R. C. Wason, George Samuel and other colleagues in preparation of the manuscript is appreciated.

References

1. Visvesvaraya, H.C. and Mullick, A.K. (1984) Developmental efforts in relation to concrete constructions in India, a state-of-art report. *ACI's Fall Convention*, New York, American Concrete Institute, Detroit, MI, pp. 9–18.
2. Gogte, B.S. (1973) An evaluation of some common Indian rocks with special reference to alkali–aggregate reaction. *Engng. Geol.* **7**, 135–153.
3. Irrigation and Power Department, Government of Orissa, India (1983) *Report of the Committee of Experts to Study and Advise Remedial Measures on Cracks in Hirakud Dam Spillway*, **1**, 67.
4. Irrigation Department, Government of Uttar Pradesh, India (1986) *Rihand Dam Experts Committee, Report*, **1**, 58.
5. Samuel, G., Mullick, A.K., Ghosh, S.P. and Wason, R.C. (1984) Alkali silica reaction in concrete—SEM and EDAX analyses. *Proceedings of the Sixth International Conference on Cement Microscopy, Albuquerque, New Mexico, USA*, pp. 276–291.
6. Regourd, M. and Horuain, H. (1986) Microstructure of reaction products. *Proceedings of the Seventh International Conference on Concrete Alkali–Aggregate Reactions*, pp. 375–380.
7. Thaulow, N. and Knudsen, T. (1975) Quantitative microanalysis of the reaction zone between cement paste and opal. *Symposium on Alkali–Aggregate Reaction, Preventive Measures, Reykjavik, Iceland*, pp. 189–203.
8. Cement Research Institute of India (1983) *Assessment of Concrete in the Spillway Blocks of Hirakud Dam*. Project Report, SP—147, p. 38.
9. Dolar-Mantuani, L. (1983) *Handbook of Concrete Aggregates, A Petrographic and Technological Evaluation*. Noyes Publication, Far Ridge, NJ, pp. 79–125.
10. Mather, B. (1975) *New Concern Over Alkali–Aggregate Reaction*. NMRCA

Publication No. 149. National Ready-Mixed Concrete Association, Silver Springs, MD, p. 20.

11. National Council for Cement and Building Materials (1985) *Assessments of Causes and Extent of Distress to Concrete in Rihand Dam and Power House Structure.* Project Report, SP-197, 38.

12. Visvesvaraya, H.C., Mullick, A.K., Samuel, G., Sinha, S.K. and Wason, R.C. (1986) Alkali reactivity of granitic rock aggregates. *Proceedings of the Eighth International Congress on the Chemistry of Cement, Rio de Janeiro, Brasil,* Vol. V, pp. 206–213.

13. Visvesvaraya, H.C., Rajmukar, C. and Mullick, A.K. (1986) Analysis of distress due to alkali–aggregate reaction in gallery structures of a concrete dam. *Proceedings of the Seventh International Conference on Concrete Alkali–Aggregate Reactions,* pp. 188–193.

14. Mullick, A. K., Samuel, G., Sinha, S.K. and Wason, R.C. (1987) Assessment of residual expansion potential in concrete structures due to alkali–silica reaction. *Proceedings of the Fourth International Conference on Durability of Building Materials and Components, Singapore,* Vol. II, pp. 793–800.

15. Mather, B. (1952) Cracking of concrete in the Tuscaloosa lock. *Proceedings of the 31st Annual Meeting of Highway Research Board, Washington DC,* pp. 218–233.

16. Mullick, A.K. and Samuel, G. (1986) Reaction products of alkali silica reaction—a microstructural study. *Proceedings of the Seventh International Conference on Concrete Alkali–Aggregate Reactions,* pp. 381–385.

17. Cole, W.F., Lancucki, C.J. and Sandy, M.J. (1981) *Cement Concr. Res.* **11**, 443–454.

18. Poole, A.B. (1975) Alkali–silica reactivity in concrete. *Symposium on Alkali–Aggregate Reaction, Preventive Measures, Reykjavik, Iceland,* pp. 101–111.

19. Jagus, P.J. and Bawa, N.S. (1957) Alkali–aggregate reaction in concrete construction. *Road Research Bulletin* **3**, 41–73.

20. Cement Research Institute of India (1982) *Assessment of Strength of Concrete in the Training Walls and Apron Slab of the Bhakra Dam,* Project Report, SP-141, p. 28.

21. Van Aardt, J.H.P. and Visser, S. (1978) Reaction of $Ca(OH)_2$ and of $Ca(OH)_2 + CaSO_4 \cdot 2H_2O$ at various temperatures with feldspars in aggregates used for concrete making. *Cement Concr. Res.* **8**, 677–682.

22. Tuthill, L.H. (1982) Alkali–silica reaction—40 years later. *Conc. Int. Design Const.* **4**, 32–36.

23. Mullick, A.K. (1987) Evaluation of ASR potential of concrete aggregates containing strained quartz. *NCB Quest* (New Delhi), **1**, 35 46.

24. Buck, A.D. (1983) Alkali reactivity of strained quartz as a constituent of concrete aggregates. *Cement Concr. Agg.* (ASTM) **5**, 131–133.

25. Mullick, A.K., Samuel, G. and Ghosh, S.N. (1985) Identification of reactive concrete aggregates containing strained quartz by SEM and IR. *Proceedings of the Seventh International Conference on Cement Microscopy, Fort Worth, TX,* pp. 316–332.

26. Visvesvaraya, H.C. and Mullick, A.K. (1986) Quality of cements in India—Results of three decadal surveys. In: Farkas, E. and Klieger, P. (Eds) *ASTM Special Technical Publication,* STP 961, pp. 66–79.

27. National Council for Cement and Building Materials (1988) *Alleviating Alkali Silica Reaction in Concrete with Particular Reference to Blended Cements.* Project Report, CDI-4, 73.

28. Bhatty, M.S.Y. (1985) Mechanism of pozzolanic reactions and control of alkali–aggregate expansion. *Cement Concr. Agg.* **7**, 69–77.

29. Chatterji, S. and Clausson-Kaas, N. F. (1984) Prevention of alkali–silica expansion by using slag-Portland cement. *Cement Concr. Res.* **14**, 816–818.

30. Raina, S.J., Ali, M.M. and Irani, D.B. (1987) Porsal, A low energy sulphate resisting cement—further studies. *NCB Quest* (New Delhi) **1**, 9 20.

Index